Cluster Algebras and Poisson Geometry

Mathematical
Surveys
and
Monographs

Volume 167

Cluster Algebras and Poisson Geometry

Michael Gekhtman
Michael Shapiro
Alek Vainshtein

American Mathematical Society
Providence, Rhode Island

2010 *Mathematics Subject Classification.* Primary 13F60, 53D17;
Secondary 05E40, 14M15, 16T30.

The first author was supported in part by NSF Grants DMS #0400484 and DMS #0801204 and BSF Grant #2002375. The second author was supported in part by NSF Grants DMS #0401178, DMS #0800671, PHY #0555346, and BSF Grant #2002375. The third author was supported in part by ISF Grant #1032/08 and BSF Grant #2002375.

Key words and phrases. Cluster algebras, Poisson structures, networks on surfaces, Grassmannians, double Bruhat cells, Coxeter–Toda lattices.

For additional information and updates on this book, visit
www.ams.org/bookpages/surv-167

Library of Congress Cataloging-in-Publication Data

Gekhtman, Michael, 1963–
Cluster algebras and Poisson geometry / Michael Gekhtman, Michael Shapiro, Alek Vainshtein.
p. cm. — (Mathematical surveys and monographs ; v. 167)
Includes bibliographical references and index.
ISBN 978-0-8218-4972-9 (alk. paper)
1. Cluster algebras. 2. Poisson algebras. I. Shapiro, Michael, 1963– II. Vainshtein, Alek, 1958– III. Title.

QA251.3.G45 2010
512′.44—dc22

2010029529

∞ The paper used in this book is acid-free and falls within the guidelines
established to ensure permanence and durability.
Visit the AMS home page at http://www.ams.org/

10 9 8 7 6 5 4 3 2 1 15 14 13 12 11 10

Dedicated to the memory of Vladimir Igorevich Arnold

Contents

Preface

Cluster algebras introduced by Fomin and Zelevinsky in [**FZ2**] are commutative rings with unit and no zero divisors equipped with a distinguished family of generators (*cluster variables*) grouped in overlapping subsets (*clusters*) of the same cardinality (*the rank of the cluster algebra*) connected by *exchange relations*. Among these algebras one finds coordinate rings of many algebraic varieties that play a prominent role in representation theory, invariant theory, the study of total positivity, etc. For instance, homogeneous coordinate rings of Grassmannians, coordinate rings of simply-connected connected semisimple groups, Schubert varieties, and other related varieties carry (possibly after a small adjustment) a cluster algebra structure. A prototypic example of exchange relations is given by the famous Plücker relations, and precursors of the cluster algebra theory one can observe in the Ptolemy theorem expressing product of diagonals of inscribed quadrilateral in terms of side lengths and in the Gauss formulae describing Pentagrama Myrificum.

Cluster algebras were introduced in an attempt to create an algebraic and combinatorial framework for the study dual canonical bases and total positivity in semisimple groups. The notion of *canonical bases* introduced by Lusztig for quantized enveloping algebras plays an important role in the representation theory of such algebras. One of the approaches to the description of canonical bases utilizes dual objects called *dual canonical bases*. Namely, elements of the enveloping algebra are considered as differential operators on the space of functions on the corresponding group. Therefore, the space of functions is considered as the dual object, whereas the pairing between a differential operator D and a function F is defined in the standard way as the value of DF at the unity. Elements of dual canonical bases possess special positivity properties. For instance, they are regular positive-valued functions on the so-called *totally positive* subvarieties in reductive Lie groups, first studied by Lusztig. In the case of GL_n the notion of total positivity coincides with the classical one, first introduced by Gantmakher and Krein: a matrix is totally positive if all of its minors are positive. Certain finite collections of elements of dual canonical bases form distinguished coordinate charts on totally positive varieties. The positivity property of the elements of dual canonical bases and explicit expressions for these collections and transformations between them were among the sources of inspiration for designing cluster algebra transformation mechanism. Transitions between distinguished charts can be accomplished via sequences of relatively simple positivity preserving transformations that served as a model for an abstract definition of a cluster transformation. Cluster algebra transformations construct new distinguished elements of the cluster algebra from the initial collection of elements. In many concrete situations all constructed distinguished elements have certain stronger positivity properties called *Laurent positivity*. It is still an open question whether Laurent positivity holds for all distinguished elements for an arbitrary cluster algebra. All Laurent positive elements of a cluster algebra form a cone. Conjecturally, for cluster algebras arising from reductive semisimple Lie groups extremal rays of this cone form a basis closely connected to the dual canonical basis mentioned above.

Since then, the theory of cluster algebras has witnessed a spectacular growth, first and foremost due to the many links that have been discovered with a wide range of subjects including

- representation theory of quivers and finite-dimensional algebras and categorification;
- discrete dynamical systems based on rational recurrences, in particular, Y-systems in the thermodynamic Bethe Ansatz;
- Teichmüller and higher Teichmüller spaces;
- combinatorics and the study of combinatorial polyhedra, such as the Stasheff associahedron and its generalizations;
- commutative and non-commutative algebraic geometry, in particular,
  - Grassmannians, projective configurations and their tropical analogues,
  - the study of stability conditions in the sense of Bridgeland,
  - Calabi-Yau algebras,
  - Donaldson-Thomas invariants,
  - moduli space of (stable) quiver representations.

In this book, however, we deal only with one aspect of the cluster algebra theory: its relations to Poisson geometry and theory of integrable systems. First of all, we show that the cluster algebra structure, which is purely algebraic in its nature, is closely related to certain Poisson (or, dually, pre-symplectic) structures. In the cases of double Bruhat cells and Grassmannians discussed below, the corresponding families of Poisson structures include, among others, standard R-matrix Poisson-Lie structures (or their push-forwards). A large part of the book is devoted to the interplay between cluster structures and Poisson/pre-symplectic structures. This leads, in particular, to revealing of cluster structure related to integrable systems called Toda lattices and to dynamical interpretation of cluster transformations, see the last chapter. Vice versa, Poisson/pre-symplectic structures turned out to be instrumental for the proof of purely algebraic results in the general theory of cluster algebras.

In Chapter 1 we introduce necessary notions and notation. Section 1.1 provides a very concise introduction to *flag varieties*, *Grassmannians* and *Plücker coordinates*. Section 1.2 treats *simple Lie algebras* and *groups*. Here we remind to the reader the standard objects and constructions used in Lie theory, including the *adjoint action*, the *Killing form*, *Cartan subalgebras*, *root systems*, and *Dynkin diagrams*. We discuss in some detail *Bruhat decompositions* of a simple Lie group, and *double Bruhat cells*, which feature prominently in the next Chapter. *Poisson–Lie groups* are introduced in Section 1.3. We start with providing basic definitions of Poisson geometry, and proceed to define main objects in Poisson–Lie theory, including the *classical R-matrix*. *Sklyanin brackets* are then defined as a particular example of an R-matrix Poisson bracket. Finally, we treat in some detail the case of the *standard Poisson–Lie structure* on a simple Lie group, which will play an important role in subsequent chapters.

Chapter 2 considers in detail two basic sources of cluster-like structures in rings of functions related to Schubert varieties: the homogeneous coordinate ring of the Grassmannian $G_2(m)$ of 2-dimensional planes and the ring of regular functions on a double Bruhat cell. The first of the two rings is studied in Section 2.1. We show that Plücker coordinates of $G_2(m)$ can be organized into a finite number of groups of the same cardinality (*clusters*), covering the set of all Plücker coordinates. Every cluster corresponds to a triangulation of a convex m-gon. The system of clusters has a natural graph structure, so that adjacent clusters differ exactly by two Plücker coordinates corresponding to a pair of crossing diagonals (and contained together in exactly one short Plücker relation). Moreover, this graph is a 1-skeleton of the Stasheff polytope of m-gon triangulations. We proceed to show that if one fixes an arbitrary cluster, any other Plücker coordinate can be expressed as a rational function in the Plücker coordinates entering this cluster. We prove that this rational function is a Laurent polynomial and find a geometric meaning for its numerator and denominator, see Proposition 2.1.

Section 2.2 starts with the formulation of Arnold's problem: find the number of connected components in the variety of real complete flags intersecting transversally a given pair of flags. We reformulate this problem as a problem of enumerating connected components in the intersection of two real open Schubert cells, and proceed to a more general problem of enumerating connected components of a real double Bruhat cell. It is proved that components in question are in a bijection with the orbits of a group generated by symplectic transvections in a vector space over the field $\mathbb{F}_2$, see Theorem 2.10. As one of the main ingredients of the proof, we provide a complete description of the ring of regular functions on the double Bruhat cell. It turns out that generators of this ring can be grouped into clusters, and that they satisfy Plücker-type exchange relations.

In Chapter 3 we introduce cluster algebras and prove two fundamental results about them. Section 3.1 contains basic definitions and examples. In this book we mainly concentrate on *cluster algebras of geometric type*, so the discussion in this Section is restricted to such algebras, and the case of general coefficients is only mentioned in Remark 3.13. We define basic notions of *cluster* and *stable variables*, *seeds*, *exchange relations*, *exchange matrices* and their *mutations*, *exchange graphs*, and provide extensive examples. The famous *Laurent phenomenon* is treated in Section 3.2. We prove both the general statement (Theorem 3.14) and its sharpening for cluster algebras of geometric type (Proposition 3.20). The second fundamental result, the classification of cluster algebras of *finite type*, is discussed in Section 3.3. We state the result (Theorem 3.26) as an equivalence of three conditions, and provide complete proofs for two of the three implications. The third implication is discussed only briefly, since a complete proof would require exploring intricate combinatorial properties of root systems for different Cartan–Killing types, which goes beyond the scope of this book. In Section 3.4 we discuss relations between cluster algebras and *rings of regular functions*, see Proposition 3.37. Finally, Section 3.5 contains a list of conjectures, some of which are treated in subsequent chapters.

Chapter 4 is central to the book. In Section 4.1 we introduce the notion of Poisson brackets *compatible* with a cluster algebra structure and provide a complete characterization of such brackets for cluster algebras with the extended exchange matrix of full rank, see Theorem 4.5. In this context, mutations of the exchange matrix are explained as transformations of the coefficient matrix of the compatible

Poisson bracket induced by a basis change. In Section 4.2 we apply this result to the study of Poisson and cluster algebra structures on Grassmannians. Starting from the Sklyanin bracket on SL_n, we define the corresponding Poisson bracket on the open Schubert cell in the Grassmannian $G_k(n)$ and construct the cluster algebra compatible with this Poisson bracket. It turns out that this cluster algebra is isomorphic to the ring of regular functions on the open cell in $G_k(n)$, see Theorem 4.14. We further investigate this construction and prove that an extension of the obtained cluster algebra is isomorphic to the homogeneous coordinate ring of the Grassmannian, see Theorem 4.17.

The smooth part of the spectrum of a cluster algebra is called the *cluster manifold* and is treated in Chapter 5. The definition of the cluster manifold is discussed in Section 5.1. In Section 5.2 we investigate the natural *Poisson toric action* on the cluster manifold and provide necessary and sufficient conditions for the extendability of the local toric action to the global one, see Lemma 5.3. We proceed to the enumeration of connected components of the regular locus of the toric action in Section 5.3 and extend Theorem 2.10 to this situation, see Theorem 5.9. In Section 5.4 we study the structure of the regular locus and prove that it is foliated into disjoint union of generic symplectic leaves of the compatible Poisson bracket, see Theorem 5.12. In Section 5.5 we apply these results to the enumeration of connected components in the intersection of n Schubert cells in general position in $G_k(n)$, see Theorem 5.15.

Note that compatible Poisson brackets are defined in Chapter 4 only for cluster algebras with the extended exchange matrix of a full rank. To overcome this restriction, we present in Chapter 6 a dual approach based on *pre-symplectic* rather than on Poisson structures. In Section 6.1 we define closed 2-forms *compatible* with a cluster algebra structure and provide a complete characterization of such forms parallel to Theorem 4.5, see Theorem 6.2. Further, we define the *secondary cluster manifold* and a compatible symplectic form on it, which we call the *Weil–Petersson form* associated to the cluster algebra. The reason for such a name is explained in Section 6.2, which treats our main example, the *Teichmüller space*. We briefly discuss *Penner coordinates* on the decorated Teichmüller space defined by fixing a triangulation of a Riemann surface Σ, and observing that Ptolemy relations for these coordinates can be considered as exchange relations. The secondary cluster manifold for the cluster algebra $\mathcal{A}(\Sigma)$ arising in this way is the Teichmüller space, and the Weil–Petersson form associated with this cluster algebra coincides with the classical Weil–Petersson form corresponding to Σ, see Theorem 6.6. We proceed with providing a geometric meaning for the degrees of variables in the denominators of Laurent polynomials expressing arbitrary cluster variables in terms of the initial cluster, see Theorem 6.7; Proposition 2.1 can be considered as a toy version of this result. Finally, we give a geometric description of $\mathcal{A}(\Sigma)$ in terms of triangulation equipped with a *spin*, see Theorem 6.9. In Section 6.3 we derive an axiomatic approach to exchange relations. We show that the class of transformations satisfying a number of natural conditions, including the compatibility with a closed 2-form, is very restricted, and that exchange transformations used in cluster algebras are simplest representatives of this class, see Theorem 6.11.

In Chapter 7 we apply the results of previous chapters to prove several conjectures about *exchange graphs* of cluster algebras listed in Section 3.5. The dependence of a general cluster algebra on the coefficients is investigated in Section 7.1.

We prove that the exchange graph of the cluster algebra with principal coefficients covers the exchange graph of any other cluster algebra with the same exchange matrix, see Theorem 7.1. In Section 7.2 we consider vertices and edges of an exchange graph and prove that distinct seeds have distinct clusters for cluster algebras of geometric type and for cluster algebras with arbitrary coefficients and a non-degenerate exchange matrix, see Theorem 7.4. Besides, we prove that if a cluster algebra has the above cluster-defines-seed property, then adjacent vertices of its exchange graph are clusters that differ only in one variable, see Theorem 7.5. Finally, in Section 7.3 we prove that the exchange graph of a cluster algebra with a non-degenerate exchange matrix does not depend on coefficients, see Theorem 7.7.

In the remaining three chapters we develop an approach to the interaction of Poisson and cluster structures based on the study of *perfect networks*—directed networks on surfaces with boundary having trivalent interior vertices and univalent boundary vertices. Perfect planar networks in the *disk* are treated in Chapter 8. Main definitions, including *weights* and *boundary measurements*, are given in Section 8.1. We prove that each boundary measurement is a rational function of the edge weights, see Proposition 8.3, and define the *boundary measurement map* from the space of edge weights of the network to the space of $k \times m$ matrices, where k and m are the numbers of sources and sinks. Section 8.2 is central to this chapter; it treats Poisson structures on the space of edge weights of the network and Poisson structures on the space of $k \times m$ matrices induced by the boundary measurement map. The former are defined axiomatically as satisfying certain natural conditions, including an analog of the Poisson–Lie property for groups. It is proved that such structures form a 6-parametric family, see Proposition 8.5. For fixed sets of sources and sinks, the induced Poisson structures on the space of $k \times m$ matrices form a 2-parametric family that does not depend on the internal structure of the network, see Theorem 8.6. Explicit expressions for this 2-parametric family are provided in Theorem 8.7. In case $k = m$ and separated sources and sinks, we recover on $k \times k$ matrices the Sklyanin bracket corresponding to a 2-parametric family of classical R-matrices including the standard R-matrix, see Theorem 8.10. In Section 8.3 we extend the boundary measurement map to the Grassmannian $G_k(k+m)$ and prove that the obtained Grassmannian boundary measurement map induces a 2-parametric family of Poisson structures on $G_k(k+m)$ that does not depend on the particular choice of k sources and m sinks; moreover, for any such choice, the family of Poisson structures on $k \times m$ matrices representing the corresponding cell of $G_k(k+m)$ coincides with the family described in Theorem 8.7 (see Theorem 8.12 for details). Next, we give an interpretation of the natural GL_{k+m} action on $G_k(k+m)$ in terms of networks and establish that every member of the above 2-parametric family of Poisson structures on $G_k(k+m)$ makes it into a Poisson homogeneous space of GL_{k+m} equipped with the Sklyanin R-matrix bracket, see Theorem 8.17. Finally, in Section 8.4 we prove that each bracket in this family is compatible with the cluster algebra constructed in Chapter 4, see Theorem 8.20. An important ingredient of the proof is the use of face weights instead of edge weights.

In Chapter 9 we extend the constructions of the previous chapter to perfect networks in an *annulus*. Section 9.1.1 is parallel to Section 8.1; the main difference is that in order to define boundary measurements we have to introduce an auxiliary parameter λ that counts intersections of paths in the network with a *cut* whose endpoints belong to distinct boundary circles. We prove that boundary measurements are rational functions in the edge weights and λ, see Corollary 9.3 and study how they depend on the choice of the cut. Besides, we provide a *cohomological description* of the space of face and trail weights, which is a higher analog of the space of edge weights used in the previous chapter. Poisson properties of the obtained boundary measurement map from the space of edge weights to the space of rational $k \times m$ matrix functions in one variable are treated in Section 9.2. We prove an analog of Theorem 8.6, saying that for fixed sets of sources and sinks, the induced Poisson structures on the space of matrix functions form a 2-parametric family that does not depend on the internal structure of the network, see Theorem 9.4. Explicit expressions for this family are much more complicated than those for the disk, see Proposition 9.6. The proof itself differs substantially from the proof of Theorem 8.6. It relies on the fact that any rational $k \times m$ matrix function belongs to the image of the boundary measurement map for an appropriated perfect network, see Theorem 9.10. The section is concluded with Theorem 9.15 claiming that for a specific choice of sinks and sources one can recover the Sklyanin bracket corresponding to the trigonometric R-matrix. In Section 9.3 we extend these results to the *Grassmannian boundary measurement map* from the space of edge weights to the space of Grassmannian loops. We define the *path reversal map* and prove that this map commutes with the Grassmannian boundary measurement map, see Theorem 9.17. Further, we prove Theorem 9.22, which is a natural extension of Theorem 8.12; once again, the proof is very different and is based on path reversal techniques and the use of face weights.

In the concluding Chapter 10 we apply techniques developed in the previous chapter to providing a cluster interpretation of *generalized Bäcklund–Darboux transformations* for *Coxeter–Toda lattices*. Section 10.1 gives an overview of the chapter and contains brief introductory information on Toda lattices, *Weyl functions* of the corresponding *Lax operators* and generalized Bäcklund–Darboux transformations between phase spaces of different lattices preserving the Weyl function. A *Coxeter double Bruhat cell* in GL_n is defined by a pair of Coxeter elements in the symmetric group S_n. In Section 10.2 we have collected and proved all the necessary technical facts about such cells and the representation of their elements via perfect networks in a disk. Section 10.3 treats the inverse problem of restoring factorization parameters of an element of a Coxeter double Bruhat cell from its Weyl function. We provide an explicit solution for this problem involving *Hankel determinants* in the coefficients of the Laurent expansion for the Weyl function, see Theorem 10.9. In Section 10.4 we build and investigate a cluster algebra on a certain space $\mathcal{R}_n$ of rational functions related to the space of Weyl functions corresponding to Coxeter double Bruhat cells. We start from defining a perfect network in an annulus corresponding to the network in a disk studied in Section 10.2. The space of face weights of this network is equipped with a particular Poisson bracket from the family studied in Chapter 9. We proceed by using results of Chapter 4 to build a cluster algebra of rank $2n-2$ compatible with this Poisson bracket. Theorem 10.27 claims that this cluster algebra does not depend on the choice of the pair of Coxeter elements, and that the ring of regular functions on $\mathcal{R}_n$ is isomorphic to the localization of this cluster algebra with respect to the stable variables. In Section 10.5 we use these results to characterize generalized Bäcklund–Darboux transformations as sequences of cluster transformations in the above cluster algebra conjugated by by a certain map defined by the solution of the inverse problem in Section 10.3, see Theorem 10.36.

We also show how one can interpret generalized Bäcklund–Darboux transformations via equivalent transformations of the corresponding perfect networks in an annulus. In conclusion, we explain that classical Darboux transformations can be also interpreted via cluster transformations, see Proposition 10.39.

The process of writing spread over several years. We would like to thank all individuals and institutions that supported us in this undertaking. The project was started in Summer 2005 during our visit to Mathematisches Forschunginstitut Oberwolfach in the framework of Research in Pairs program. Since then we continued to work on the book during our joint visits to Institut des Hautes Etudes Scientifiques (Bures-sur-Yvette), Institut Mittag-Leffler (Djursholm), Max-Planck-Institut fur Mathematik (Bonn), University of Stockholm, Royal School of Technology (Stockholm), University of Haifa, Michigan State University, and University of Michigan. We thank all these institutions for warm hospitality and excellent working conditions. During the process of writing, we had numerous discussions with A. Berenstein, L. Chekhov, V. Fock, S. Fomin, R. Kulkarni, A. Postnikov, N. Reshetikhin, J. Scott, B. Shapiro, M. Yakimov, A. Zelevinsky and others that helped us to understand many different facets of cluster algebras. It is our pleasure to thank all of them for their help.

CHAPTER 1

Preliminaries

In this chapter we collect necessary terms and notation that will be used throughout the book.

1.1. Flag manifolds, Grassmannians, Plücker coordinates and Plücker relations

1.1.1. Recall that the *Grassmannian* $G_k(m)$ is the set of all k-dimensional subspaces in an m-dimensional vector space V over a field $\mathbb{F}$. Any k-dimensional subspace W of V can be described by a choice of k independent vectors $w_1, \dots, w_k \in W$, and, vice versa, any choice of k independent vectors $w_1, \dots, w_k \in V$ determines a k-dimensional subspace $W = \operatorname{span}\{w_1, \dots, w_k\}$. Fix a basis in V. Any k independent vectors in V give rise to a $k \times m$ matrix of rank k whose rows are coordinates of these vectors. Hence any element X of the Grassmannian can be represented (non-uniquely) as a $k \times m$ matrix $\bar{X}$ of rank k. Two $k \times m$ matrices represent the same element of the Grassmannian $G_k(m)$ if one of them can be obtained from the other one by the left multiplication with a nondegenerate $k \times k$ matrix.

Let I be a k-element subset of $[1, m] = \{1, \dots, m\}$. The *Plücker coordinate* x_I is a function on the set of $k \times m$ matrices which is equal to the value of the minor formed by the columns of the matrix indexed by the elements of I. Note that for any $k \times m$ matrix M and any nondegenerate $k \times k$ matrix A, the Plücker coordinate $x_I(AM)$ is equal to $\det A \cdot x_I(M)$. The *Plücker embedding* is an embedding of the Grassmannian into the projective space of dimension $\binom{m}{k} - 1$ that sends $X \in G_k(m)$ to a point with homogeneous coordinates

$$x_{1,2,\dots,k}(\bar{X}) : \cdots : x_{i_1,\dots,i_k}(\bar{X}) : \cdots : x_{m-k+1,m-k+2,\dots,m}(\bar{X}).$$

Plücker coordinates are subject to quadratic constraints called *Plücker relations.* Fix two k-element subsets of the set $[1, m]$, $I = \{i_1, \dots, i_k\}$ and $J = \{j_1, \dots, j_k\}$. In what follows, we will sometimes use notation

$$I(i_\alpha \to l) = \{i_1, \dots, i_{\alpha-1}, l, i_{\alpha+1}, \dots, i_k\} \tag{1.1}$$

for $\alpha \in [1, k]$ and $l \in [1, m]$.

Then for every I, J and α Plücker coordinates satisfy a quadratic equation

$$x_I x_J = \sum_{\beta=1}^{k} x_{I(i_\alpha \to j_\beta)} x_{J(j_\beta \to i_\alpha)}. \tag{1.2}$$

If I, J are of the form $I = \{I', i, j\}$, $J = \{I', p, q\}$, where I' is a $(k-2)$-subset of $[1, m]$ and i, j, p, q are distinct indices not contained in I', then (1.2) reduces to a 3-term relation

$$x_{I'ij} x_{I'pq} = x_{I'pj} x_{I'iq} + x_{I'qj} x_{I'pi} \tag{1.3}$$

called the *short Plücker relation*, which will be of special interest to us.

The Plücker embedding shows that the Grassmannian $G_k(m)$ is a smooth algebraic projective variety. Its *homogeneous coordinate ring* $\mathbb{C}[G_k(m)]$ is the quotient of the ring of polynomials in Plücker coordinates by the ideal of homogeneous polynomials vanishing on $G_k(m)$. The latter ideal is generated by Plücker relations.

1.1.2. Grassmannians $G_k(m)$ are particular examples of *(partial) flag varieties*. More generally, the *(partial) flag variety* $\mathrm{Fl}(V, \mathbf{d})$ associated to the data $\mathbf{d} = (d_1, \ldots, d_n)$ with $1 \le d_1 < \cdots < d_n \le m = \dim V$ is defined as the set of all *flags* $f = (\{0\} \subset f^1 \subset \cdots \subset f^n)$ of linear subspaces of V with $\dim f^j = d_j$; it can be naturally embedded into the product of Grassmannians $G_{d_1}(m) \times \cdots \times G_{d_n}(m)$. The image of this embedding is closed, thus $\mathrm{Fl}(V, \mathbf{d})$ is a projective variety. If $\mathbf{d} = (1, 2, \ldots, m-1)$, then $\mathrm{Fl}(V, \mathbf{d})$ is called the *complete flag variety* and denoted $\mathrm{Fl}(V)$.

Let us choose a basis $(e_1, \ldots, e_m)$ in V and a *standard* flag

$$f_0(\mathbf{d}) = \{\{0\} \subset \mathrm{span}\{e_1, \ldots, e_{d_1}\} \subset \cdots \subset \mathrm{span}\{e_1, \ldots, e_{d_n}\}\}$$

in $\mathrm{Fl}(V, \mathbf{d})$. The group $SL(V)$ acts transitively on $\mathrm{Fl}(V, \mathbf{d})$. We can use the chosen basis to identify V with $\mathbb{F}^m$. The stabilizer of the point $f_0(\mathbf{d})$ under this action is a parabolic subgroup $\mathcal{P} \subset SL_m(\mathbb{F})$ of elements of the form

$$\begin{pmatrix} A_1 & \star & \cdots & \star \\ 0 & A_2 & \ddots & \vdots \\ \vdots & \ddots & \ddots & \star \\ 0 & \cdots & 0 & A_n \end{pmatrix},$$

where A_i are $d_i \times d_i$ invertible matrices. Thus, as a homogeneous space, $\mathrm{Fl}(V, \mathbf{d}) \cong SL_m(\mathbb{F})/\mathcal{P}$. In particular, the complete flag variety is isomorphic to

$$\mathrm{Fl}_m(\mathbb{F}) := SL_m(\mathbb{F})/\mathcal{B}_+,$$

where $\mathcal{B}_+$ is the subgroup of invertible upper triangular matrices over $\mathbb{F}$.

Two complete flags f and g in $\mathrm{Fl}_m(\mathbb{F})$ are called *transversal* if

$$\dim(f^i \cap g^j) = \max(i + j - m, 0)$$

for any $i, j \in [1, m]$. When two flags are transversal, we often say that they are in *general position*.

Fix two transversal flags f and g and consider the set $U^m_{f,g}(\mathbb{F}) \subset \mathrm{Fl}_m(\mathbb{F})$ of all flags that are transversal to both f and g. The set $U^m_{f,g}(\mathbb{R})$ serves as one of our main motivational examples and will be studied in great detail in the next chapter. For now, note that, since $SL_m(\mathbb{F})$ acts transitively on the set of pairs of transverse flags, we may assume without loss of generality that $g^j = \mathrm{span}\{e_1, \ldots, e_j\}$ and $f^i = \mathrm{span}\{e_{m-i+1}, \ldots, e_m\}$. Then any flag h transversal to f can be naturally identified with a unipotent upper triangular matrix H whose first i rows span the subspace h^i. In order for h to be also transversal to g, the matrix H must satisfy an additional condition: for every $1 \le k \le m$, the minor of H formed by the first k rows and the last k columns must be non-zero.

1.2. Simple Lie algebras and groups

1.2.1. Let $\mathcal{G}$ be a connected simply-connected complex Lie group with an identity element e and let $\mathfrak{g} \cong T_e\mathcal{G}$ be the corresponding Lie algebra. We will often use notation $\mathcal{G} = \exp \mathfrak{g}$ and $\mathfrak{g} = \mathrm{Lie}\,\mathcal{G}$. The dual space to $\mathfrak{g}$ will be denoted by $\mathfrak{g}^*$ and the value of $\ell \in \mathfrak{g}^*$ at $\xi \in \mathfrak{g}$ will be denoted by $\langle \ell, \xi \rangle$.

Recall the definitions of the adjoint (resp. co-adjoint) actions of $\mathcal{G}$ and $\mathfrak{g}$ on $\mathfrak{g}$ (resp. $\mathfrak{g}^*$):

$$\mathrm{Ad}_x \, \xi = \frac{d}{dt} \left(x^{-1} \exp(\xi t) x \right) \Big|_{t=0}, \qquad \mathrm{ad}_\eta \, \xi = [\eta, \xi]$$

and

$$\langle \mathrm{Ad}^*_x \, \ell, \xi \rangle = \langle \ell, \mathrm{Ad}_x \, \xi \rangle, \qquad \langle \mathrm{ad}^*_\eta \, \ell, \xi \rangle = \langle \ell, \mathrm{ad}_\eta \, \xi \rangle$$

for any $\xi \in \mathfrak{g}$.

The *Killing form* on $\mathfrak{g}$ is a bilinear symmetric form $(\cdot,\cdot)_{\mathfrak{g}}$ defined by

$$(\xi, \eta)_{\mathfrak{g}} = \mathrm{tr}(\mathrm{ad}_\xi \, \mathrm{ad}_\eta), \tag{1.4}$$

where $\mathrm{ad}_\xi \, \mathrm{ad}_\eta$ is viewed as a linear operator acting in the vector space $\mathfrak{g}$.

1.2.2. Let now $\mathfrak{g}$ be a complex simple Lie algebra of rank r with a Cartan subalgebra $\mathfrak{h} = \mathfrak{a} \oplus i\mathfrak{a}$. Recall that $\mathfrak{h}$ is a maximal commutative subalgebra of $\mathfrak{g}$ of dimension r, and that the adjoint action of $\mathfrak{h}$ on $\mathfrak{g}$ can be diagonalized. More precisely, for $\alpha \in \mathfrak{h}^*$, define

$$\mathfrak{g}_\alpha = \{\xi \in \mathfrak{g} \; : \; [h, \xi] = \alpha(h)\xi \quad \text{for any } h \in \mathfrak{h}\}.$$

Clearly, $\mathfrak{g}_0 = \mathfrak{h}$. A nonzero α such that $\mathfrak{g}_\alpha \neq 0$ is called a *root* of $\mathfrak{g}$, and a collection Φ of all roots is called the *root system* of $\mathfrak{g}$. Then:

(i) For any $\alpha \in \Phi$, $\dim \mathfrak{g}_\alpha = 1$, and

(ii) $\mathfrak{g}$ has a direct sum decomposition

$$\mathfrak{g} = \mathfrak{h} \oplus \left(\oplus_{\alpha \in \Phi} \mathfrak{g}_\alpha \right),$$

which is graded by Φ, that is, $[\mathfrak{g}_\alpha, \mathfrak{g}_\beta] = \mathfrak{g}_{\alpha+\beta}$, where the right hand side is zero if $\alpha + \beta$ is not a root.

The Killing form (1.4) is nondegenerate on the simple algebra $\mathfrak{g}$, and so is its restriction to $\mathfrak{h}$, which we will denote by $(\cdot,\cdot)$. Thus $\mathfrak{h}^*$ and $\mathfrak{h}$ can be identified via $(\cdot,\cdot)$:

$$\mathfrak{h}^* \ni \alpha \;\mapsto\; h_\alpha \in \mathfrak{h} \; : \alpha(h) = (h_\alpha, h)$$

for all $h \in \mathfrak{h}$. Moreover, the restriction of $(\cdot,\cdot)$ to the *real* vector space spanned by $h_\alpha, \alpha \in \Phi$ is real-valued and positive definite. Thus, the real vector space $\mathfrak{h}_0^*$ spanned by $\alpha \in \Phi$ is a real Euclidean vector space with the inner product $(\cdot,\cdot)$. In particular, a reflection with respect to the hyperplane normal to $\alpha \in \mathfrak{h}_0^*$ can be defined by

$$s_\alpha \beta = \beta - \langle \beta | \alpha \rangle \alpha, \tag{1.5}$$

where $\langle \beta | \alpha \rangle = 2\frac{(\beta,\alpha)}{(\alpha,\alpha)}$.

We can now summarize the properties of the root system Φ:

(i) Φ contains finitely many nonzero vectors that span $\mathfrak{h}_0^*$;

(ii) for any $\alpha \in \Phi$, the only multiples of $\alpha \in \Phi$ are $\pm\alpha$;

(iii) Φ is invariant under reflections s_α, $\alpha \in \Phi$;

(iv) for any $\alpha, \beta \in \Phi$, the number $\langle \beta | \alpha \rangle$ is an integer.

The second property above allows one to choose a *polarization* of Φ, i.e., a decomposition $\Phi = \Phi_+ \cup \Phi_-$, where $\Phi_- = -\Phi_+$. Elements of Φ_+ (resp. Φ_-) are called *positive* (resp. *negative*) roots. The set $\Pi = \{\alpha_1, \dots, \alpha_r\}$ of *positive simple roots* is a maximal collection of positive roots that can not be represented as a positive linear combination of other roots. Every positive root is a nonnegative linear combination of elements of Π with integer coefficients. The sum of these coefficients is called the *height* of the root.

The *Cartan matrix* of $\mathfrak{g}$ is defined by

$$A = (a_{ij})_{i,j=1}^r = (\langle \alpha_j | \alpha_i \rangle)_{i,j=1}^r$$

and is characterized by the following properties:

(i) A is an integral matrix with 2's on the diagonal and non-positive off-diagonal entries.

(ii) A is symmetrizable, i.e. there exists an invertible diagonal matrix D (e.g. $D = \operatorname{diag}((\alpha_i, \alpha_i))_{i=1}^r$) such that DA is symmetric. In particular, A is *sign-symmetric*, that is, $a_{ij}a_{ji} \geq 0$ and $a_{ij}a_{ji} = 0$ implies that both a_{ij} and a_{ji} are zero.

(iii) DA defined as above is positive definite. This implies, in particular, that $a_{ij}a_{ji} < 3$.

The information contained in the Cartan matrix can also be encoded in the corresponding *Dynkin diagram*, defined as a graph on r vertices with ith and jth vertices joined by an edge of multiplicity $a_{ij}a_{ji}$. If $a_{ij}a_{ji} > 1$ then an arrow is added to the edge. It points to j if $(\alpha_i, \alpha_i) > (\alpha_j, \alpha_j)$.

A simple Lie algebra is determined, up to an isomorphism, by its Cartan matrix (or, equivalently, its Dynkin diagram). Namely, it has a presentation with the so-called *Chevalley generators* $\{h_i, e_{\pm i}\}_{i=1}^r$ and relations

$$[h_i, h_j] = 0, \quad [h_i, e_{\pm j}] = \pm c_{ji} e_{\pm j}, \quad [e_j, e_{-j}] = \delta_{ij} h_j, \tag{1.6}$$

and

$$\operatorname{ad}_{e_{\pm i}}^{1-a_{ij}} e_{\pm j} = 0.$$

The latter are called the *Serre relations.* In terms of positive simple roots and Chevalley generators, the elements of the Cartan matrix can be rewritten as $a_{ij} = \alpha_j(h_i)$, $i, j \in [1, r]$.

The classification theorem for simple Lie algebras states that every Dynkin diagram is of one of the types described in Fig. 1.1.

It is clear from relations (1.6) that for every $i \in [1, r]$, one can construct an embedding ρ_i of the Lie algebra sl_2 into $\mathfrak{g}$ by assigning

$$\rho_i \begin{pmatrix} 1 & 0 \\ 0 & 1 \end{pmatrix} = h_i, \quad \rho_i \begin{pmatrix} 0 & 1 \\ 0 & 0 \end{pmatrix} = e_i, \quad \rho_i \begin{pmatrix} 0 & 0 \\ 1 & 0 \end{pmatrix} = e_{-i},$$

which can be integrated to an embedding (also denoted by ρ_i) of the group SL_2 into the group $\mathcal{G}$. In particular,

$$\rho_i \begin{pmatrix} 1 & t \\ 0 & 1 \end{pmatrix} = \exp(te_i), \quad \rho_i \begin{pmatrix} 1 & 0 \\ t & 1 \end{pmatrix} = \exp(te_{-i}) \tag{1.7}$$

for any $t \in \mathbb{C}$. One-parameter subgroups

$$x_i^{\pm}(t) = \exp(te_{\pm i}) \tag{1.8}$$

generated by $e_{\pm i}$ will play an important role in what follows.

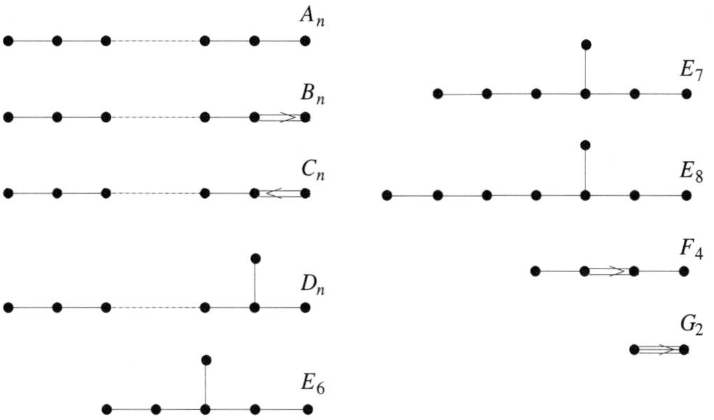

FIGURE 1.1. The list of Dynkin diagrams

1.2.3. The *Borel* (resp., *opposite Borel*) subalgebra $\mathfrak{b}_+$ (resp., $\mathfrak{b}_-$) of $\mathfrak{g}$ relative to $\mathfrak{h}$ is spanned by $\mathfrak{h}$ and vectors e_α, $\alpha \in \Phi_+$ (resp., e_α, $\alpha \in \Phi_-$). Let $\mathfrak{n}_+ = [\mathfrak{b}_+, \mathfrak{b}_+]$ and $\mathfrak{n}_- = [\mathfrak{b}_-, \mathfrak{b}_-]$ be the corresponding maximal nilpotent subalgebras of $\mathfrak{g}$. The connected subgroups that correspond to $\mathfrak{g}$, $\mathfrak{h}$, $\mathfrak{b}_+$, $\mathfrak{b}_-$, $\mathfrak{n}_+$, $\mathfrak{n}_-$ will be denoted by $\mathcal{G}$, $\mathcal{H}$, $\mathcal{B}_+$, $\mathcal{B}_-$, $\mathcal{N}_+$, $\mathcal{N}_-$.

The *weight lattice* P consists of elements $\gamma \in \mathfrak{h}^*$ such that $\gamma(h_i) \in \mathbb{Z}$ for $i \in [1, r]$. The basis of *fundamental weights* $\omega_1, \dots, \omega_r$ in P is defined by $\omega_j(h_i) = \langle \omega_j | \alpha_i \rangle = \delta_{ij}$, $i, j \in [1, r]$. Every weight $\gamma \in P$ defines a multiplicative character $a \mapsto a^\gamma$ of $\mathcal{H}$ as follows: let $a = \exp(h)$ for some $h \in \mathfrak{h}$, then $a^\gamma = e^{\gamma(h)}$. Positive simple roots, fundamental weights and entries of the Cartan matrix are related via

$$\alpha_i = \sum_{j=1}^{r} a_{ji}\omega_j = 2\omega_i + \sum_{\substack{j=1 \\ j \neq i}}^{r} a_{ji}\omega_j. \tag{1.9}$$

The *Weyl group* W is defined as a quotient $W = \mathrm{Norm}_{\mathcal{G}}\mathcal{H}/\mathcal{H}$. For each of the embeddings ρ_i: $SL_2 \to \mathcal{G}$, $i \in [1, r]$, defined in (1.7), put

$$\bar{s}_i = \rho_i \begin{pmatrix} 0 & -1 \\ 1 & 0 \end{pmatrix} \in \mathrm{Norm}_{\mathcal{G}}\,\mathcal{H}$$

and $s_i = \bar{s}_i\mathcal{H} \in W$. All s_i are of order 2, and together they generate W. We can identify s_i with the *elementary reflection* s_{α_i}, $\alpha_i \in \Pi$, in $\mathfrak{h}_0^*$ defined by (1.5). Consequently, W can be identified with the *Coxeter group* generated by reflections $s_1, \dots, s_r$. The Weyl group acts on the weight lattice. For any $w \in W$ and $\gamma \in P$ the action $\gamma \mapsto w\gamma$ is defined by expanding w into a product of elementary reflections and applying the reflections one by one to γ.

A *reduced decomposition* of an element $w \in W$ is a representation of w as a product $w = s_{i_1} \cdots s_{i_l}$ of the smallest possible length. A reduced decomposition is not unique, but the number l of reflections in the product depends only on w and is called the *length* of w and denoted by $l(w)$. The sequence of indices $(i_1, \dots, i_l)$ that corresponds to a given reduced decomposition of w is called a *reduced word* for w. The unique element of W of maximal length (also called *the longest element* of W) will be denoted by w_0. Braid relations for $\bar{s}_i$ guarantee that the representative $\bar{w} \in \mathrm{Norm}_{\mathcal{G}}\,\mathcal{H}$ can be unambiguously defined for any $w \in W$ by requiring that

$\overline{uv} = \bar{u}\bar{v}$ whenever $l(uv) = l(u)+l(v)$. In what follows we will slightly abuse notation by denoting an element of the Weyl group and its representative in $\mathrm{Norm}_{\mathcal{G}} \mathcal{H}$ with the same letter.

We will also need a notion of a reduced word for an ordered pair (u, v) of elements in W. It is defined as follows: if $(i_1, \dots, i_{l(u)})$ is a reduced word for u and $(i'_1, \dots, i'_{l(v)})$ is a reduced word for v, then any shuffle of sequences $(i_1, \dots, i_{l(u)})$ and $(-i'_1, \dots, -i'_{l(v)})$ is called a reduced word for (u, v).

Every $\xi \in \mathfrak{g}$ can be uniquely decomposed

$$\xi = \xi_- + \xi_0 + \xi_+, \tag{1.10}$$

where $\xi_+ \in \mathfrak{n}_+$, $\xi_- \in \mathfrak{n}_-$ and $\xi_0 \in \mathfrak{h}$. Consequently, for every x in an open Zarisky dense subset

$$\mathcal{G}^0 = \mathcal{N}_- \mathcal{H} \mathcal{N}_+$$

of $\mathcal{G}$ there exists a unique *Gauss factorization*

$$x = x_- x_0 x_+, \quad x_+ \in \mathcal{N}_+, \ x_- \in \mathcal{N}_-, \ x_0 \in \mathcal{H}.$$

For any $x \in \mathcal{G}^0$ and a fundamental weight ω_i define

$$\Delta_i(x) = x_0^{\omega_i};$$

this function can be extended to a regular function on the whole $\mathcal{G}$. In particular, for $\mathcal{G} = SL_{r+1}$, $\Delta_i(x)$ is just the principal $i \times i$ minor of a matrix x. For any pair $u, v \in W$, the corresponding *generalized minor* is a regular function on $\mathcal{G}$ given by

$$\Delta_{u\omega_i, v\omega_i}(x) = \Delta_i(u^{-1} x v). \tag{1.11}$$

These functions depend only on the weights $u\omega_i$ and $v\omega_i$, and do not depend on the particular choice of u and v.

1.2.4. The *Bruhat decompositions* of $\mathcal{G}$ with respect to $\mathcal{B}_+$ and $\mathcal{B}_-$ are defined, resp., by

$$\mathcal{G} = \cup_{u \in W} \mathcal{B}_+ u \mathcal{B}_+, \qquad \mathcal{G} = \cup_{v \in W} \mathcal{B}_- v \mathcal{B}_-.$$

The sets $\mathcal{B}_+ u \mathcal{B}_+$ (resp. $\mathcal{B}_- v \mathcal{B}_-$) are disjoint. They are called *Bruhat cells* (resp. *opposite Bruhat cells*). The *generalized flag variety* is defined as a homogeneous space

$$\mathcal{G}/\mathcal{B}_+ = \cup_{u \in W} \mathcal{B}_+ u.$$

Clearly, this is a generalization of the complete flag variety in a finite dimensional vector space.

Finally, for any $u, v \in W$, the *double Bruhat cell* is defined as

$$\mathcal{G}^{u,v} = \mathcal{B}_+ u \mathcal{B}_+ \cap \mathcal{B}_- v \mathcal{B}_-,$$

and the *reduced double Bruhat cell*, as

$$L^{u,v} = \mathcal{N}_+ u \mathcal{N}_+ \cap \mathcal{B}_- v \mathcal{B}_-.$$

Double Bruhat cells and reduced double Bruhat cells will be the subject of our discussion in the next chapter. Here we will only mention that the variety $\mathcal{G}^{u,v}$ is biregularly isomorphic to a Zariski open subset of $\mathbb{C}^{r+l(u)+l(v)}$, where $r = \operatorname{rank} \mathcal{G}$. A corresponding birational map from $\mathbb{C}^{r+l(u)+l(v)}$ to $\mathcal{G}^{u,v}$ can be constructed quite explicitly, though not in a unique way.

Namely, let $\mathbf{j} = (j_1, \dots, j_{l(u)+l(v)+r})$ be a shuffle of a reduced word for (u, v) and any re-arrangement of numbers $\mathrm{i}, \dots, \mathrm{i}r$ with $\mathrm{i}^2 = -1$; we define $\theta(l) = +$ if

$j_l > 0$, $\theta(l) = -$ if $j_l < 0$ and $\theta(l) = 0$ if $j_l \in i\mathbb{Z}$. Besides, for $i \in [1, r]$ and $t \in \mathbb{C}$, put

$$x_i^0(t) = \rho_i(\operatorname{diag}(t, t^{-1})).$$

Then one defines a map $x_{\mathbf{j}}\colon \mathbb{C}^{l(u)+l(v)+r} \to \mathcal{G}^{u,v}$ by

$$x_{\mathbf{j}}(\mathbf{t}) = \prod_{l=1}^{l(u)+l(v)+r} x_{|j_l|}^{\theta(l)}(t_l), \tag{1.12}$$

where elements $x_i^{\pm}(t)$ are defined as in (1.8). This map is a biregular isomorphism between $(\mathbb{C}^*)^{l(u)+l(v)+r}$ and a Zariski open subset of $\mathcal{G}^{u,v}$. Parameters $t_1, \dots, t_{l(u)+l(v)+r}$ constituting $\mathbf{t}$ are called *factorization parameters.* Note that setting $t_l = 1$ in the definition of $x_{\mathbf{j}}$ for all l such that $\theta(l) = 0$, one gets a parametrization for a Zariski open set in $L^{u,v}$, and so $L^{u,v}$ is biregularly isomorphic to a Zariski open subset in $\mathbb{C}^{l(u)+l(v)}$. In this case, we replace the factorization map above by

$$x_{\mathbf{j}}(\mathbf{t}) = \prod_{l=1}^{l(u)+l(v)} x_{|j_l|}^{\theta(l)}(t_l), \tag{1.13}$$

where now $\mathbf{j}$ is a reduced word for (u, v) and $\mathbf{t} = (t_1, \dots, t_{l(u)+l(v)})$. We retain notations $\mathbf{j}$ and $\mathbf{t}$, since it will be always clear from context which of the factorizations (1.12), (1.13) is being used.

REMARK 1.1. It will be sometimes convenient to interpret θ as taking values ± 1 instead of $\pm$. The proper interpretation will be clear from the context.

It is possible to find explicit formulae for the inverse of the map (1.12) in terms of the so-called *twisted generalized minors*. We will not review them here, but instead conclude the subsection with an example.

EXAMPLE 1.2. Let $\mathcal{G} = SL_n$ and $(u, v) = (w_0, w_0)$, where w_0 is the longest element of the Weyl group S_n. Then $\mathcal{G}^{w_0,w_0}$ is open and dense in SL_n and consists of elements such that all minors formed by several first rows and last columns as well as all minors formed by several last rows and first columns are nonzero. A generic element of $\mathcal{G}^{w_0,w_0}$ has nonzero leading principal minors, and by right multiplying by an invertible diagonal matrix D^{-1} can be reduced to a matrix $A \in \mathcal{G}^{w_0,w_0}$ with all leading principal minors equal to 1. Subtracting a multiple of the $(n-1)$st row of A from its nth row we can ensure that the $(n, 1)$-entry of the resulting matrix is 0, while (generically) all other entries are not. Applying similar elementary transformations to rows $n-2$ and $n-1$, then $n-3$ and $n-2$, and so on up to 1 and 2, we obtain a matrix with all off-diagonal entries in the first column equal to 0. Starting from the bottom row again and proceeding in a similar manner, we will eventually reduce A to a unipotent upper triangular matrix B. Applying (in the same order) the corresponding column transformations to B, we will eventually reduce it to the identity matrix. If one now translates operations above into the matrix multiplication, they can be summarized as a factorization of A into

$$A = (\mathbf{1} + t_1 e_{n,n-1}) \cdots (\mathbf{1} + t_{n-1} e_{2,1})(\mathbf{1} + t_n e_{n,n-1}) \cdots (\mathbf{1} + t_{2n-3} e_{3,2}) \cdots$$
$$(\mathbf{1} + t_{\frac{n(n-1)}{2}} e_{n,n-1})(\mathbf{1} + t_{\frac{n(n-1)}{2}+1} e_{n-1,n}) \cdots (\mathbf{1} + t_{n(n-1)} e_{n-1,n}).$$

Furthermore, denote $m = n(n-1)$ and factor D as

$$D = \operatorname{diag}(t_{m+1}, t_{m+1}^{-1}, 1, \dots, 1) \cdots \operatorname{diag}(1, \dots, 1, t_{m+n-1}, t_{m+n-1}^{-1}).$$

The resulting factorization for $X = AD$ is exactly formula (1.12) corresponding to the word $\mathbf{j} = (-(n-1), -(n-2), \dots, -1, -(n-1), -(n-2), \dots, -2, \dots, -(n-1), n-1, n-2, n-1, \dots, 1, \dots, n-1, \mathrm{i}, \dots, \mathrm{i}(n-1))$.

1.3. Poisson-Lie groups

The theory of Poisson-Lie groups and Poisson homogeneous spaces serves as a source of many important examples that will be considered below. This section contains a brief review of the theory.

1.3.1. We first recall basic definitions of Poisson geometry.

A *Poisson algebra* is a commutative associative algebra $\mathfrak{F}$ equipped with a *Poisson bracket* defined as a skew-symmetric bilinear map $\{\cdot,\cdot\} : \mathfrak{F} \times \mathfrak{F} \to \mathfrak{F}$ that satisfies, for any $f_1, f_2, f_3 \in \mathfrak{F}$, the *Leibniz identity*

$$\{f_1 f_2, f_3\} = f_1\{f_2, f_3\} + \{f_1, f_3\} f_2$$

and the *Jacobi identity*

$$\{f_1, \{f_2, f_3\}\} + \{f_2, \{f_3, f_1\}\} + \{f_3, \{f_1, f_2\}\} = 0.$$

An element $c \in \mathfrak{F}$ such that $\{c, f\} = 0$ for any $f \in \mathfrak{F}$ is called a *Casimir element.*

A smooth real manifold M is called a *Poisson manifold* if the algebra $C^\infty(M)$ of smooth functions on M is a Poisson algebra. In this case we say that M is equipped with a *Poisson structure.*

Let $(M_1, \{\cdot,\cdot\}_1)$ and $(M_2, \{\cdot,\cdot\}_2)$ be two Poisson manifolds. Then $M_1 \times M_2$ is equipped with a natural Poisson structure: for any $f_1, f_2 \in C^\infty(M_1 \times M_2)$ and any $x \in M_1,\ y \in M_2$

$$\{f_1, f_2\}(x, y) = \{f_1(\cdot, y), f_2(\cdot, y)\}_1(x) + \{f_1(x, \cdot), f_2(x, \cdot)\}_2(y).$$

Let $F : M_1 \to M_2$ be a smooth map; it is called a *Poisson map* if

$$\{f_1 \circ F, f_2 \circ F\}_1 = \{f_1, f_2\}_2 \circ F$$

for every $f_1, f_2 \in C^\infty(M_2)$. A submanifold N of a Poisson manifold $(M, \{\cdot,\cdot\})$ is called a *Poisson submanifold* if it is equipped with a Poisson bracket $\{\cdot,\cdot\}_N$ such that the inclusion map $\mathrm{i} : N \to M$ is Poisson.

A bracket $\{\cdot,\cdot\}$ on $C^\infty(M)$ is called *non-degenerate* if there are no non-constant Casimir elements in $C^\infty(M)$ for $\{\cdot,\cdot\}$. Otherwise, $\{\cdot,\cdot\}$ is called *degenerate.* Every *symplectic* manifold, that is, an even-dimensional manifold M equipped with a non-degenerate closed 2-form ω^2, gives rise to a nondegenerate Poisson bracket

$$\{f_1, f_2\} = \omega^2(Idf_1, Idf_2), \tag{1.14}$$

where $I\ : T^*M \ni \omega^1 \mapsto I\omega^1 \in TM$ is an isomorphism defined by

$$\langle \omega^1, \cdot \rangle = \omega^2(\ \cdot\ , I\omega^1).$$

Conversely, if a Poisson bracket $\{\cdot,\cdot\}$ on M is nondegenerate, then M is a symplectic manifold and the Poisson bracket is defined by (1.14).

Fix a smooth function φ on a Poisson manifold M. By the Leibniz rule, the map $\{\varphi, \cdot\} : C^\infty(M) \to C^\infty(M)$ is a differentiation. This implies an equation

$$\{\varphi, f\} = \langle df, V_\varphi \rangle,$$

which defines a vector field V_φ even if the Poisson bracket on M is degenerate. V_φ is called a *Hamiltonian vector field* generated by φ. If M is a symplectic manifold, then $V_\varphi = Id\varphi$.

The *Poisson bivector field* π is a section of T^2M defined by

$$\{f_1, f_2\} = (df_1 \wedge df_2)(\pi).$$

In terms of the Poisson bivector field, a Poisson submanifold can be defined as a submanifold $N \subset M$ such that $\pi|_N \in T^2N$. The *rank* of the Poisson structure at a point of M is defined to be the rank of π at that point.

Now, we are ready to review the notion of a symplectic leaf. First, introduce an equivalence relation $\sim$ on M: for $x, y \in M$, $x \sim y$ if x and y can be connected by a piecewise smooth curve whose every segment is an integral curve of a Hamiltonian vector field. Then it can be shown that

(i) equivalence classes of $\sim$ are Poisson submanifolds of M;

(ii) the rank of the Poisson structure at every point of such submanifold N is equal to the dimension of N.

An equivalence class, M_x, of a point $x \in M$ is called a *symplectic leaf through* x. It is clear that M_x is a symplectic manifold with respect to the restriction of $\{\cdot,\cdot\}$ to M_x. Casimir functions of $\{\cdot,\cdot\}$ are constant on symplectic leaves.

EXAMPLE 1.3. An important example of a degenerate Poisson bracket is the *Lie-Poisson bracket* on a dual space $\mathfrak{g}^*$ to a Lie algebra $\mathfrak{g}$ defined by

$$\{f_1, f_2\}(a) = \langle a, [df_1(a), df_2(a)]\rangle$$

for any $f_1, f_2 \in C^\infty(\mathfrak{g}^*)$ and $a \in \mathfrak{g}^*$. Casimir functions for the Lie-Poisson bracket are functions invariant under the co-adjoint action on $\mathfrak{g}^*$ of the Lie group $\mathcal{G} = \exp \mathfrak{g}$. Symplectic leaves of the Lie-Poisson bracket are co-adjoint orbits of $\mathcal{G}$.

1.3.2. Let $\mathcal{G}$ be a Lie group equipped with a Poisson bracket $\{\cdot,\cdot\}$. $\mathcal{G}$ is called a *Poisson-Lie group* if the multiplication map

$$\mathfrak{m} : \mathcal{G} \times \mathcal{G} \ni (x, y) \mapsto xy \in \mathcal{G}$$

is Poisson. This condition can be re-written as

$$\{f_1, f_2\}(xy) = \{\rho_y f_1, \rho_y f_2\}(x) + \{\lambda_x f_1, \lambda_x f_2\}(y), \tag{1.15}$$

where ρ_y and λ_x are, respectively, right and left translation operators on $\mathcal{G}$:

$$(\rho_y f)(x) = (\lambda_x f)(y) = f(xy).$$

EXAMPLE 1.4. A dual space $\mathfrak{g}^*$ to a Lie algebra $\mathfrak{g}$ equipped with a Lie-Poisson bracket (Example 1.3) is an additive Poisson-Lie group.

To construct examples of non-abelian Poisson-Lie groups, one needs a closer look at the additional structure induced by property (1.15) on the tangent Lie algebra $\mathfrak{g}$ of $\mathcal{G}$. First, we introduce *right-* and *left-invariant* differentials, D and D', on $\mathcal{G}$:

$$\langle Df(x), \xi\rangle = \frac{d}{dt} f(\exp(t\xi)x)\Big|_{t=0}, \qquad \langle D'f(x), \xi\rangle = \frac{d}{dt} f(x\exp(t\xi))\Big|_{t=0}, \tag{1.16}$$

where $f \in C^\infty(\mathcal{G})$, $\xi \in \mathfrak{g}$. The Poisson bracket $\{\cdot,\cdot\}$ on $\mathcal{G}$ can then be written as

$$\{f_1, f_2\}(x) = \langle \pi(x) Df_1(x), Df_2(x)\rangle = \langle \pi'(x) D'f_1(x), D'f_2(x)\rangle,$$

where π, π' map $\mathcal{G}$ into $\operatorname{Hom}(\mathfrak{g}^*, \mathfrak{g})$. Then property (1.15) translates into the 1-*cocycle* condition for π and π' :

$$\pi(xy) = \operatorname{Ad}_x \pi(y) \operatorname{Ad}^*_{x^{-1}} + \pi(x), \qquad \pi'(xy) = \operatorname{Ad}_{y^{-1}} \pi'(x) \operatorname{Ad}^*_y + \pi'(y). \tag{1.17}$$

Using the identification $\mathrm{Hom}(\mathfrak{g}^*, \mathfrak{g}) \cong \mathfrak{g} \otimes \mathfrak{g}$, we conclude from (1.17) that the map $\delta : \mathfrak{g} \to \mathfrak{g} \otimes \mathfrak{g}$ defined by

$$\delta(\xi) = \frac{d}{dt}\pi(\exp(t\xi))\Big|_{t=0} \tag{1.18}$$

satisfies

$$\delta([\xi, \eta]) = [\xi \otimes \mathbf{1} + \mathbf{1} \otimes \xi, \delta(\eta)] - [\eta \otimes \mathbf{1} + \mathbf{1} \otimes \eta, \delta(\xi)].$$

In other words, δ is a 1-cocycle on $\mathfrak{g}$ with values in the $\mathfrak{g}$-module $\mathfrak{g} \otimes \mathfrak{g}$.

Another consequence of (1.17) is that the Poisson operators π, π' vanish at the identity element e of $\mathcal{G}$ and, thus, any Poisson-Lie bracket is degenerate at the identity. The linearization of $\{\cdot, \cdot\}$ at the identity equips $\mathfrak{g}^* \cong T_e^*\mathcal{G}$ with a Lie algebra structure:

$$[a_1, a_2]_* = d_e\{\varphi_1, \varphi_2\},$$

where φ_i, $i = 1, 2$ are any functions such that $d_e\varphi_i = a_i$. Comparison with (1.18) gives

$$\langle [a_1, a_2]_*, \xi \rangle = \langle a_1 \wedge a_2, \delta(\xi) \rangle.$$

To summarize, if $\mathcal{G}$ is a Poisson-Lie group then the pair $(\mathfrak{g}, \mathfrak{g}^*)$ satisfies the following conditions:

(i) $\mathfrak{g}$ and $\mathfrak{g}^*$ are Lie algebras,

(ii) the map δ dual to the commutator $[\cdot, \cdot]_* : \mathfrak{g}^* \otimes \mathfrak{g}^* \to \mathfrak{g}^*$ is a 1-cocycle on $\mathfrak{g}$ with values in $\mathfrak{g} \otimes \mathfrak{g}$.

A pair $(\mathfrak{g}, \mathfrak{g}^*)$ satisfying the two conditions above is called a *Lie bialgebra* and the corresponding map δ is called a *cobracket*.

THEOREM 1.5. (*Drinfeld*) *If $(\mathfrak{g}, \mathfrak{g}^*)$ is a Lie bialgebra and $\mathcal{G}$ is a connected simply-connected Lie group with the Lie algebra $\mathfrak{g}$, then there exists a unique Poisson bracket on $\mathcal{G}$ that makes $\mathcal{G}$ into a Poisson-Lie group with the tangent Lie bialgebra $(\mathfrak{g}, \mathfrak{g}^*)$.*

If $\mathcal{H}$ is a Lie subgroup of $\mathcal{G}$, it is natural to ask if it is also a Poisson-Lie subgroup, i. e. if $\mathcal{H}$ is a Poisson submanifold of $\mathcal{G}$ such that property (1.15) remains valid for the restriction of the Poisson-Lie bracket on $\mathcal{G}$ to $\mathcal{H}$. The answer to this question is also conveniently described in terms of the tangent Lie bialgebra of $\mathcal{G}$. Let $\mathfrak{h} \subset \mathfrak{g}$ be the Lie algebra that corresponds to $\mathcal{H}$.

PROPOSITION 1.6. *$\mathcal{H}$ is a Poisson-Lie subgroup of $\mathcal{G}$ if and only if the annihilator of $\mathfrak{h}$ in $\mathfrak{g}^*$ is an ideal in the Lie algebra $\mathfrak{g}^*$.*

1.3.3. Next, we discuss an important class of Lie bialgebras, called *factorizable Lie bialgebras*.

A Lie bialgebra $(\mathfrak{g}, \mathfrak{g}^*)$ is called factorizable if the following two conditions hold:

(i) $\mathfrak{g}$ is equipped with an invariant bilinear form $(\cdot, \cdot)$, so that $\mathfrak{g}^*$ can be identified with $\mathfrak{g}$ via

$$\mathfrak{g}^* \ni a \mapsto \xi_a \in \mathfrak{g} \ : \ \langle a, \ \cdot \rangle = (\xi_a, \cdot);$$

(ii) the Lie bracket on $\mathfrak{g}^* \cong \mathfrak{g}$ is given by

$$[\xi, \eta]_* = \frac{1}{2}\left([R(\xi), \eta] + [\xi, R(\eta)]\right), \tag{1.19}$$

where $R \in \mathrm{End}(\mathfrak{g})$ is a skew-symmetric operator satisfying the *modified classical Yang-Baxter equation (MCYBE)*

$$[R(\xi), R(\eta)] - R\left([R(\xi), \eta] \ + \ [\xi, R(\eta)]\right) = -[\xi, \eta]; \tag{1.20}$$

R is called a *classical R-matrix.*

The invariant bilinear form $(\cdot,\cdot)$ can be represented by a Casimir element $\mathfrak{t} \in \mathfrak{g}\otimes\mathfrak{g}$. Let r be the image of R under the identification $\mathfrak{g}\otimes\mathfrak{g} \cong \mathfrak{g}\otimes\mathfrak{g}^* \cong \mathrm{End}(\mathfrak{g})$ and let r_{ij} $(1 \le i < j \le 3)$ denote the image of r under the embedding $\mathfrak{g}\otimes\mathfrak{g} \to \mathfrak{g}\otimes\mathfrak{g}\otimes\mathfrak{g}$ such that the first (resp. second) factor in $\mathfrak{g}\otimes\mathfrak{g}$ is mapped into the ith (resp. jth) factor in $\mathfrak{g}\otimes\mathfrak{g}\otimes\mathfrak{g}$. Define

$$[[r,r]] := [r_{12}, r_{13}] \; + \; [r_{12}, r_{23}] \; + \; [r_{13}, r_{23}].$$

Then condition (1.20) in the definition of a factorizable Lie bialgebra can be equivalently re-phrased as a condition on the cobracket δ:

$$\delta(\xi) = [\xi\otimes\mathbf{1} + \mathbf{1}\otimes\xi, r_\pm],$$

where $r_\pm = \frac{1}{2}(r \pm \mathfrak{t}) \in \mathfrak{g}\otimes\mathfrak{g}$ satisfy

$$[[r_\pm, r_\pm]] = 0.$$

Let $\mathcal{G}$ be a Poisson-Lie group with a factorizable tangent Lie bialgebra $(\mathfrak{g},\mathfrak{g}^*)$. The corresponding Poisson bracket whose existence is guaranteed by Theorem 1.5 is called the *Sklyanin bracket.* To provide an explicit formula for the Sklyanin bracket, we first define, in analogy to (1.16), right and left gradients for a function $f \in C^\infty(\mathcal{G})$:

$$(1.21)\quad (\nabla f(x),\xi) = \frac{d}{dt} f(\exp(t\xi)x)\Big|_{t=0}, \qquad (\nabla' f(x),\xi) = \frac{d}{dt} f(x\exp(t\xi))\Big|_{t=0}.$$

Then the Sklyanin bracket has a form

$$(1.22)\qquad \{f_1,f_2\} = \frac{1}{2}(R(\nabla' f_1),\nabla' f_2) - \frac{1}{2}(R(\nabla f_1),\nabla f_2).$$

1.3.4. Standard Poisson-Lie structure on a simple Lie group. We now turn to the main example of a Poisson-Lie group to be used in this exposition. Let $\mathcal{G}$ be a connected simply-connected Lie group with the Lie algebra $\mathfrak{g}$. The Killing form is a bilinear nondegenerate form on $\mathfrak{g}\cong\mathfrak{g}^*$. Define $R \in \mathrm{End}(\mathfrak{g})$ by

$$(1.23)\qquad R(\xi) = \xi_+ - \xi_-,$$

where $\xi_\pm$ are defined as in (1.10). It is easy to check that R satisfies (1.20) and thus (1.22), (1.23) endow $\mathcal{G}$ with a Poisson-Lie structure, called the *standard* Poisson-Lie structure and denoted by $\{\cdot,\cdot\}_{\mathcal{G}}$. Furthermore, the Lie bracket (1.19) on $\mathfrak{g}^*\cong\mathfrak{g}$ in this case is given by

$$(1.24)\qquad [\xi,\eta]_* = [\xi_+ + \frac{1}{2}\xi_0, \eta_+ + \frac{1}{2}\eta_0] - [\xi_- + \frac{1}{2}\xi_0, \eta_- + \frac{1}{2}\eta_0].$$

It follows from (1.24), that subalgebras $\mathfrak{n}_\pm$ of $\mathfrak{g}$ are ideals in $\mathfrak{g}^*$. Since $\mathfrak{n}_\pm = \mathfrak{b}_\pm^\perp$ with respect to the Killing form, Proposition 1.6 implies that Borel subgroups $\mathcal{B}_\pm$ are Poisson-Lie subgroups of $\mathcal{G}$.

EXAMPLE 1.7. Let us equip $\mathcal{G} = SL_n$ and $\mathfrak{g} = sl_n$ with the trace-form

$$(\xi,\eta) = \operatorname{tr}\xi\eta.$$

Then the right and left gradients (1.21) are

$$\nabla f(x) = x\ \mathrm{grad} f(x), \qquad \nabla' f(x) = \mathrm{grad} f(x)\ x,$$

where

$$\operatorname{grad} f(x) = \left(\frac{\partial f}{\partial x_{ji}}\right)_{i,j=1}^{n}.$$

Thus the standard Poisson-Lie bracket becomes

(1.25)

$$\{f_1, f_2\}_{SL_n}(x) = \frac{1}{2}(R(\operatorname{grad} f_1(x)\, x), \operatorname{grad} f_2(x)\, x) - \frac{1}{2}(R(x \operatorname{grad} f_1(x)), x \operatorname{grad} f_2(x)),$$

where the action of the R-matrix (1.23) on $\xi = (\xi_{ij})_{i,j=1}^n \in sl_n$ is given by

$$R(\xi) = (\operatorname{sign}(j-i)\xi_{ij})_{i,j=1}^n.$$

Substituting into (1.25) coordinate functions x_{ij}, x_{kl}, we obtain

$$\{x_{ij}, x_{kl}\}_{SL_n} = \frac{1}{2}\left(\operatorname{sign}(k-i) + \operatorname{sign}(l-j)\right) x_{il} x_{kj}.$$

In particular, the standard Poisson-Lie structure on

$$SL_2 = \left\{ \begin{pmatrix} a & b \\ c & d \end{pmatrix} : ad - bc = 1 \right\}$$

is described by the relations

$$\{a,b\}_{SL_2} = \tfrac{1}{2}ab, \quad \{a,c\}_{SL_2} = \tfrac{1}{2}ac, \quad \{a,d\}_{SL_2} = bc,$$
$$\{c,d\}_{SL_2} = \tfrac{1}{2}cd, \quad \{b,d\}_{SL_2} = \tfrac{1}{2}bd, \quad \{b,c\}_{SL_2} = 0.$$

Note that the Poisson bracket induced by $\{\cdot,\cdot\}_{SL_2}$ on upper and lower Borel subgroups of SL_2

$$\mathcal{B}_+ = \left\{ \begin{pmatrix} p & q \\ 0 & p^{-1} \end{pmatrix} \right\}, \qquad \mathcal{B}_- = \left\{ \begin{pmatrix} p & 0 \\ q & p^{-1} \end{pmatrix} \right\}$$

has an especially simple form:

$$\{p,q\} = \frac{1}{2}pq.$$

The example of SL_2 is instrumental in an alternative characterization of the standard Poisson-Lie structure on $\mathcal{G}$. Fix $i \in [1,l]$, $l = \operatorname{rank} \mathfrak{g}$, and consider an annihilator of the subalgebra $\mathfrak{g}^i = \operatorname{span}\{e_i, e_{-i}, h_i\}$ with respect to the Killing form:

$$(\mathfrak{g}^i)^\perp = h_i^\perp \oplus \left(\oplus_{\alpha \in \Phi \setminus \{\alpha_{\pm i}\}} \mathfrak{g}_\alpha\right),$$

where $h_i^\perp$ is the orthogonal complement of h_i in $\mathfrak{h}$ with respect to the restriction of the Killing form. Though $(\mathfrak{g}^i)^\perp$ is not even a subalgebra in $\mathfrak{g}$, it is an ideal in $\mathfrak{g}^*$ equipped with the Lie bracket (1.24). It follows that for every i, the image of the embedding $\rho_i : SL_2 \to \mathcal{G}$ defined in (1.7) is a Poisson-Lie subgroup of $\mathcal{G}$. The restriction $\{\cdot,\cdot\}_{\rho_i(SL_2)}$ of the Sklyanin bracket (1.22), (1.23) to $\rho_i(SL_2)$ involves only the restriction of the standard R-matrix to $\mathfrak{g}^i = \rho_i(sl_2)$. It is not difficult to check then, that

$$\{\cdot,\cdot\}_{\rho_i(SL_2)} = (\alpha_i, \alpha_i)\{\cdot,\cdot\}_{SL_2}.$$

Thus the standard Poisson-Lie structure is characterized by the condition that for every $i \in [1,l]$, the map

(1.26) $$\rho_i : \left(SL_2, (\alpha_i,\alpha_i)\{\cdot,\cdot\}_{SL_2}\right) \to \left(\mathcal{G}, \{\cdot,\cdot\}_{\mathcal{G}}\right)$$

is Poisson.

Bibliographical notes

1.1. For a detailed treatment of Grassmannians and Plücker relations we address the reader to [**HdP**] or [**GrH**].

1.2. The standard material about simple Lie algebras and groups can be found in [**Hu2**] and [**OV**]. For a good introduction to Coxeter groups, including reduced words for pairs of elements, Bruhat decompositions and Bruhat cells, see [**Hu3**]. Double Bruhat cells were introduced and studied in [**FZ1**], as well as generalized minors and their twisted counterparts.

1.3. The bulk of the material reviewed in this section, including Drinfeld's theorem and Proposition 1.6, can be found in [**ES**] and [**ReST**]. For an introduction to symplectic manifolds and Hamilton vector fields, see [**A2**]. The description of the standard Poisson–Lie structure on a simple Lie group follows [**HKKR**] and [**KoZ**].

CHAPTER 2

Basic examples: Rings of functions on Schubert varieties

We start with a few basic examples which motivate the notion of a cluster algebra.

2.1. The homogeneous coordinate ring of $G_2(m)$

In this section we consider the problem of constructing an additive monomial basis in the homogeneous coordinate ring of the Grassmannian of two-dimensional planes in the m-dimensional space, and introduce the cluster structure of Plücker coordinates.

2.1.1. Cluster structure of Plücker coordinates. While Plücker coordinates provide a convenient system of generators for the homogeneous coordinate ring of $G_2(m)$, these generators are not independent: Plücker coordinates satisfy *short Plücker relations* (1.3). In $G_2(m)$ Plücker relations take the form

$$x_{ij}x_{kl} = x_{ik}x_{jl} + x_{il}x_{kj} \qquad \text{for } 1 \leq i < k < j < l \leq m. \tag{2.1}$$

A classical question of the invariant theory is to choose an additive basis of the homogeneous coordinate ring $\mathbb{C}[G_2(m)]$ formed by monomials in Plücker coordinates.

More exactly, all homogeneous degree d polynomials in Plücker coordinates form a vector space, and the question is to describe explicitly a subsystem of monomials of degree d that forms a basis in this vector space.

The answer to this question has been known for a long time. Consider a convex m-gon R_m. Enumerate its vertices counterclockwise from 1 to m. A pair (ij), $i < j$, is called a *diagonal* of R_m; it is a proper diagonal if $j - i > 1$ and a side otherwise. Any diagonal (ij) encodes the Plücker coordinate x_{ij}. We call two diagonals (ij) and (kl) (and the corresponding Plücker coordinates x_{ij} and x_{kl}) *crossing* if $i < k < j < l$. In other words, both diagonals should be proper and should intersect inside R_m.

Note that by definition, the pair of Plücker coordinates in the left hand side of Plücker relations (2.1) is crossing, and for any pair of crossing diagonals (ij) and (kl), the Plücker relation (2.1) expresses the product of these crossing Plücker coordinates as a sum of monomials containing only non-crossing Plücker coordinates. We see therefore that non-crossing monomials form a linear generating set in the homogeneous coordinate ring $\mathbb{C}[G_2(m)]$. Let us check now that all such monomials are linearly independent.

Given a linear relation between non-crossing monomials, we say that a Plücker coordinate is labeled if it enters at least one of the monomials involved in the relation. Next, if x_{ij} is labeled, we say that both indices i and j are labeled.

From all the linear relations between the non-crossing monomials choose those with the minimal number of distinct labeled indices. Among the obtained set of relations, choose those with the minimal number of distinct labeled Plücker coordinates (observe that the left hand side of such a relation is an irreducible polynomial). Take an arbitrary relation as above, and let i and j be two labeled indices such that any k, $i < k < j$, is not labeled. Consider a new relation obtained from the chosen one by replacing each x_{il} by x_{jl}, each x_{li} by x_{lj}, and x_{ij} by 0. By the non-crossing property, this is indeed a relation, and by the irreducibility, it includes monomials with non-zero coefficients. It is easy to see that this relation is nontrivial, and hence we obtained a relation with a smaller number of labeled indices, a contradiction. Therefore the set S of all monomials each containing only non-crossing Plücker coordinates forms a basis in $\mathbb{C}[G_2(m)]$.

Let us study the maximal families of non-crossing Plücker coordinates. Any such family gives rise to an infinite set of monomials from S formed by products of powers of Plücker coordinates from this family. The basis S is covered by the union of these families.

We call a maximal family of non-crossing Plücker coordinates an *extended cluster*. Adding any extra Plücker coordinate to such an extended cluster violates the non-crossing property.

Below we describe all extended clusters for $G_2(4)$ and $G_2(5)$. Note that each maximal non-crossing subset contains all sides of the m-gon R_m. The sides of R_m (or, equivalently, *cyclically dense* minors $x_{12}, \dots, x_{m-1,m}, x_{1,m}$) play a special role in the construction. We call them *stable* Plücker coordinates, and consider $\mathbb{C}[G_2(m)]$ as an algebra over $\mathbb{K} = \mathbb{C}[x_{12}, x_{23}, \dots, x_{m-1,m}, x_{1,m}]$.

For the sake of simplicity we will discuss below only how the proper diagonals are distributed among extended clusters, keeping in mind that all sides of R_m are common for all extended clusters. A maximal set of non-crossing proper diagonals of R_m we call a *cluster*. The corresponding extended cluster is obtained by adding to the cluster all sides of R_m.

2.1.2. Grassmannian $G_2(4)$. If $m = 4$ then R_m becomes a quadrilateral with two proper diagonals. We have two clusters $\{x_{13}\}$ and $\{x_{24}\}$. Each cluster contains exactly one proper diagonal. The only Plücker relation is $x_{13}x_{24} = x_{12}x_{34} + x_{14}x_{23}$. The homogeneous coordinate ring $\mathbb{C}[G_2(4)]$ has an additive basis over $\mathbb{K} = \mathbb{C}[x_{12}, x_{23}, x_{34}, x_{14}]$ formed by powers of either one of the variables x_{13} and x_{24}. As a vector space, $\mathbb{C}[G_2(4)] = \mathbb{K}[x_{13}] + \mathbb{K}[x_{24}]$, with $\mathbb{K}[x_{13}] \cap \mathbb{K}[x_{24}] = \mathbb{K}$.

2.1.3. Grassmannian $G_2(5)$ and the Stasheff pentagon. In a convex pentagon we have five proper diagonals: $(13), (14), (24), (25), (35)$. It is easy to check that there are five clusters:

$$\{x_{13}, x_{14}\}, \{x_{24}, x_{25}\}, \{x_{13}, x_{35}\}, \{x_{24}, x_{14}\}, \{x_{25}, x_{35}\},$$

and five Plücker relations. These clusters exhibit nice combinatorial properties. Each of them describes a triangulation of the pentagon into three triangles. Each one contains exactly two proper diagonals. It is well known that any triangulation can be transformed into another one by a sequence of *Whitehead moves* (see Fig. 2.1).

If we represent triangulations as a vertices of a graph and connect two vertices by an edge when the corresponding triangulations are connected by a Whitehead move, then we obtain the *Stasheff pentagon* (see Figure 2.2).

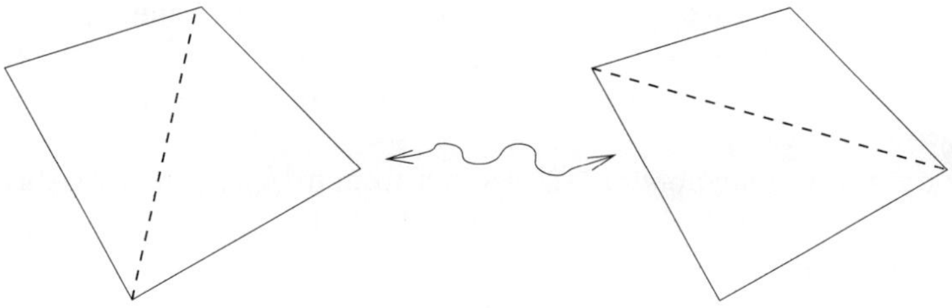

FIGURE 2.1. Whitehead move

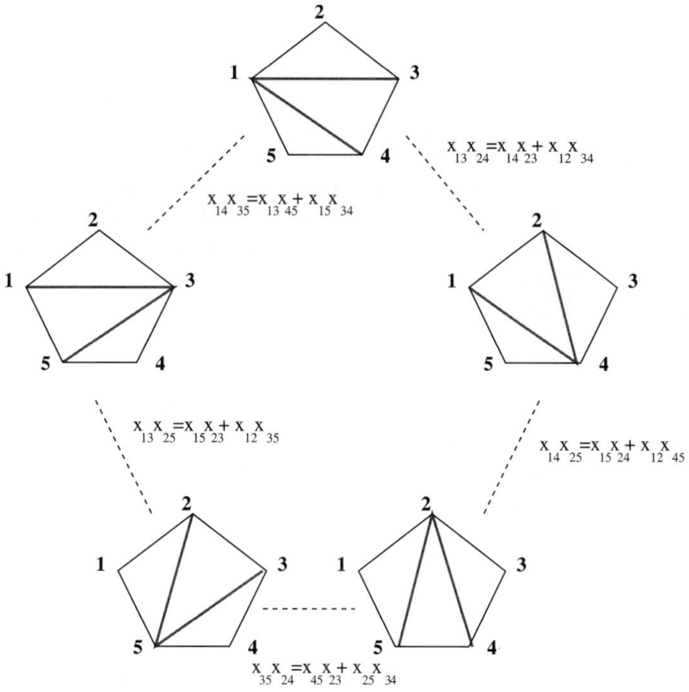

FIGURE 2.2. Stasheff pentagon

Edges of the Stasheff pentagon correspond to Plücker relations. Note also that adjacent clusters always contain one common proper diagonal and the second proper diagonal is replaced according to the Whitehead move.

2.1.4. Grassmannian $G_2(m)$ and Stasheff polytopes. A convex m-gon has $m(m-3)/2$ proper diagonals $\{(ij) : 1 \le i < j \le m, |j-i| > 1\}$. The number of clusters coincides with the number of all possible triangulations and equals to the Catalan number $C_{m-2} = \frac{1}{m-1}\binom{2(m-2)}{m-2}$. Each cluster contains exactly $m-3$ proper diagonals (since each triangulation contains $m-3$ proper diagonals).

The Stasheff pentagon is replaced by the *Stasheff polytope*, also known as the *associahedron*. To define the latter we need some preparations.

Recall that an *abstract polytope* is a partially ordered set $(\mathfrak{P}, \prec)$ with the following properties:

(i) $\mathfrak{P}$ contains the least element p_{-1} and the greatest element p_∞;

(ii) each maximal chain between p_{-1} and p_∞ has the same finite length, and hence the *rank* of each element $p \in \mathfrak{P}$ can be defined as the length of the maximal chain between p_{-1} and p;

(iii) for any two elements $p, q \in \mathfrak{P}$ distinct from p_{-1}, p_∞, there exists a sequence $p = p_0, p_1, \dots, p_k = q$ such that for each $i \in [0, k-1]$,

a) either $p_i \prec p_{i+1}$ or $p_{i+1} \prec p_i$,

b) $p_i \prec r$ for any r such that $p \prec r$, $q \prec r$, and

c) $r \prec p_i$ for any r such that $r \prec p$ and $r \prec q$;

(iv) if $p \prec q$ and the ranks of p and q differ by two, then there exist exactly two elements $r \in \mathfrak{P}$ such that $p \prec r \prec q$ and $\operatorname{rank} p < \operatorname{rank} r < \operatorname{rank} q$.

The elements in $\mathfrak{P}$ of rank i are called its $(i-1)$-dimensional faces; faces of dimension 0 are known as vertices, those of dimension 1, as edges.

We now define the associahedron S_m as an abstract polytope. The ground set of S_m is formed by sets of noncrossing diagonals containing all the sides of R_m; we will call them *subtriangulations*. The order relation is given by inverse inclusion. Clearly, the set of sides of R_m is the greatest element p_∞ of this partial order; the least element p_{-1} should be added artificially, to satisfy condition (i) above. Any maximal chain between p_{-1} and p_∞ corresponds to a sequence of deletions of proper diagonals one by one from a triangulation, until only the sides of R_m remain; clearly, the length of such a chain equals the number of proper diagonals in a triangulation, and is therefore the same for all chains. Condition (iii) for triangulations follows immediately from the fact that any two triangulations can be connected by a sequence of Whitehead moves. Verifying this condition for subtriangulations is left to the the reader. Finally, condition (iv) just says that if one subtriangulation is obtained from another one by deleting two proper diagonals, there are exactly two ways to perform this deletion.

Vertices of the associahedron thus defined correspond to triangulations of R_m, and its edges correspond to subtriangulations having exactly one quadrilateral; therefore, each edge corresponds to the Plücker relation defined by this quadrilateral. In other words, adjacent clusters contain $m-2$ common proper diagonals and differ in only one proper diagonal. These two diagonals are the only crossing diagonals in the corresponding Plücker relation. Two-dimensional faces of S_m correspond to subtriangulations obtained from a triangulation by deleting a pair of its proper diagonals. These diagonals are either disjoint, or have a common endpoint. In the former case the subtriangulation has exactly two quadrilaterals, while in the latter case it has exactly one pentagon. Consequently, there are two types of 2-dimensional faces: those corresponding to disjoint diagonals are quadrilaterals, and those corresponding to diagonals with a common endpoint are Stasheff pentagons. Finally, the facets of the associahedron correspond to subtriangulations with only one proper diagonal left; therefore, there is a bijection between the set of facets of S_m and the set of proper diagonals of R_m.

It is worth to note for future applications that the associahedron S_m can be realized as a simple convex polytope in $\mathbb{R}^{m-3}$; the proof of this fact is elementary but rather tedious, and falls beyond the scope of this book. The convex polytope S_6 in $\mathbb{R}^3$ corresponding to triangulations of the hexagon is shown in Fig. 2.3.

Each triangulation T of R_m defines a directed graph $\Gamma(T)$ in the following way. The vertices of $\Gamma(T)$ are the proper diagonals in T. The vertices d_i and d_j are joined

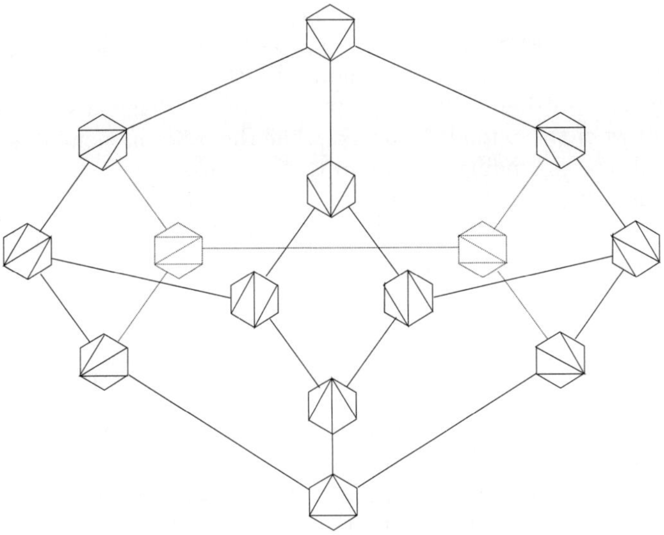

FIGURE 2.3. Stasheff polytope S_6

by an edge if d_i and d_j share a common endpoint v and d_i immediately precedes d_j in the counterclockwise order around v. For example, "snake-like" triangulations (see Fig. 2.4) lead to an orientation of the Dynkin diagram A_{m-3}.

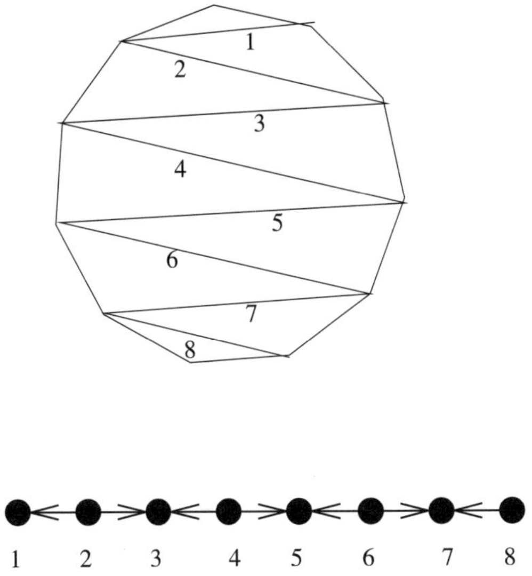

FIGURE 2.4. Snake-like triangulation and its graph

Consider the $(m-3) \times (m-3)$ incidence matrix of $\Gamma(T)$. Extend it to a $(m-3) \times (2m-3)$ matrix by adding an additional column for each side of R_m describing the counterclockwise adjacency between this side and proper diagonals. It is easy to see that the rows of this extended matrix correspond to the Plücker relations (2.1) in the following sense. For an arbitrary diagonal $d \in T$, let d' denote the unique proper diagonal that intersects d and is disjoint from all the other diagonals in T. Then the left hand side of the corresponding Plücker relation is $x_d x_{d'}$. One of the monomials in the right hand side is the product of the diagonals corresponding to 1's in the dth row of the extended matrix, while the other monomial is the product of the diagonals corresponding to -1's in the same row.

Consider for example the uppermost pentagon in Fig. 2.2. Assume that the diagonals are ordered as follows: $\{13, 14, 12, 23, 34, 45, 15\}$. Then the extended counterclockwise adjacency matrix is

$$\begin{matrix} 0 & -1 & 1 & -1 & 1 & 0 & 0 \\ 1 & 0 & 0 & 0 & -1 & 1 & -1 \end{matrix}$$

and we immediately read off both Plucker relations shown on Fig. 2.2.

2.1.5. Transformation rules. It is known that any two triangulations are connected by a sequence of Whitehead moves, and each Whitehead move replaces one of the Plücker coordinates by a new one, which is expressed as a rational function of the replaced coordinate and the remaining ones. Therefore, fixing some initial cluster, one can express all other Plücker coordinates (and hence, any element in $\mathbb{C}[G_2(m)]$) as rational functions in the Plücker coordinates from this cluster. We will describe how to compute these rational functions.

Let us introduce the following notation. Let us fix a triangulation T of our m-gon consisting of non-crossing proper diagonals $d_1, \dots, d_{m-3}$. The corresponding cluster contains Plücker coordinates $x_{d_1}, \dots, x_{d_{m-3}}$; together with the "sides" $x_{i,i+1}$ they form the extended cluster.

Given an arbitrary proper diagonal $d \notin T$, let us express x_d as a rational function of $\{x_t : t \in T\}$ (slightly abusing notation we assume that T also includes all sides of the m-gon).

Let T' denote the subset of diagonals from T intersecting d, and let $p_{t'}$, $t' \in T'$, denote the corresponding intersection point. In a neighborhood of $p_{t'}$ one can resolve the intersection in two different ways to produce smooth nonintersecting curves (see Figure 2.5).

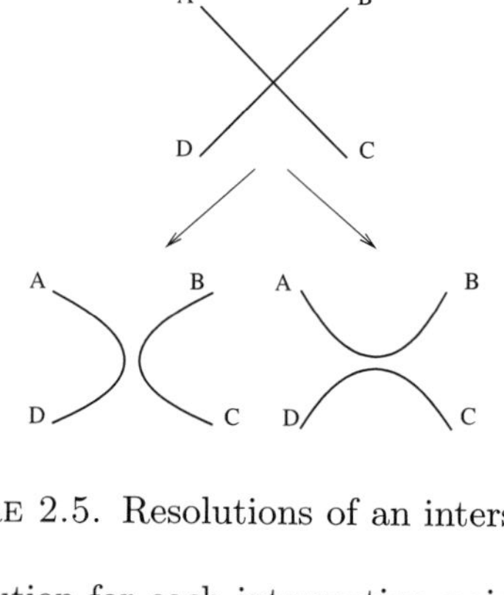

FIGURE 2.5. Resolutions of an intersection

Let us choose a resolution for each intersection point; we obtain a number of non-intersecting curves inside the m-gon. Replace each curve with distinct endpoints by the diagonal (or the side) connecting the same endpoints. If a curve is closed, we replace it by $\varnothing$. Denote the set of all possible resolutions of $T \cup \{d\}$

by $\mathrm{Res}(T, d)$, and the set of sides and proper diagonals (including $\varnothing$ if necessary) obtained from a resolution $z \in \mathrm{Res}(T, d)$ by $E(z)$. For any subset A of $T \cup \{\varnothing\}$ denote by x_A the product $\prod_{t \in A} x_t$, where $x_\varnothing = 0$. In particular, the monomial $x_{E(z)} = \prod_{t \in E(z)} x_t$ vanishes if $E(z)$ contains $\varnothing$; we call such a resolution z *degenerate*.

PROPOSITION 2.1. *All diagonals obtained by a resolution of $T \cup \{d\}$ belong to T. The Plücker coordinate x_d equals to the following rational expression in x_t, $t \in T$:*

$$x_d = \frac{1}{x_{T'}} \sum_{z \in \mathrm{Res}(T,d)} x_{E(z)}.$$

EXAMPLE 2.2. Take a triangulation T of a pentagon as in Figure 2.6.

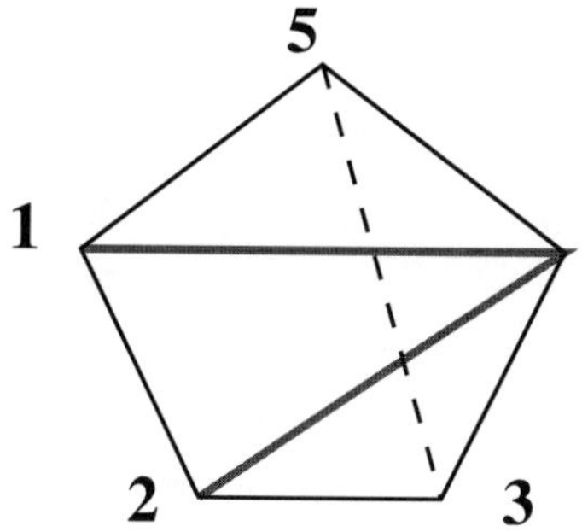

FIGURE 2.6. Transformation rules

Let us express x_{35} as a rational function of x_{12}, x_{23}, x_{34}, x_{45}, x_{15}, x_{14}, x_{24}. By Plücker relations,

$$x_{35} = \frac{1}{x_{24}} \left(x_{23} x_{45} + x_{25} x_{34} \right), \qquad x_{25} = \frac{1}{x_{14}} \left(x_{12} x_{45} + x_{15} x_{24} \right).$$

After substitution we obtain

$$x_{35} = \frac{x_{14} x_{23} x_{45} + x_{12} x_{45} x_{34} + x_{15} x_{24} x_{34}}{x_{24} x_{14}}$$

The monomial in the denominator is determined by the intersection. The numerator coincides with the decomposition into the sum over all possible resolutions, as seen in the Figure 2.7.

PROOF. The proof goes by induction. The case $m = 4$ follows immediately from the Plücker relations. Assume that the statement holds true for any n between 4 and m, and consider the case $n = m + 1$. Without loss of generality one may assume that the new diagonal d connects the vertices $m + 1$ and $k \in [2, m - 1]$. If there exists at least one initial diagonal d' not intersecting properly with d, we are done, since we can cut our $(m + 1)$-gon along d' and thus reduce the problem to the case $n \leq m$. Therefore, we assume that all initial diagonals are intersected by d. In particular, this implies that the first diagonal met by d upon leaving vertex $m + 1$ is $\bar{d}$ connecting 1 and m. The corresponding Plücker relation reads

$$x_{1,m} x_{k,m+1} = x_{1,k} x_{m,m+1} + x_{1,m+1} x_{k,m}. \tag{2.2}$$

Consider other initial diagonals incident to 1 or m. Since all such diagonals intersect d and do not intersect each other, we immediately infer that either all of

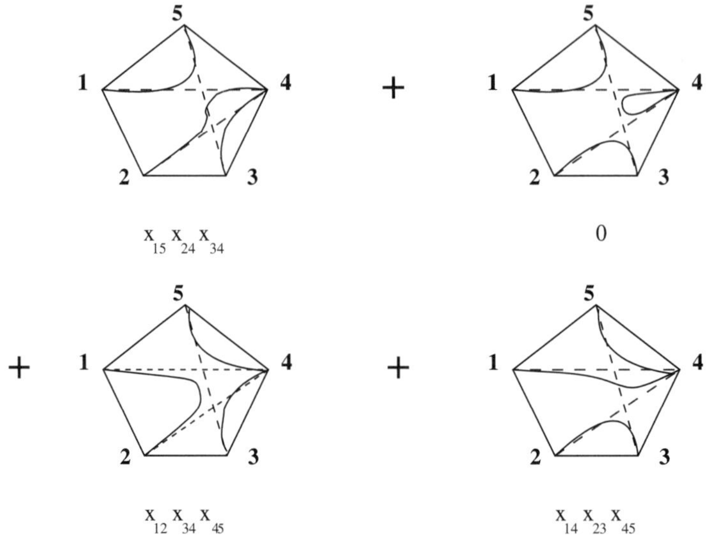

FIGURE 2.7. Resolutions for the pentagon

them are incident to 1 or all of them are incident to m. Assume without loss of generality that they all are incident to m.

There are two possible cases: $k = m - 1$ and $2 \leq k \leq m - 2$.

In the first case, all the initial diagonals are incident to m. Relation (2.2) implies

$$x_{1,m}x_{m-1,m+1}x_{\bar{T}} = x_{1,m-1}x_{m,m+1}x_{\bar{T}} + x_{1,m+1}x_{m-1,m}x_{\bar{T}},$$

where $\bar{T} = T \backslash \{\bar{d}\}$. The decomposition into two summands in this formula evidently corresponds to the two possible resolutions of the intersection of d and $\bar{d}$. The first summand corresponds to the resolution in which the side $(m, m+1)$ is cut off. All the $m - 3$ remaining intersections can be resolved independently, and each $(m-3)$-tuple of such resolutions corresponds bijectively to a certain resolution of all intersections of the diagonal $(1, m-1)$ with the diagonals (i, m), $2 \leq i \leq m-2$ in an m-gon spanned by the vertices $1, \ldots, m$ (see Fig. 2.8).

The second summand corresponds to the resolution in which the side $(1, m+1)$ is cut off. It is easy to see that in this case there exists a unique nondegenerate resolution of the remaining $m - 3$ intersections (see Fig. 2.9).

In the second case, relation (2.2) implies

$$x_{1,m}x_{k,m+1}x_{\bar{T}} = x_{1,k}x_{m,m+1}x_{\bar{T}} + x_{1,m+1}x_{m-1,m}x_{\bar{T}}.$$

Here again the two summands correspond to the two different resolutions of the intersection of d and $\bar{d}$. The first summand correspond to the resolution in which the side $(m, m+1)$ is cut off. All the $m-3$ remaining intersections can be resolved independently, and each $(n-3)$-tuple of such resolutions corresponds bijectively to a certain resolution of all intersections of the diagonal $(1, k)$ with the corresponding diagonals in an m-gon spanned by the vertices $1, \ldots, m$. Similarly, the second summand corresponds to the resolution in which the side $(1, m+1)$ is cut off. All the $m - 4$ remaining intersections can be resolved independently, and each $(m - 4)$-tuple of such resolutions corresponds bijectively to a certain resolution

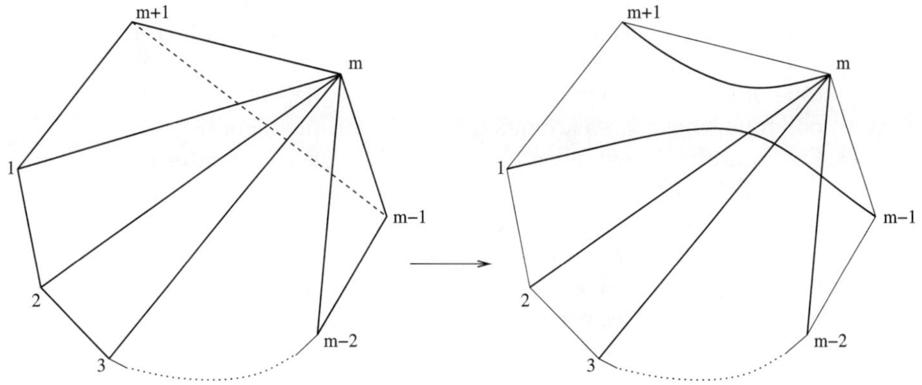

FIGURE 2.8. Cutting off $(m, m+1)$

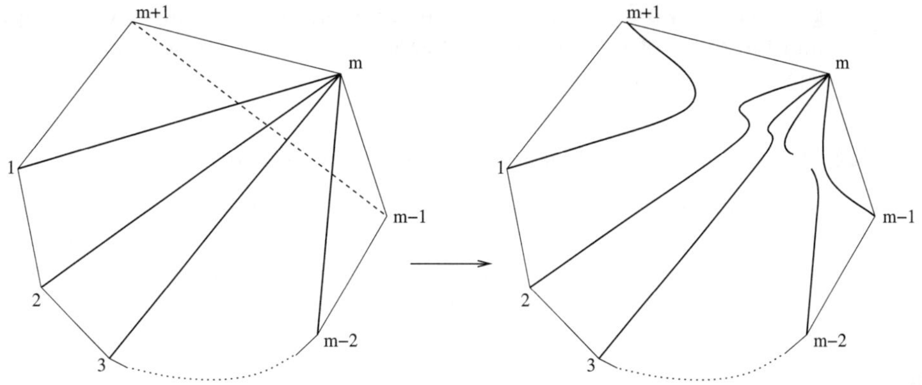

FIGURE 2.9. Cutting off $(1, m+1)$

of all intersections of the diagonal (k, m) with the corresponding diagonals in an $(m-1)$-gon spanned by the vertices $2, \ldots, m$. $\square$

2.1.6. Summary. Let us summarize what we have learned about $\mathbb{C}[G_2(m)]$.

- Plücker coordinates of $G_2(m)$ can be organized into a system of C_{m-2} extended clusters (where $C_m = \frac{1}{m+1}\binom{2m}{m}$ is the mth Catalan number) covering the set of all Plücker coordinates. Any extended cluster contains all "cyclically dense" Plücker coordinates (corresponding to the sides of a convex m-gon) and exactly $m-3$ other Plücker coordinates encoding a maximum set of non-crossing diagonals and forming the corresponding cluster. In other words, every (extended) cluster corresponds to a triangulation of a convex m-gon.
- The system of clusters has a natural graph structure. Namely, two clusters are connected by an edge if and only if they have exactly $2m-4$ common Plücker coordinates. Adjacent clusters exchange exactly two Plücker

coordinates corresponding to the crossing diagonals. (These Plücker coordinates are contained together in exactly one short Plücker relation.)
- The "exchange" graph described above is a 1-skeleton of the Stasheff polytope of m-gon triangulations. In particular, it is a 1-skeleton of a convex polytope whose faces are quadrilaterals and pentagons.
- Fix an arbitrary extended cluster. Its Plücker coordinates are generators of the field of meromorphic functions $\mathbb{C}(G_2(m))$. In particular, any other Plücker coordinate can be expressed as a rational function in the Plücker coordinates of the initial extended cluster. This rational function is a Laurent polynomial whose denominator equals to the product of initial Plücker coordinates corresponding to the diagonals crossing the diagonal of the coordinate in question. Moreover, the numerator is the sum of nonnegative monomials over all possible resolutions of the diagonal diagram.
- We can assign a directed graph $\Gamma(T)$ to each triangulation of the convex m-gon. The graph corresponding to a "snake-like" triangulation is an orientation of the Dynkin diagram A_{m-3}. Note that there is only a finite number (a Catalan number) of different triangulations of the m-gon. These two facts (the existence of a diagram of a finite Dynkin type and finiteness of the number of clusters) are related to each other in the general theory of cluster algebras, see Chapter 3.

2.2. Rings of regular functions on reduced double Bruhat cells

In this Section we describe a geometric problem of counting connected components of a double Bruhat cell. Our solution to this problem is based on the construction of cluster-like structures in the ring of regular functions on the double Bruhat cell. This example was one of the initial motivations for Fomin and Zelevinsky to introduce the cluster algebra formalism.

2.2.1. Intersection of Schubert cells. In early 1980's, V.I.Arnold asked the following question. Given two real n-dimensional complete flags f and g in general position, what can one say about the variety $U^n_{f,g}(\mathbb{R})$ of real complete flags intersecting these two fixed flags transversally? In particular, what is the topology of this variety and how many connected components this variety has? The latter question is equivalent to the following problem: "What is the number of connected components of the complement of the train of a complete flag in $\mathbb{R}^n$ in a neighborhood of this flag?"

This problem can be reformulated in terms of Schubert calculus. Namely, the variety of all real complete flags intersecting transversally a fixed flag is the maximal *Schubert cell* $\mathcal{B}_+ w_0 \mathcal{B}_+$. The variety of all real flags intersecting transversally two fixed flags in general position is the intersection of two generic maximal Schubert cells $\mathcal{B}_+ w_0 \mathcal{B}_+ \cap w_0 \mathcal{B}_+ w_0 \mathcal{B}_+$. If two flags are not in general position then we obtain the intersection $\mathcal{B}_+ w \mathcal{B}_+ \cap w_0 \mathcal{B}_+ w_0 \mathcal{B}_+$. Hence, the problem is to compute the number of connected components of $\mathcal{B}_+ w \mathcal{B}_+ \cap w_0 \mathcal{B}_+ w_0 \mathcal{B}_+$.

EXAMPLE 2.3. Let the dimension n be equal to 2. We use projective spaces instead of vector ones, to decrease the dimensions and to facilitate visualization. The vector space $\mathbb{R}^2$ becomes $\mathbb{R}P^1$, complete flags become points on $\mathbb{R}P^1$. Flags that intersect transversally two other fixed flags in general position become points on $\mathbb{R}P^1$ avoiding two fixed points on $\mathbb{R}P^1$. The variety of such flags consists of two disjoint intervals and has two connected components.

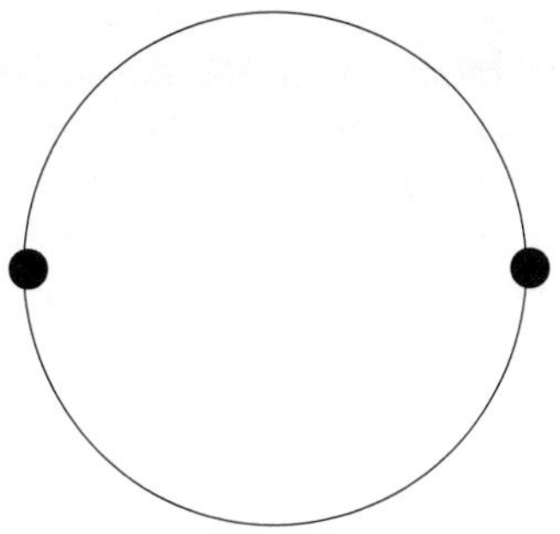

FIGURE 2.10. Intersection of opposite Schubert cells in complete flags in $\mathbb{R}^2$

EXAMPLE 2.4. Consider the case $n = 3$. The three-dimensional vector space becomes $\mathbb{R}P^2$, a complete flag is represented as a pair consisting of a projective line in $\mathbb{R}P^2$ and a point on it.

It is easy to see that the are six different pairs "point $\in$ line" that are disjoint (belong to different connected components) (see Figure 2.11) in the class of complete flags transversal to the two fixed flags.

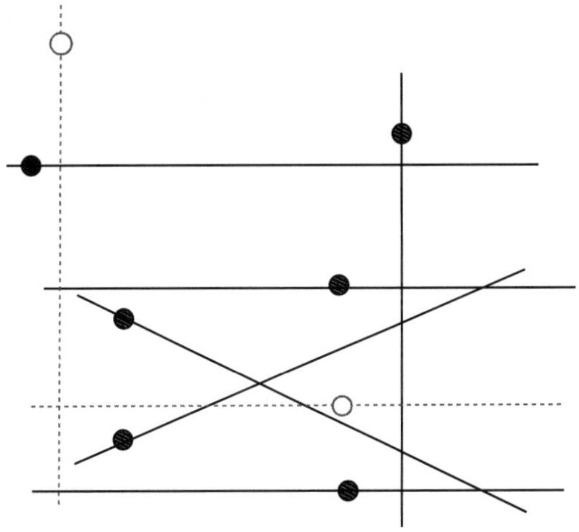

FIGURE 2.11. Six flags in $\mathbb{R}P^2$ that can not be deformed one into another in the class of flags transversal to the two fixed flags. Fixed flags are represented by dashed lines and empty circles.

To prove that the number of connected components is indeed 6, consider the following general setting. Let e_i be a standard basis in $\mathbb{R}^n$. Following our explanation in Section 1.1.2, denote by f the complete flag

$$\{\text{span}\{e_n\} \subset \text{span}\{e_{n-1}, e_n\} \dots \text{span}\{e_2, \dots, e_{n-1}, e_n\} \subset \mathbb{R}^n\},$$

and by g the "opposite" flag

$$\{\mathrm{span}\{e_1\} \subset \mathrm{span}\{e_1, e_2\} \ldots \mathrm{span}\{e_1, \ldots, e_{n-2}, e_{n-1}\} \subset \mathbb{R}^n\}.$$

Recall that the maximal Schubert cell of flags transversal to f can be identified with the set of unipotent upper triangular matrices (we consider the rows of a matrix as vectors in $\mathbb{R}^n$, the corresponding k-dimensional subspace is generated by the first k rows). The condition that flags are transversal to g is transformed into inequalities $\Delta_k \neq 0$, where Δ_k is the dense upper-right $k \times k$ minor. Hence, for the dimension three the intersection of the maximal generic Schubert cells is

$$\left\{ \begin{pmatrix} 1 & x & z \\ 0 & 1 & y \\ 0 & 0 & 1 \end{pmatrix} \; : \; z \neq 0, xy - z \neq 0 \right\}. \tag{2.3}$$

In other words, we have to find the number of connected components of the complement to the union of two surfaces $z = 0$ and $z = xy$ in the standard three-dimensional space with coordinates x, y, z.

From Figure 2.12 we see that there are totally six components with representatives as in Figure 2.11.

In dimension 4, the number of connected components was computed by V.I.Arnold; the answer is 20. The case $n = 5$ was studied by B. Shapiro and A. Vainshtein, and the number of connected components is equal to 52. However, ad hoc computations did not seem to work in dimensions higher than 5. The progress in this problem was made only several years later; the same approach applies also to double Bruhat cells.

2.2.2. Double Bruhat cells. The notion of a double Bruhat cell in a semisimple Lie group $\mathcal{G}$ was introduced as a geometrical framework for the study of total positivity. Recall that given a pair (u, v) of elements in the Weyl group of $\mathcal{G}$, the double Bruhat cell $\mathcal{G}^{u,v}$ is the intersection $\mathcal{G}^{u,v} = \mathcal{B}_+ u \mathcal{B}_+ \cap \mathcal{B}_- v \mathcal{B}_-$, see Section 1.2.4 for details. Besides, the corresponding reduced double Bruhat cell $L^{u,v}$ is the intersection $L^{u,v} = \mathcal{N}_+ u \mathcal{N}_+ \cap \mathcal{B}_- v \mathcal{B}_-$. The maximal torus acts on $\mathcal{G}^{u,v}$ by left multiplication. This action makes $\mathcal{G}^{u,v}$ into a trivial fiber bundle over $L^{u,v}$ with the fiber $(\mathbb{C}^*)^r$, where r is the rank of $\mathcal{G}$. Taking real parts, we get

$$\mathcal{G}^{u,v}(\mathbb{R}) \simeq L^{u,v}(\mathbb{R}) \times (\mathbb{R}^*)^r.$$

Therefore, the number of connected components in the real part of $\mathcal{G}^{u,v}$ equals 2^r times the number of components in the real part of $L^{u,v}$.

We will explain how to compute the number of connected components in the real part $L^{u,v}(\mathbb{R})$ of a reduced double Bruhat cell $L^{u,v}$. In particular, this solves the initial Arnold's problem, since L^{e,w_0} is isomorphic to the intersection of the two opposite Schubert cells $\mathcal{B}_+ w_0 \mathcal{B}_+ \cap w_0 \mathcal{B}_+ w_0 \mathcal{B}_+$ in the flag variety $SL_n/\mathcal{B}_+$.

The answer is given in the following terms. The reduced double Bruhat cell $L^{u,v}$ is biregularly isomorphic to an open subset of an affine space of dimension $m = l(u)+l(v)$. For example, we have already seen above that L^{e,w_0} in the case $\mathcal{G} = SL_n$ is an open subset of the set of unipotent upper triangular matrices; the latter set is an affine space of dimension $l(w_0) = n(n-1)/2$. In the general case, every reduced word $\mathbf{i}$ of $(u, v) \in W \times W$ gives rise to a subgroup $\Gamma_{\mathbf{i}}(\mathbb{F}_2) \subset GL_m(\mathbb{F}_2)$ generated by symplectic transvections (to be defined below). Connected components of $L^{u,v}(\mathbb{R})$ are in a natural bijection with the $\Gamma_{\mathbf{i}}(\mathbb{F}_2)$-orbits in $\mathbb{F}_2^m$.

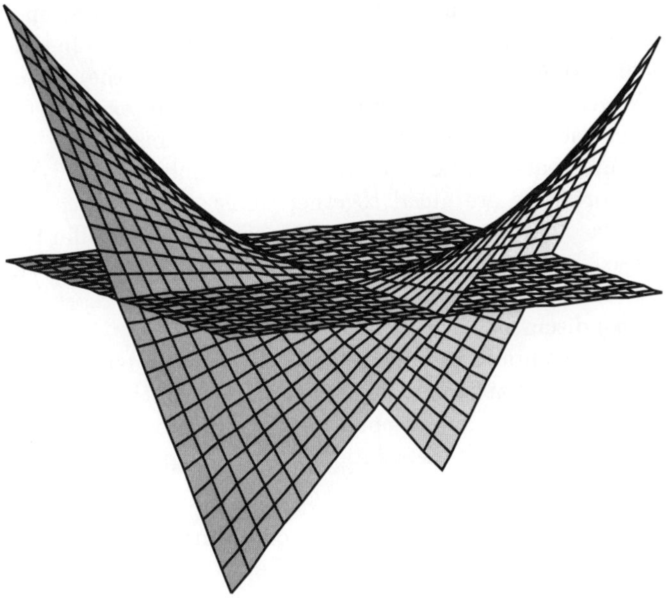

FIGURE 2.12. Connected components in the intersection of generic Schubert cells in 3D flags

The idea of the proof is as follows. We construct a set of *toric charts* (that is, isomorphic to $(\mathbb{R}^*)^m$) covering $L^{u,v}(\mathbb{R})$ except for a codimension 2 locus. More exactly, given a reduced word $\mathbf{i}$ for (u, v), we consider the corresponding factorization of the elements in $L^{u,v}$ to the product of elementary matrices, as in Section 1.2.4. The toric chart associated with $\mathbf{i}$ is the subset $\mathcal{T}_\mathbf{i}$ of $L^{u,v}(\mathbb{R})$ defined as the image of the product map (1.13) for all nonzero values of factorization parameters. The analysis of the inverse problem (restoring factorization parameters t_i from the product x) shows that t_i are reconstructed uniquely from any $x \in \mathcal{T}_\mathbf{i}$. Therefore, $\mathcal{T}_\mathbf{i} \simeq (\mathbb{R} \setminus 0)^m$ and t_i are coordinates for the toric chart $\mathcal{T}_\mathbf{i}$. Note that we have constructed a toric chart for every reduced word of (u, v). Moreover, elementary transformation of reduced words lead to the corresponding transformations of factorization parameters. Even more convenient to make a monomial change of variables and replace factorization parameters t_i by collection of monomial functions M_i, which have the following nice properties: M_i are regular functions on the whole $L^{u,v}$, and transformations of M_i's induced by elementary transformations of reduced words have an especially simple form.

Clearly, the union $U(\mathbb{R})$ of the toric charts has the same number of connected components as $L^{u,v}(\mathbb{R})$. Each chart is dense in $U(\mathbb{R})$ and consists of 2^m disjoint connected pieces called *orthants.* The orthants belonging to the same chart are naturally encoded by vectors in $\mathbb{F}_2^m$. Consider two orthants whose encoding vectors differ only in one coordinate. We prove that they are connected in $U(\mathbb{R})$ if and only if there exists an orthant of a different chart that intersects the above two orthants nontrivially. Moreover, this happens if and only if there exists a transvection in $\Gamma_{\mathbf{i}}(\mathbb{F}_2)$ that maps one of the encoding vectors into the other. Therefore the number of connected components can be computed as the number of $\Gamma_{\mathbf{i}}(\mathbb{F}_2)$-orbits.

To see how orthants are glued together, it is enough to consider all charts obtained from a fixed one by one "exchange" transformation similar to short Plücker relations discussed in Section 2.1. Note that even in the case of SL_n, the coordinates in these charts are no longer minors of the standard matrix representation, unlike the case of $G_2(m)$ discussed above.

It is possible to compute the number of $\Gamma_{\mathbf{i}}(\mathbb{F}_2)$-orbits explicitly for a vast majority of pairs (u, v). In particular, for dimension $n \geq 6$, the number of connected components in the intersection of open opposite Schubert cells equals $3 \cdot 2^{n-1}$. This observation accomplishes the solution of Arnold's problem discussed above in Section 2.2.1.

2.2.3. Computation of the number of connected components. Let Π be the Dynkin diagram of $\mathcal{G}$, and $A = (a_{kl})$ be its Cartan matrix. We associate with $\mathbf{i}$ an $m \times m$ matrix $C_{\mathbf{i}} = C = (c_{kl})$ in the following way: set $c_{kl} = 1$ if $|i_k| = |i_l|$ and $c_{kl} = -a_{|i_k|,|i_l|}$ if $|i_k| \neq |i_l|$. The diagonal entries c_{ii} do not play any role in what follows.

For $l \in [1, m]$, we denote by $l^- = l^-_{\mathbf{i}}$ the maximal index k such that $k \in [1, l]$ and $|i_k| = |i_l|$. If $|i_k| \neq |i_l|$ for all $k \in [1, l]$ then $l^- = 0$.

DEFINITION 2.5. We call an index $q \in [1, m]$ **i**-*bounded* if $q^- > 0$.

EXAMPLE 2.6. Let $W = S_4$ and $\mathbf{i} = (1\ 3\ 2\ -3\ -2\ 1)$, hence $u = s_1 s_3 s_2 s_1 = (3\ 2\ 4\ 1)$, $v = s_3 s_2 = (1\ 3\ 4\ 2)$, whereas $l(u) = 4$, $l(v) = 2$. Next, $1^- = 0$, $2^- = 0$, $3^- = 0$, $4^- = 2$, $5^- = 3$, $6^- = 1$, and so 4, 5 and 6 are **i**-bounded. Finally,

$$C = \begin{pmatrix} * & 0 & 1 & 0 & 1 & 1 \\ 0 & * & 1 & 1 & 1 & 0 \\ 1 & 1 & * & 1 & 1 & 1 \\ 0 & 1 & 1 & * & 1 & 0 \\ 1 & 1 & 1 & 1 & * & 1 \\ 1 & 0 & 1 & 0 & 1 & * \end{pmatrix}.$$

Let us associate to $\mathbf{i}$ a directed graph $\Sigma_{\mathbf{i}}$ on the set of vertices $[1, m]$. The edges of $\Sigma_{\mathbf{i}}$ are defined as follows.

DEFINITION 2.7. A pair $\{k, l\} \subset [1, m]$ with $k < l$ is an edge of $\Sigma_{\mathbf{i}}$ if it satisfies one of the following three conditions:

(i) $k = l^-$;

(ii) $k^- < l^- < k$, $\{|i_k|, |i_l|\} \in \Pi$, and $\theta(l^-) = \theta(k)$;

(iii) $l^- < k^- < k$, $\{|i_k|, |i_l|\} \in \Pi$, and $\theta(k^-) \neq \theta(k)$.

The edges of type (i) are called *horizontal*, and those of type (ii) and (iii) *inclined.* A horizontal (resp. inclined) edge $\{k, l\}$ with $k < l$ is directed from k to l if and only if $\theta(k) = +$ (resp. $\theta(k) = -$). We write $(k \to l) \in \Sigma_{\mathbf{i}}$ if $k \to l$ is a directed edge of $\Sigma_{\mathbf{i}}$.

EXAMPLE 2.8. The reduced word $\mathbf{i} = (1\ 3\ 2\ \ -3\ \ -2\ 1)$ from the previous example gives rise to the graph in Figure 2.13. The corresponding Dynkin diagram is shown on the left.

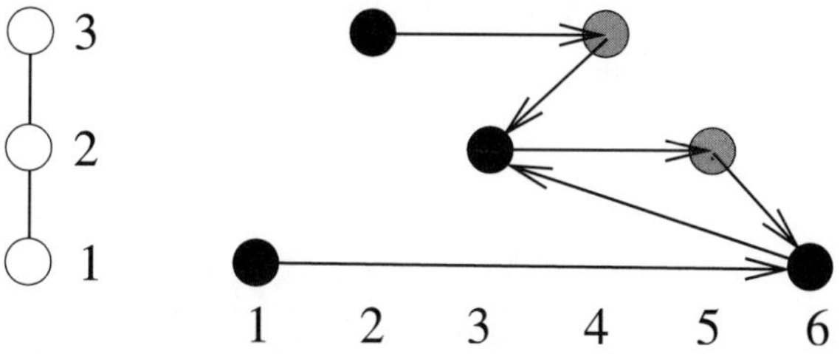

FIGURE 2.13. The graph associated to reduced word $\mathbf{i} = (1\ 3\ 2\ \ -3\ \ -2\ 1)$. Indices k for which $\theta(k) = +$ correspond to bold circles, those for which $\theta(k) = -$ correspond to gray circles.

To each $q \in [1, m]$ we associate a *transvection* $\tau_q = \tau_{q,\mathbf{i}} : \mathbb{Z}^m \to \mathbb{Z}^m$ defined as follows. Let $\tau_q(\xi_1, \ldots, \xi_m) = (\xi'_1, \ldots, \xi'_m)$, then $\xi'_k = \xi_k$ for all $k \neq q$, and

$$\xi'_q = \xi_q - \sum_{(k \to q) \in \Sigma_{\mathbf{i}}} c_{kq}\xi_k + \sum_{(q \to l) \in \Sigma_{\mathbf{i}}} c_{lq}\xi_l. \tag{2.4}$$

Let $\Gamma_{\mathbf{i}}$ denote the group of linear transformations of $\mathbb{Z}^m$ generated by the transvections τ_q for all $\mathbf{i}$-bounded indices $q \in [1, m]$. Similarly, $\Gamma_{\mathbf{i}}(\mathbb{F}_2)$ denotes the group of linear transformations of the $\mathbb{F}_2$-vector space $\mathbb{F}_2^m$ obtained from $\Gamma_{\mathbf{i}}$ by reduction modulo 2.

EXAMPLE 2.9. For $W = S_3$ the pair (w_0, e) gives rise to the reduced word $\mathbf{i} = (1\ 2\ 1)$, which corresponds to the following graph:

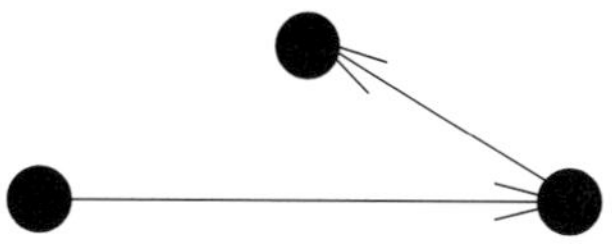

FIGURE 2.14. The graph associated to reduced word $\mathbf{i} = (1\ 2\ 1)$

The only bounded index is 3. The group $\Gamma_{\mathbf{i}}(\mathbb{F}_2)$ is generated by the unique symplectic transvection $\tau_3 : (\xi_1, \xi_2, \xi_3) \mapsto (\xi_1, \xi_2, \xi_3 + \xi_2 - \xi_3)$. The linear space $\mathbb{F}_2^3$ is subdivided into six $\Gamma_{\mathbf{i}}(\mathbb{F}_2)$-orbits:

$$\{(0,0,0)\}, \{(0,0,1)\}, \{(1,1,0)\}, \{(1,1,1)\}, \{(0,1,0),(0,1,1)\}, \{(1,0,0),(1,0,1)\}.$$

In the above example, the number of orbits coincides with the number of connected components in the intersection of two open opposite Schubert cells in SL_3. Let us formulate a general result.

THEOREM 2.10. *For every reduced word* $\mathbf{i}$ *of* $(u,v) \in W \times W$, *the connected components of* $L^{u,v}(\mathbb{R})$ *are in a natural bijection with the* $\Gamma_{\mathbf{i}}(\mathbb{F}_2)$*-orbits in* $\mathbb{F}_2^m$.

To prove this theorem we need the following lemma.

LEMMA 2.11. *For any fixed* $\mathbf{i}$ *there exist regular functions* $M_1, \dots M_m$ *on* $L^{u,v}$ *with the following properties:*

(i) *If* $k \in [1,m]$ *is not* $\mathbf{i}$*-bounded then* M_k *vanishes nowhere on* $L^{u,v}$.

(ii) *The map* $(M_1, \dots, M_m) : L^{u,v} \to \mathbb{C}^m$ *restricts to a biregular isomorphism* $U_{\mathbf{i}} \to (\mathbb{C}^*)^m$, *where* $U_{\mathbf{i}}$ *is the locus of all* $x \in L^{u,v}$ *such that* $M_k(x) \neq 0$ *for all* $k \in [1,m]$.

(iii) *For every* $\mathbf{i}$*-bounded* $q \in [1,m]$, *the rational function* M'_q *defined by*

$$M'_q M_q = \prod_{(k \to q) \in \Sigma_{\mathbf{i}}} M_k^{c_{kq}} + \prod_{(q \to l) \in \Sigma_{\mathbf{i}}} M_l^{c_{lq}} \tag{2.5}$$

is regular on $L^{u,v}$.

(iv) *For every* $\mathbf{i}$*-bounded* $n \in [1,m]$, *the map*

$$(M_1, \dots, M_{n-1}, M'_n, M_{n+1}, \dots M_m) : L^{u,v} \to \mathbb{C}^m$$

restricts to a biregular isomorphism $U_{n,\mathbf{i}} \to (\mathbb{C}^*)^m$, *where* $U_{n,\mathbf{i}}$ *is the locus of all* $x \in L^{u,v}$ *such that* $M'_n \neq 0$ *and* $M_k(x) \neq 0$ *for all* $k \in [1,m] \setminus \{n\}$.

(v) *The functions* M_k *and* M'_n *take real values on* $L^{u,v}(\mathbb{R})$, *and the biregular isomorphisms in* (2) *and* (4) *restrict to biregular isomorphisms* $U_{\mathbf{i}}(\mathbb{R}) \to (\mathbb{R}^*)^m$ *and* $U_{n,\mathbf{i}}(\mathbb{R}) \to (\mathbb{R}^*)^m$.

Regular functions $M_{r_1}, \dots, M_{r_s}$ for unbounded indices r_j play a special role in what follows.

The proof of Lemma 2.11 is rather technical, and will be omitted. Instead, we would like to illustrate the lemma by providing the construction of M_k in SL_n-case.

Recall that a reduced word gives rise to a *pseudoline arrangement* as in the following example (each entry i of $\mathbf{i}$ produces a crossing in the ith level counting from the bottom).

EXAMPLE 2.12. The reduced decomposition $v = s_1 s_3 s_2 s_1$ produces the pseudoline arrangement shown in Fig. 2.15.

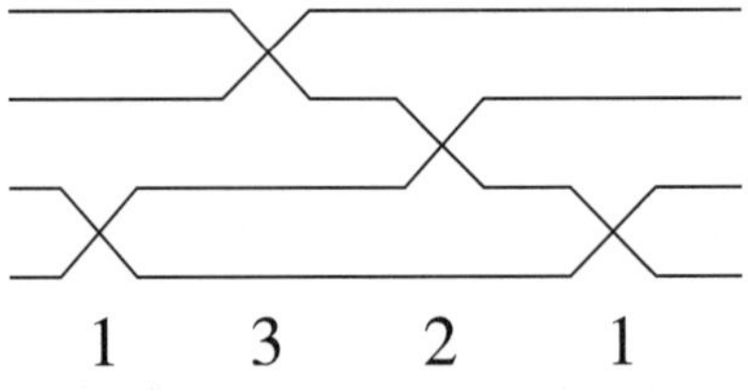

FIGURE 2.15. The pseudoline arrangement associated to the reduced word $\mathbf{i} = (1\ 3\ 2\ 1)$

Given a reduced word $\mathbf{i}$ for $(u, v) \in S_n \times S_n$, with the subword E corresponding to u and the subword F corresponding to v, we construct a double pseudoline arrangement, superimposing pseudoline arrangements for E and F, aligning them closely in the vertical direction, and placing the intersections so that tracing them left-to-right would produce the same shuffle of E and F that appears in $\mathbf{i}$.

Say, the reduced word $\mathbf{i} = (2\ -1\ 3\ -3\ -2\ 1\ 2\ -1\ 1)$ for the pair of permutations (u, v), where $u = (3\ 2\ 4\ 1)$ and $v = (3\ 4\ 2\ 1)$, gives the pseudoline arrangement shown in Fig. 2.16.

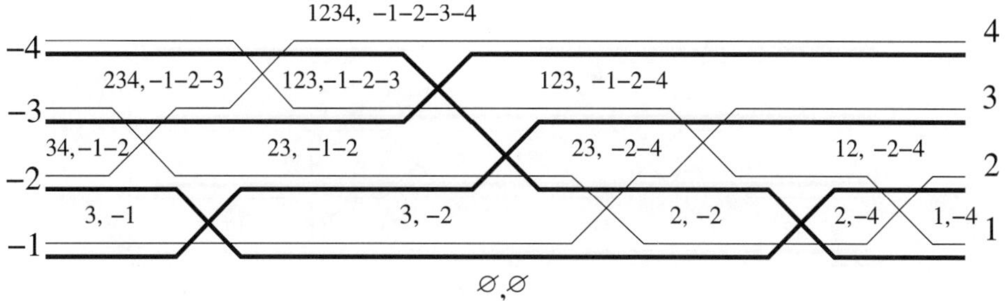

FIGURE 2.16. Double pseudoline arrangement associated to the reduced word $\mathbf{i} = (2\ -1\ 3\ -3\ -2\ 1\ 2\ -1\ 1)$

The pseudoline arrangement for $E = (-1\ \ -3\ \ -2\ \ -1)$ is shown in bold lines, the one for $F = (2\ 3\ 1\ 2\ 1)$, in thin lines. Each E–E (F–F) intersection corresponds to a letter of E (F) in $\mathbf{i}$; E–F intersections are of no interest. The pseudolines of the E-part (F-part) are numbered from 1 to n bottom-up at the left (right) end. Note that reading off the labels of E bottom-up at the right end we obtain u^{-1}, while reading off the labels of F bottom-up at the left end we obtain v.

The arrangement has clearly visible horizontal strips of levels 1 to $n - 1$. (The bottom strip is numbered 1, the top strip is $n - 1$.) Each strip of the pseudoline arrangement is divided into closed chambers. We take into account also chambers at the ends of the strips. The leftmost (rightmost) endpoint of a chamber correspond to an E–E or an F–F intersection. Given $i \in [1, m]$, $C(i)$ stands for the chamber whose rightmost point is the ith intersection point counting from left to right.

We assign to each chamber two sets of indices: indices of pseudolines of the E-part (F-part) below the chamber (see Figure 2.16). For a chamber C, we denote these sets by $I(C)$ and $J(C)$, respectively. The cardinalities of $I(C)$ and in $J(C)$ coincide and are equal to the level of C.

For an $n \times n$ matrix X, denote by $\Delta_{I(C)}^{J(C)}(X)$ the minor of X consisting of rows $I(C)$ and columns $J(C)$. Finally, put $M_i(X) = \Delta_{I(C(i))}^{J(C(i))}(X)$.

EXAMPLE 2.13. Consider the reduced double Bruhat cell L^{e,w_0} in SL_5. As we have discussed above, it is isomorphic to the following set of unipotent upper triangular matrices:

(2.6)
$$\left\{ \begin{pmatrix} 1 & x_{12} & x_{13} & x_{14} & x_{15} \\ 0 & 1 & x_{23} & x_{24} & x_{25} \\ 0 & 0 & 1 & x_{34} & x_{35} \\ 0 & 0 & 0 & 1 & x_{45} \\ 0 & 0 & 0 & 0 & 1 \end{pmatrix} : \Delta_1^5 \neq 0,\ \Delta_{12}^{45} \neq 0,\ \Delta_{123}^{345} \neq 0,\ \Delta_{1234}^{2345} \neq 0, \right\}$$

Take $\mathbf{i} = (1\ 2\ 3\ 4\ 1\ 2\ 3\ 1\ 2\ \ 1)$. Figure 2.17 presents the corresponding double pseudoline arrangement. Note that since the E-part of this arrangement is trivial, we omit the corresponding E-indices.

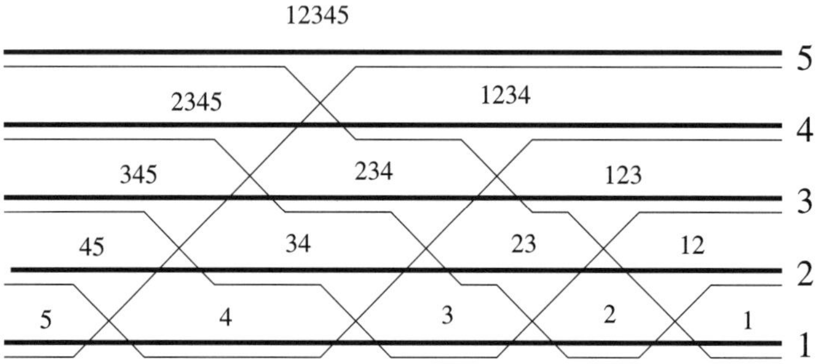

FIGURE 2.17. Double pseudoline arrangement associated to reduced word $\mathbf{i} = (\ 1\ 2\ 3\ 4\ 1\ 2\ 3\ 1\ 2\ 1)$

It follows immediately from the definition that all M_i's are regular functions on SL_5. Moreover, M_4, M_7, M_9, M_{10} do not vanish on L^{e,w_0} (see inequalities (2.6)).

Let us check the behavior of M_i'. The graph induced by the reduced decomposition $\mathbf{i}$ is shown in Fig. 2.18.

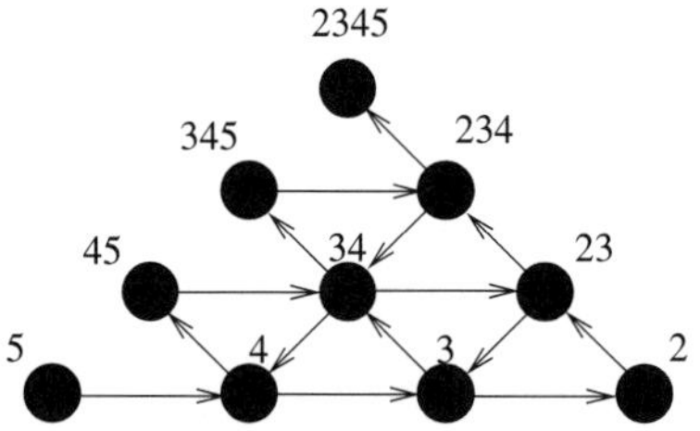

FIGURE 2.18. Oriented graph associated to reduced word $\mathbf{i} = (\ 1\ 2\ 3\ 4\ 1\ 2\ 3\ 1\ 2\ 1)$

Consider for instance $M_9' = \frac{M_8+M_6}{M_9}$. Since M_9 does not vanish on L^{e,w_0}, the function M_9' is regular. The function

$$M_6' = \frac{M_2M_7M_8 + M_4M_5M_9}{M_6} = x_{13}\Delta_{123}^{245} - x_{12}\Delta_{123}^{345}$$

is evidently regular on the whole SL_5. Finally, the function

$$M_2' = \frac{M_3M_5 + M_1M_6}{M_2} = \Delta_{12}^{24}$$

is also regular on SL_5.

REMARK 2.14. Functions M_i arise in a natural way from the factorization problem discussed in Section 1.2.4. Namely, factorization parameters t_i are meromorphic functions on $L^{u,v}$. Functions M_i are regular functions on $L^{u,v}$ and are monomials in t_i. Vice versa, it can be shown that factorization parameters are expressed monomially in terms of M_is. Factorization parameters are naturally associated with intersection points of pseudolines, while functions M_i are labeled by chambers formed by pseudolines. Another collection of functions τ_i monomial in t_i and M_i can be defined as follows. Functions τ_i are associated with chambers as well. If a chamber i is bounded by an intersection point p_r of pseudolines from the right and by an intersection point p_l from the left then $\tau_i = t(p_r)/t(p_l)$. Generalizations of these functions in a more general context of cluster algebras will be discussed in Section 4.1.

REMARK 2.15. It is not true that functions M_i are minors of the matrix X for an arbitrary reduced decomposition. For a generic reduced decomposition, M_i are generalized minors of the so-called twisted matrix X', where the *twist* is a birational map between $L^{u,v}$ and $L^{u^{-1},v^{-1}}$.

Now we derive Theorem 2.10 from Lemma 2.11. In fact, we provide an explicit bijection between connected components and orbits.

For every $\xi = (\xi_1, \dots, \xi_m) \in \mathbb{F}_2^m$, denote by $U_{\mathbf{i}}(\xi)$ the set of all $x \in U_{\mathbf{i}}(\mathbb{R})$ such that $(-1)^{\xi_k} M_k(x) > 0$ for all k, and by $\overline{U_{\mathbf{i}}(\xi)}$ its closure in $U_{\mathbf{i}}(\mathbb{R})$.

THEOREM 2.16. *For any reduced word* $\mathbf{i}$ *for* (u,v)*, the correspondence* $\mathcal{Q} \mapsto \cup_{\xi\in\mathcal{Q}}\overline{U_{\mathbf{i}}(\xi)}$*, where* $\mathcal{Q}$ *is a* $\Gamma_{\mathbf{i}}(\mathbb{F}_2)$*-orbit, is a bijection between* $\Gamma_{\mathbf{i}}(\mathbb{F}_2)$*-orbits in* $\mathbb{F}_2^m$ *and connected components of* $L^{u,v}(\mathbb{R})$.

PROOF. Let $U = U_{\mathbf{i}} \bigcup (\cup_n U_{n,\mathbf{i}}) \subset L^{u,v}$, where the inner union is taken over all $\mathbf{i}$-bounded n. We are going to prove first that the number of connected components of $L^{u,v}(\mathbb{R})$ coincides with the number of connected components of $U(\mathbb{R})$ by showing that the complement of $U(\mathbb{R})$ in $L^{u,v}(\mathbb{R})$ has codimension two. As a second step, we are going to label connected components in $U(\mathbb{R})$ by the orbits of $\Gamma_{\mathbf{i}}(\mathbb{F}_2)$.

Let us start with the first step. We show that the complement of $U(\mathbb{R})$ can be described as a finite union of common zero sets of at least two irreducible polynomial functions. Then the codimension two statement follows immediately.

From Lemma 2.11 we know that all M_n and M'_n are regular functions on $L^{u,v}$. Let $x \in L^{u,v}$ be a common zero of M_n and M_k (or of M_n and M'_n), where n and k are $\mathbf{i}$-bounded. Then $x \notin U$.

Conversely, suppose $x \in L^{u,v} \setminus U$. Since $x \notin U_{\mathbf{i}}$, property (i) in Lemma 2.11 implies that $M_n(x) = 0$ for some $\mathbf{i}$-bounded n. Since $x \notin U_{n,\mathbf{i}}$, it follows that either $M'_n(x) = 0$, or $M_k(x) = 0$ for some $\mathbf{i}$-bounded $k \neq n$.

We see that that $L^{u,v}(\mathbb{R}) \setminus U(\mathbb{R})$ is the union of finitely many subvarieties, each given by two (distinct) equations. Since $L^{u,v}$ is a Zariski open subset in $\mathbb{C}^m$, it remains to prove that these equations are irreducible, and the codimension 2 result will follow.

More exactly, we need to show that M_n are irreducible functions for all $\mathbf{i}$-bounded n, whereas each M'_n is equal to an irreducible element $M''_n \in \mathcal{O}(L^{u,v})$ times a Laurent monomial in $M_1, \dots, \widehat{M_n}, \dots, M_m$.

We start with the irreducibility of M_n. Assume that $M_n = P \cdot Q$, where P and Q are some regular functions on $L^{u,v}$. Restrict all functions to $U_{\mathbf{i}}$. Since P and Q are regular functions on $U_{\mathbf{i}}$, they are Laurent polynomials in $M_1, \dots, M_m$. Equality $M_n = P \cdot Q$ implies that both P and Q are in fact monomials. However, a monomial $M_1^{d_1} \times \cdots \times M_m^{d_m}$ is regular on $L^{u,v}$ if and only if all d_i corresponding to $\mathbf{i}$-bounded i are nonnegative. Indeed, the product $M_1^{d_1} \times \cdots \times M_m^{d_m}$ must be a regular function on $U_{k,\mathbf{i}}$, i.e., it must be a Laurent polynomial in $M_1, \dots, M'_k, \dots, M_m$, which is possible in view of (2.5) only for a nonnegative value of d_k. In particular, it means, that either P or Q is a monomial containing only unbounded variables M_i, i.e., it is an invertible element of $\mathcal{O}(L^{u,v})$. Hence, M_n is irreducible.

Assume now that $M'_n = P \cdot Q$, where $P, Q \in \mathcal{O}(L^{u,v})$ are as above. Restriction to $U_{n,\mathbf{i}}$ shows that both P and Q must be Laurent monomials in $M_1, \dots, M_{n-1}, M'_n, M_{n+1}, \dots, M_m$. Moreover, since both P and Q may contain only positive powers of M'_n, we can conclude that one of them, say P, contains M'_n in some nonnegative power times a Laurent monomial in $M_1, \dots, \widehat{M_n}, \dots, M_m$, whereas Q is a Laurent monomial in $M_1, \dots, \widehat{M_n}, \dots, M_m$.

This accomplishes the first step in the proof of Theorem 2.16.

We illustrate this on the following example.

EXAMPLE 2.17. Let $W = S_3$. Then L^{e,w_0} is given by (2.3). Choosing $\mathbf{i} = (121)$ we obtain $U_{\mathbf{i}} = (\mathbb{C}^*)^3$ with toric coordinates $M_1 = z \neq 0$, $M_2 = xy - z \neq 0$, $M_3 = x \neq 0$. As we have seen above, the only bounded index is 3. The open set $U_{3,\mathbf{i}}$ is $(\mathbb{C}^*)^3$ with coordinates $M_1 = z \neq 0$, $M_2 = xy - z \neq 0$, $M'_3 = y \neq 0$. The complement

$$L^{e,w_0} \setminus (U_{\mathbf{i}} \cup U_{3,\mathbf{i}}) = \left\{ \begin{pmatrix} 1 & 0 & z \\ 0 & 1 & 0 \\ 0 & 0 & 1 \end{pmatrix} : z \neq 0 \right\}$$

has codimension 2 in L^{e,w_0}.

In particular, the above codimension 2 result gives rise to the following statement.

COROLLARY 2.18. *For any reduced word* $\mathbf{i}$ *for* (u, v)*, the ring* $\mathcal{O}(L^{u,v})$ *of regular functions on the reduced double Bruhat cell* $L^{u,v}$ *coincides with* $\mathcal{O}(U)$*, where* $U = U_{\mathbf{i}} \bigcup (\cup_n U_{n,\mathbf{i}})$.

PROOF. Indeed, the inclusion $\mathcal{O}(L^{u,v}) \hookrightarrow \mathcal{O}(U)$ follows immediately from the embedding $U \hookrightarrow L^{u,v}$. On the other hand, let f be a regular function on U that is not regular on $L^{u,v}$. Recall that $L^{u,v}$ is biregularly isomorphic to the complement of a finite union of irreducible hypersurfaces D_i in $\mathbb{C}^m$. Therefore, the divisor of poles of f has codimension 1 in $\mathbb{C}^m$. Since f is not regular on $L^{u,v}$, this latter divisor does not lie entirely in the union of D_i, and hence its intersection with $L^{u,v}$ has codimension 1 in $L^{u,v}$. Therefore, it intersects U nontrivially, a contradiction. $\square$

To accomplish the second step of the proof of Theorem 2.16, let us note that $U_{\mathbf{i}}$ and all the $U_{n,\mathbf{i}}$ are dense in U. Moreover, $U_{\mathbf{i}}(\mathbb{R})$ consists of 2^m connected orthants $U_{\mathbf{i}}(\xi)$, $\xi \in \mathbb{F}_2^m$. To finish the proof it is enough to find out which of the orthants $U_{\mathbf{i}}(\xi)$ are connected in $U(\mathbb{R})$. We say that two orthants $U_{\mathbf{i}}(\xi)$ and $U_{\mathbf{i}}(\xi')$ are adjacent if $\overline{U_{\mathbf{i}}(\xi)} \cap \overline{U_{\mathbf{i}}(\xi')} \neq \emptyset$.

Let us prove that two orthants are connected in $U(\mathbb{R})$ if and only if there exists a sequence of adjacent orthants starting at one of them and ending at the other one. The if statement is evident. Let $U_{\mathbf{i}}(\xi)$ and $U_{\mathbf{i}}(\xi')$ be adjacent, and let $x \in \overline{U_{\mathbf{i}}(\xi)} \cap \overline{U_{\mathbf{i}}(\xi')}$. Recall that all M_i are regular functions on $L^{u,v}$. Therefore, there exists $k \in [1, m]$ such that $M_k(x) = 0$, while $M_k(U_{\mathbf{i}}(\xi))$ and $M_k(U_{\mathbf{i}}(\xi'))$ have different signs, and hence $\xi_k \neq \xi'_k$. Observe that it follows from the first part of the proof that such k is unique, that is, $\xi_j = \xi'_j$ for $j \neq k$.

Since $x \in U(\mathbb{R})$ and all $U_{\mathbf{i}}(\xi)$ are disjoint for distinct ξ, we see that $x \in U_{k,\mathbf{i}}$ and $x \notin U_{\mathbf{i}}(\xi)$ as well as $x \notin U_{\mathbf{i}}(\xi')$. In particular, $M'_k(x) \neq 0$, which means that $U_{\mathbf{i}}(\xi)$ and $U_{\mathbf{i}}(\xi')$ are indeed connected in $U(\mathbb{R})$. Moreover, it follows that two monomials on the right hand side of transformation formula (2.5) must have opposite signs at x. We thus have

$$\xi'_k - \xi_k = 1 = \sum_{(k \to r) \in \Sigma_{\mathbf{i}}} c_{rk}\xi_r - \sum_{(s \to k) \in \Sigma_{\mathbf{i}}} c_{sk}\xi_s$$

(since all the expressions in the latter formula are defined over $\mathbb{F}_2$, subtraction signs are used for illustrative purposes only). Comparing this with (2.4), we conclude that $\xi' = \tau_k(\xi)$, and hence ξ' and ξ belong to the same $\Gamma_{\mathbf{i}}(\mathbb{F}_2)$-orbit. Conversely, if ξ and ξ' are connected by a transvection, we see that the corresponding orthants are adjacent. □

2.2.4. Summary. Let us summarize what we had learned about $\mathcal{O}(L^{u,v})$.

- The ring of regular functions on a reduced double Bruhat cell $\mathcal{O}(L^{u,v})$ coincides with the ring $\mathcal{O}\left(U_{\mathbf{i}} \cup (\cup_n U_{n,\mathbf{i}})\right)$, where $U_{\mathbf{i}}, U_{n,\mathbf{i}}$ are toric charts isomorphic to $(\mathbb{C}^*)^m$.
- The latter ring is an algebra over $\mathbb{K}$ generated by $2b$ generators $M_1, \dots, M_b$, $M'_1, \dots, M'_b$, where b is the number of bounded indices in $\mathbf{i}$.
- For a bounded index q, generators M_q and M'_q satisfy Plücker-type relations (2.5).
- The generators of $\mathcal{O}(L^{u,v})$ are organized into $b+1$ groups ("clusters") numbered 0 to b. Group 0 corresponds to the toric chart $U_{\mathbf{i}}$, group q is associated with the toric chart $U_{q,\mathbf{i}}$. Cluster 0 contains $M_1, \dots, M_b$, cluster q contains $M_1, \dots, M_{q-1}, M'_q, M_{q+1}, \dots, M_b$. These clusters form a star-like graph with $b+1$ labeled vertices and edges connecting vertex 0 with all other vertices (see Fig. 2.19). The edge number q corresponds to the unique Plücker-type relation (2.5).

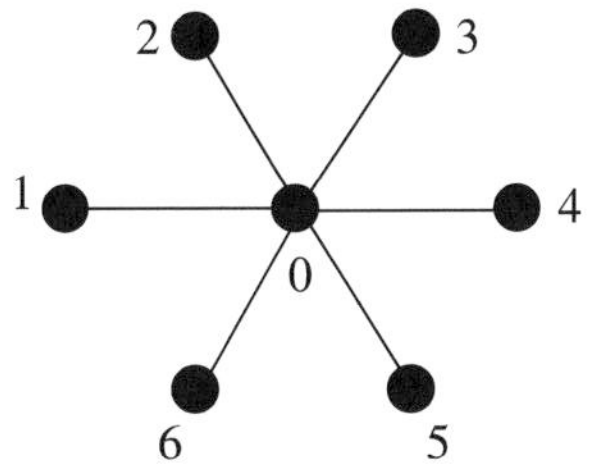

FIGURE 2.19. Star on 7 clusters

An algebra with such a special structure of generators is an example of the so called *upper cluster algebra*.

Bibliographical notes

2.1. For the description of an additive basis in the homogeneous coordinate ring of $G_2(M)$, and other related questions, see [**KR**]. Stasheff's pentagon, and, in general, Stasheff's polytopes (associahedra) were first introduced in [**St**] as CW-complexes. The realization theorem for associahedra is proved, e.g., in [**Le**]. Our definition of abstract polytopes follows [**McMS**].

Triangulations of polygons and Whitehead moves (also called diagonal flips) are discussed, e.g., in [**SlTTh**]. Proposition 2.1 is a particular case of an unpublished result due to D. Thurston (see also [**GSV3**]).

2.2. The problem stated at the beginning of Section 2.2.1 appears as Problem 1987-7 in [**A1**]. The solution for $n = 5$ was obtained in [**SV**]. The complete solution in any dimension mentioned at the end of Section 2.2.2 is given in [**SSV1, SSV2**]. For the generalization of the same approach to double Bruhat cells, see [**SSVZ, Z, GSV1**].

Double Bruhat cells for semisimple Lie groups are introduced in [**FZ1**], reduced double Bruhat cells and the inverse problem of restoring factorization parameters are studied in [**BZ2**]. In Section 2.2.3 we follow [**SSVZ, Z**]. Lemma 2.11 is proved in [**Z**]. The construction of functions M_i in the SL_n-case based on pseudoline arrangements is borrowed from [**BFZ1, FZ1**].

Monomial expressions for factorization parameters via M_i mentioned in Remark 2.14 are obtained in [**FZ1**]. Functions τ_i are introduced and studied in [**SSV2**]. Change of τ_i under elementary transformations of reduced decompositions was computed in the same paper.

For a more detailed definition of the twist mentioned in Remark 2.15, see [**BZ2**] and [**FZ1**].

CHAPTER 3

Cluster algebras

3.1. Basic definitions and examples

The goal of this chapter is to introduce cluster algebras and to study their basic properties. To make the exposition more transparent, and to emphasize the connections to geometry, we decided to restrict our attention to the so-called cluster algebras of geometric type. However, in the text we have indicated possible extensions to general cluster algebras.

Let us start from several technical definitions.

DEFINITION 3.1. Let B be an $n \times n$ integer matrix. We say that

B is *skew-symmetric* if $b_{ij} = -b_{ji}$ for any $i, j \in [1, n]$;

B is *skew-symmetrizable* if there exists a positive integer diagonal matrix D such that DB is skew-symmetric; in this case D is called the *skew-symmetrizer* of B, and B is called *D-skew-symmetrizable*;

B is *sign-skew-symmetric* if $b_{ij}b_{ji} \leq 0$ for any $i, j \in [1, n]$ and $b_{ij}b_{ji} = 0$ implies $b_{ij} = b_{ji} = 0$.

Evidently, each skew-symmetric matrix is skew-symmetrizable, and each skew-symmetrizable matrix is sign-skew-symmetric.

The *coefficient group* $\mathcal{P}$ is a free multiplicative abelian group of a finite rank m with generators $g_1, \dots, g_m$. An *ambient field* in our setting is the field $\mathcal{F}$ of rational functions in n independent variables with coefficients in the field of fractions of the integer group ring $\mathbb{Z}\mathcal{P} = \mathbb{Z}[g_1^{\pm 1}, \dots, g_m^{\pm 1}]$ (here and in what follows we write $x^{\pm 1}$ instead of x, x^{-1}).

DEFINITION 3.2. A *seed* (of *geometric type*) in $\mathcal{F}$ is a pair $\Sigma = (\mathbf{x}, \widetilde{B})$, where $\mathbf{x} = (x_1, \dots, x_n)$ is a transcendence basis of $\mathcal{F}$ over the field of fractions of $\mathbb{Z}\mathcal{P}$ and $\widetilde{B}$ is an $n \times (n+m)$ integer matrix whose principal part B (that is, the $n \times n$ submatrix formed by the columns $1, \dots, n$) is sign-skew-symmetric.

In what follows $\mathbf{x}$ is called a *cluster*, and its elements $x_1, \dots, x_n$ are called *cluster variables*. Denote $x_{n+i} = g_i$ for $i \in [1, m]$. We say that $\widetilde{\mathbf{x}} = (x_1, \dots, x_{n+m})$ is an *extended cluster*, and $x_{n+1}, \dots, x_{n+m}$ are *stable variables*. It will be sometimes convenient to think of $\mathcal{F}$ as of the field of rational functions in $n + m$ independent variables with rational coefficients. The square matrix B is called the *exchange matrix*, and $\widetilde{B}$, the *extended exchange matrix*. Slightly abusing notation, we denote the entries of $\widetilde{B}$ by b_{ij}, $i \in [1, n]$, $j \in [1, n+m]$; besides, we say that $\widetilde{B}$ is skew-symmetric (D-skew-symmetrizable, sign-skew-symmetric) whenever B possesses this property.

DEFINITION 3.3. Given a seed as above, the *adjacent cluster* in direction $k \in [1, n]$ is defined by

$$\mathbf{x}_k = (\mathbf{x} \setminus \{x_k\}) \cup \{x'_k\},$$

where the new cluster variable x'_k is defined by the *exchange relation*

$$x_k x'_k = \prod_{\substack{1\le i\le n+m\\ b_{ki}>0}} x_i^{b_{ki}} + \prod_{\substack{1\le i\le n+m\\ b_{ki}<0}} x_i^{-b_{ki}}; \tag{3.1}$$

here, as usual, the product over the empty set is assumed to be equal to 1.

REMARK 3.4. It is important to note that most papers use a different correspondence between the (extended) exchange matrix and the exchange relations. Namely, their exchange relations correspond to the *columns* of the matrix, and not to its *rows* as above; respectively, the extended exchange matrix has n columns and $n+m$ rows. Clearly, these two approaches are equivalent up to taking the transpose.

Occasionally we will be interested in separating cluster and stable variables in (3.1) by rewriting it in the form

$$x_k x'_k = p_k^+ \prod_{\substack{1\le i\le n\\ b_{ki}>0}} x_i^{b_{ki}} + p_k^- \prod_{\substack{1\le i\le n\\ b_{ki}<0}} x_i^{-b_{ki}}, \tag{3.2}$$

where the coefficients $p_k^\pm \in \mathcal{P}$ are given by

$$p_k^+ = \prod_{\substack{1\le i\le m\\ b_{kn+i}>0}} x_{n+i}^{b_{kn+i}}, \qquad p_k^- = \prod_{\substack{1\le i\le m\\ b_{kn+i}<0}} x_{n+i}^{-b_{kn+i}}. \tag{3.3}$$

DEFINITION 3.5. Let A and A' be two matrices of the same size. We say that A' is obtained from A by a *matrix mutation* in direction k and write $A' = \mu_k(A)$ if

$$a'_{ij} = \begin{cases} -a_{ij}, & \text{if } i = k \text{ or } j = k;\\ a_{ij} + \dfrac{|a_{ik}|a_{kj} + a_{ik}|a_{kj}|}{2}, & \text{otherwise.}\end{cases}$$

It can be easily verified that $\mu_k(\mu_k(A)) = A$.

We say that two matrices $\widetilde{B}$ and $\widetilde{B}'$ are *mutation equivalent*, and write $\widetilde{B} \simeq \widetilde{B}'$, if each of them can be obtained from the other by a sequence of matrix mutations. It is easy to see that the property of a matrix to be sign-skew-symmetric is not necessarily preserved under mutation equivalence. For example, the matrices

$$\begin{pmatrix} 0 & 1 & -2\\ -3 & 0 & 1\\ 1 & -2 & 0\end{pmatrix} \quad \text{and} \quad \begin{pmatrix} 0 & -1 & 2\\ 3 & 0 & -5\\ -1 & -1 & 0\end{pmatrix}$$

are connected by a matrix mutation in direction 1. However, the first of these matrices is sign-skew-symmetric, while the second is not. This brings us to the following definition: a matrix is said to be *totally* sign-skew-symmetric if any matrix that is mutation equivalent to it is sign-skew-symmetric.

An important class of totally sign-skew-symmetric matrices is furnished by the following proposition.

PROPOSITION 3.6. *Skew-symmetrizable matrices are totally sign-skew-symmetric.*

PROOF. The proof follows immediately from Definition 3.1 and Definition 3.5. Moreover, it is easy to see that if $\widetilde{B} \simeq \widetilde{B}'$ and DB is skew-symmetric for some integer diagonal matrix D, then DB' is skew-symmetric as well. □

Given a seed $\Sigma = (\mathbf{x}, \widetilde{B})$, we say that a seed $\Sigma' = (\mathbf{x}', \widetilde{B}')$ is *adjacent* to Σ (in direction k) if $\mathbf{x}'$ is adjacent to $\mathbf{x}$ in direction k and $\widetilde{B}' = \mu_k(\widetilde{B})$. Two seeds are *mutation equivalent* if they can be connected by a sequence of pairwise adjacent seeds.

To get better acquainted with the introduced objects, let us study the change of the $2n$-tuple of coefficients $\mathbf{p} = (p_1^{\pm}, \dots, p_n^{\pm})$ after a transition to an adjacent seed.

PROPOSITION 3.7. *Let the seeds Σ and Σ' be adjacent in direction k. Then the coefficients $(p_j^{\pm})'$ in the exchange relations of the seed Σ' satisfy relations $(p_k^{\pm})' = p_k^{\mp}$ and*

$$\frac{(p_j^+)'}{(p_j^-)'} = \begin{cases} \dfrac{p_j^+}{p_j^-}(p_k^+)^{b_{jk}}, & \textit{if } b_{jk} \geq 0, \\ \dfrac{p_j^+}{p_j^-}(p_k^-)^{b_{jk}}, & \textit{if } b_{jk} \leq 0. \end{cases} \tag{3.4}$$

PROOF. The case $j = k$ follows immediately from (3.3) and the definition of matrix mutations. Let $j \neq k$, then by (3.3) and Definition 3.5 one has

$$\frac{(p_j^+)'}{(p_j^-)'} = \prod_{1 \leq i \leq m} (x'_{n+i})^{b'_{jn+i}} = \prod_{1 \leq i \leq m} x_{n+i}^{b_{jn+i}} \prod_{1 \leq i \leq m} x_{n+i}^{(|b_{jk}|b_{kn+i} + b_{jk}|b_{kn+i}|)/2},$$

since stable variables in the cluster Σ' remain the same. The first factor in the latter product equals p_j^+/p_j^-. Let $b_{jk} \geq 0$, then the second factor equals

$$\prod_{1 \leq i \leq m} x_{n+i}^{b_{jk}(b_{kn+i} + |b_{kn+i}|)/2} = \left(\prod_{\substack{1 \leq i \leq m \\ b_{kn+i} > 0}} x_{n+i}^{b_{kn+i}} \right)^{b_{jk}} = (p_k^+)^{b_{jk}}.$$

Similarly, for $b_{jk} \leq 0$ the second factor equals $(p_k^-)^{b_{jk}}$, and the result follows. □

Now we are ready to define the main object of our studies.

DEFINITION 3.8. Let $\Sigma = (\mathbf{x}, \widetilde{B})$ be a seed with a skew-symmetrizable matrix $\widetilde{B}$, $\mathbb{A}$ be a subring with unity in $\mathbb{Z}\mathcal{P}$ containing all the coefficients $p_k^{\pm}$ for all seeds mutation equivalent to Σ. The *cluster algebra* (of *geometric type*) $\mathcal{A} = \mathcal{A}(\widetilde{B})$ over $\mathbb{A}$ associated with Σ is the $\mathbb{A}$-subalgebra of $\mathcal{F}$ generated by all cluster variables in all seeds mutation equivalent to Σ. The number n of rows in the matrix $\widetilde{B}$ is said to be the *rank* of $\mathcal{A}$. The ring $\mathbb{A}$ is said to be the *ground ring* of $\mathcal{A}$.

In what follows we usually assume that for cluster algebras of geometric type the ground ring is the polynomial ring $\mathbb{Z}[x_{n+1}, \dots, x_{n+m}]$, and do not mention the ground ring explicitly.

A convenient tool in dealing with cluster algebras is the n-regular tree $\mathbb{T}_n$. Its vertices correspond to seeds, and two vertices are connected by an edge labeled by k if and only if the corresponding seeds are adjacent in direction k. The edges of $\mathbb{T}_n$ are thus labeled by the numbers $1, \dots, n$ so that the n edges emanating from each vertex receive different labels. The *distance* $d(\Sigma, \Sigma')$ between the seeds Σ and Σ' is the length of the path between the corresponding vertices in $\mathbb{T}_n$; when this does not lead to a confusion, we will write $d(\mathbf{x}, \mathbf{x}')$ instead of $d(\Sigma, \Sigma')$. Two seeds obtained one from the other by an arbitrary permutation of cluster (or stable) variables and the corresponding permutation of the rows and columns of $\widetilde{B}$ are called *equivalent*. The *exchange graph* of a cluster algebra is defined as the quotient of the tree $\mathbb{T}_n$ modulo this equivalence relation.

EXAMPLE 3.9. Let $n = 1$. The matrix $\widetilde{B}$ is a vector $(b_1, \dots, b_{m+1})$, and the skew-symmetrizability condition boils down to $b_1 = 0$. The initial cluster $\mathbf{x}$ contains one cluster variable x_1, and the unique exchange relation reads $x_1 x_1' = p_1^+ + p_1^-$, where $p_1^\pm$ are monomials in stable variables $x_2, \dots, x_{m+1}$ with exponents prescribed by $\widetilde{B}$. The adjacent cluster consists of x_1', and the corresponding exchange relation coincides with the above one, with p_1^+ and p_1^- interchanged. The two adjacent clusters exhaust the class of mutation equivalence of $\mathbf{x}$. The exchange graph coincides with the 1-regular tree $\mathbb{T}_1$ and is the path of length 1.

Though trivial, this example serves for the description of the homogeneous coordinate ring of $G_2(4)$ and of the ring of regular functions on L^{e,w_0} in SL_3. In the former case one takes $m = 4$ and $\widetilde{B} = (0, 1, 1, -1, -1)$, which leads to the Plücker exchange relation from Section 2.1.2. The ring $\mathbb{C}[G_2(4)]$ is obtained by tensoring the corresponding cluster algebra with $\mathbb{C}$. In the latter case one takes $m = 2$ and $\widetilde{B} = (0, 1, -1)$, which leads to relation $M_3 M_3' = M_1 + M_2$ from Example 2.17. The ring $\mathcal{O}(L^{e,w_0})$ is obtained by tensoring the corresponding cluster algebra over the ground ring $\mathbb{Z}\mathcal{P}$ with $\mathbb{C}$. Instead of changing the ground ring, one can take the localization of the cluster algebra (over the standard polynomial ring) with respect to the stable variables.

Let $n = 2$. Since $\mathbb{T}_2$ is an infinite path, its quotient can be either an infinite path or a cycle. Instead of trying to provide a complete classification of cluster algebras of rank 2, we give several examples.

EXAMPLE 3.10. Let us start from the matrix

$$\widetilde{B} = \begin{pmatrix} 0 & 1 & 0 & -1 & 1 & -1 & 0 \\ -1 & 0 & 1 & 0 & 0 & 1 & -1 \end{pmatrix}.$$

The two possible exchange relations are $x_1 x_1' = x_2 x_5 + x_4 x_6$ and $x_2 x_2' = x_1 x_7 + x_3 x_6$; recall that the variables $x_3, \dots, x_7$ are stable, so $p_1^+ = x_5$, $p_1^- = x_4 x_6$, $p_2^+ = x_3 x_6$, $p_2^- = x_7$. Applying matrix mutation in direction 1 we get

$$\widetilde{B}' = \begin{pmatrix} 0 & -1 & 0 & 1 & -1 & 1 & 0 \\ 1 & 0 & 1 & -1 & 0 & 0 & -1 \end{pmatrix}.$$

The corresponding cluster variables are $(x_2 x_5 + x_4 x_6)/x_1$ and x_2.

On the next step we mutate the obtained matrix in direction 2, then again in direction 1, and so on. The matrices obtained on the four consecutive steps are

$$\begin{pmatrix} 0 & 1 & 0 & 0 & -1 & 1 & -1 \\ -1 & 0 & -1 & 1 & 0 & 0 & 1 \end{pmatrix}, \quad \begin{pmatrix} 0 & -1 & 0 & 0 & 1 & -1 & 1 \\ 1 & 0 & -1 & 1 & -1 & 0 & 0 \end{pmatrix},$$
$$\begin{pmatrix} 0 & 1 & -1 & 0 & 0 & -1 & 1 \\ -1 & 0 & 1 & -1 & 1 & 0 & 0 \end{pmatrix}, \quad \begin{pmatrix} 0 & -1 & 1 & 0 & 0 & 1 & -1 \\ 1 & 0 & 0 & -1 & 1 & -1 & 0 \end{pmatrix},$$

the corresponding pairs of cluster variables are

$$\left(\frac{x_2x_5+x_4x_6}{x_1}, \frac{x_1x_4x_7+x_2x_3x_5+x_3x_4x_6}{x_1x_2}\right),$$
$$\left(\frac{x_1x_7+x_3x_6}{x_2}, \frac{x_1x_4x_7+x_2x_3x_5+x_3x_4x_6}{x_1x_2}\right),$$
$$\left(\frac{x_1x_7+x_3x_6}{x_2}, x_1\right),$$
$$(x_2, x_1).$$

Observe that the last matrix coincides with the original matrix $\widetilde{B}$ up to the transposition of rows and columns, and that the same is true for the elements of the corresponding cluster. Therefore, the sixth cluster is equivalent to the initial one, and hence the total number of nonequivalent clusters in this case is equal to 5. Comparing the exchange relations with the Plücker relations (2.1), one concludes that the arising cluster algebra tensored with $\mathbb{C}$ is isomorphic to the homogeneous coordinate ring of the Grassmannian $G_2(5)$. The corresponding exchange graph is the Stasheff pentagon described in Section 2.1.3. It is convenient to identify the initial cluster with the triangulation in the lower left corner of the pentagon, and the variables $x_1, \dots, x_7$ with the Plücker coordinates $x_{24}, x_{25}, x_{12}, x_{23}, x_{34}, x_{45}, x_{15}$.

Our next example shows that the number of nonequivalent clusters in a cluster algebra is not necessarily finite.

EXAMPLE 3.11. Let again $n = 2$ and let

$$\widetilde{B} = B = \begin{pmatrix} 0 & 3 \\ -3 & 0 \end{pmatrix}.$$

Here $m = 0$, so we are dealing with a *coefficient-free* cluster algebra (that is, all the coefficients in the exchange relations are equal to 1). The two possible exchange relations are $x_1x_1' = x_2^3 + 1$ and $x_2x_2' = 1 + x_1^3$. Applying matrix mutation in direction 1 we get $B' = -B$. The corresponding cluster variables are $(x_2^3+1)/x_1$ and x_2.

On the next step we mutate the obtained matrix in direction 2, then again in direction 1, and so on. The matrices obtained this way are equal to $\pm B$, so the exchange relations remain the same all the time. The new cluster variables are rational functions in x_1 and x_2. Given a rational function f, write it as $f = p/q$ where p and q are polynomials and define $\delta(f) = \deg p - \deg q$ (evidently, $\delta(f)$ depends only on f and does not depend on the specific choice of p and q). Let now $(\bar{x}_1, \bar{x}_2)$ be one of the clusters obtained by the above process. Assume that $\delta(\bar{x}_1) \ge \delta(\bar{x}_2) > 0$; it is easy to see that for $\bar{x}_2' = (1+\bar{x}_1^3)/\bar{x}_2$ one has $\delta(\bar{x}_2') = 3\delta(\bar{x}_1) - \delta(\bar{x}_2) > \delta(\bar{x}_1)$. Since $\delta(x_1) = \delta(x_2) = 1$, this means that δ is not bounded from above on the set of cluster variables, and hence the number of nonequivalent clusters in the cluster algebra $\mathcal{A}(B)$ is infinite. The corresponding exchange graph is a path infinite in both directions.

Though the cluster algebra in the previous example has an infinite number of nonequivalent clusters, it is not difficult to prove that it is finitely generated; in fact, each cluster algebra of rank 2 is generated by four cluster variables subject to two relations, see Lemma 3.17 below. Our next example shows that already cluster algebras of rank 3 can be much more complicated.

EXAMPLE 3.12. Consider the coefficient-free cluster algebra of rank 3 defined by the matrix

$$B = \begin{pmatrix} 0 & 2 & -2 \\ -2 & 0 & 2 \\ 2 & -2 & 0 \end{pmatrix}.$$

Similarly to the previous example, the only other matrix that is mutation equivalent to B is $-B$. Therefore, all the exchange relations are the same as in the initial cluster, namely $x_i x_i' = x_j^2 + x_k^2$, where $\{i, j, k\} = \{1, 2, 3\}$.

All cluster variables are rational functions in x_1, x_2, x_3. Let f be an arbitrary rational function in x_1, x_2, x_3, and $\alpha_1, \alpha_2, \alpha_3$ be arbitrary integers. Define the *valuation* $\nu_{\alpha_1\alpha_2\alpha_3}(f)$ as the multiplicity at point 0 of f restricted to the curve $\{x_i = t^{\alpha_i}\}_{i=1,2,3}$. It is easy to see that $\nu_{\alpha_1\alpha_2\alpha_3}$ has the following properties:

$$(3.5)\qquad \begin{aligned} \nu_{\alpha_1\alpha_2\alpha_3}(fg) &= \nu_{\alpha_1\alpha_2\alpha_3}(f) + \nu_{\alpha_1\alpha_2\alpha_3}(g), \\ \nu_{\alpha_1\alpha_2\alpha_3}(f+g) &\geq \min\{\nu_{\alpha_1\alpha_2\alpha_3}(f), \nu_{\alpha_1\alpha_2\alpha_3}(g)\}, \end{aligned}$$

provided $f + g$ does not vanish identically; the latter relation turns to an equality if both f and g can be represented as ratios of polynomials with nonnegative coefficients.

Let $\bar{x}_i, \bar{x}_j, \bar{x}_k$ form a cluster $\bar{\mathbf{x}}$, and let $\bar{x}_i'$ belong to the cluster adjacent to $\bar{\mathbf{x}}$ in direction i; we denote $\nu_{\alpha_1\alpha_2\alpha_3}(\bar{\mathbf{x}}) = \min\{\nu_{\alpha_1\alpha_2\alpha_3}(\bar{x}_i)\colon i = 1, 2, 3\}$. For a particular choice of $\alpha_1 = \alpha_2 = 1$, $\alpha_3 = 2$ one infers from (3.5) by induction on the distance $d(\mathbf{x}, \bar{\mathbf{x}})$ that

$$(3.6)\qquad \nu_{1,1,2}(\bar{x}_i') = 2\min\{\nu_{1,1,2}(\bar{x}_j), \nu_{1,1,2}(\bar{x}_k)\} - \nu_{1,1,2}(\bar{x}_i) \leq \nu_{1,1,2}(\bar{x}_i),$$

provided $d(\mathbf{x}, \bar{\mathbf{x}}') > d(\mathbf{x}, \bar{\mathbf{x}})$, and the inequality is strict if the path between $\mathbf{x}$ and $\bar{\mathbf{x}}'$ in $\mathbb{T}_3$ contains at least one mutation in direction 3. Define

$$\nu_r = \min\{\nu_{1,1,2}(\bar{\mathbf{x}})\colon d(\mathbf{x}, \bar{\mathbf{x}}) = r\}.$$

It follows from the above inequality that the sequence $\nu_0, \nu_1, \dots$ is strictly decreasing.

Let us first prove that the cluster algebra $\mathcal{A}(B)$ has an infinite number of clusters. To do this, consider the so-called *Markov equation*

$$t_1^2 + t_2^2 + t_3^2 = 3t_1t_2t_3;$$

positive integer triples (t_1, t_2, t_3) satisfying this equation are called *Markov triples.* [1] It is easy to check that if t_i, t_j and t_k form a Markov triple, then so do t_i, t_k and $3t_it_k - t_j$. Moreover, if $t_i \geq t_j \geq t_k$ then $3t_it_k - t_j > t_i \geq t_k$, and hence proceeding in this way we can get an infinite number of different Markov triples. Finally, Markov equation stipulates $3t_it_k - t_j = (t_i^2 + t_k^2)/t_j$.

Let us assign to each cluster $\bar{\mathbf{x}}$ a triple of integer numbers by expressing cluster variables $\bar{x}_1, \bar{x}_2, \bar{x}_3$ via the initial cluster variables x_1, x_2, x_3 and substituting $x_1 = x_2 = x_3 = 1$. Evidently, the triple $(1, 1, 1)$ assigned to the initial cluster is a Markov triple. It follows from the above discussion that the triple of integers assigned to

[1] Markov triples appear mysteriously in different areas of mathematics, such as mathematical physics, algebraic geometry, theory of triangulated categories, etc. Connections of these areas with the cluster algebra theory became a subject of active study in recent years. One of the most intriguing problems involving Markov triples is a Markov conjecture stating that any triple is determined uniquely by its largest element.

any cluster is a Markov triple, and that in such a way we get the same infinite set of Markov triples as above. Therefore, the number of clusters in $\mathcal{A}(B)$ is infinite.

Assume now that $\mathcal{A}(B)$ is finitely generated, and choose r to be the maximum among the distances from $\mathbf{x}$ to generators of $\mathcal{A}(B)$. Then the valuation $\nu_{1-\nu_r,1-\nu_r,2-\nu_r}$ possesses the following properties:

(i) $\nu_{1-\nu_r,1-\nu_r,2-\nu_r}(\bar{\mathbf{x}}) \geq 0$ for any $\bar{\mathbf{x}}$ such that $d(\mathbf{x},\bar{\mathbf{x}}) \leq r$;

(ii) $\nu_{1-\nu_r,1-\nu_r,2-\nu_r}(\bar{\mathbf{x}}) < 0$ for some $\bar{\mathbf{x}}$ such that $d(\mathbf{x},\bar{\mathbf{x}}) = r+1$.

Property (i) together with the inequality in (3.5) implies, in particular, that $\nu_{1-\nu_r,1-\nu_r,2-\nu_r}$ is nonnegative on the whole cluster algebra $\mathcal{A}(B)$, which contradicts property (ii).

Let $\mathcal{A} = \mathcal{A}(\widetilde{B})$ be a cluster algebra of rank n. We can declare some of the cluster variables to be stable. The corresponding exchange relations are discarded. Assuming without loss of generality that only the variables $x_1, \dots, x_k$ remain in the initial cluster for some $k < n$, we get a subalgebra of the original cluster algebra, which is called the *restriction* of $\mathcal{A}$. The corresponding extended exchange matrix is the submatrix of $\widetilde{B}$ in the rows $1, \dots, k$ and columns $1, \dots, n+m$. The tree $\mathbb{T}_k$ is obtained from $\mathbb{T}_n$ by deleting all the edges labeled by $k+1, \dots, n$ and chosing the connected component containing the initial seed. For example, restricting the algebra from Example 3.10 to any one of its cluster variables we get an algebra from Example 3.9.

In a similar way, we can declare some of the stable variables to be equal to one. Assuming without loss of generality that only the stable variables $x_{n+1}, \dots, x_{n+k}$ remain in the initial extended cluster for some $k < m$, we get a quotient algebra of the original cluster algebra; we say that it is obtained from $\mathcal{A}$ by *freezing* variables $x_{n+k+1}, \dots, x_{n+m}$. The corresponding extended exchange matrix is the submatrix of $\widetilde{B}$ in the rows $1, \dots, n$ and columns $1, \dots, n+k$. For example, the algebra related to L^{e,w_0} in SL_3 is obtained from the algebra related to $G_2(4)$ by freezing stable variables x_3 and x_4, see Example 3.9.

REMARK 3.13. To define a *general* cluster algebra, we have to consider, as a coefficient group, an arbitrary multiplicative abelian group without torsion. A seed is defined as a triple $\Sigma = (\mathbf{x}, \mathbf{p}, B)$, where $\mathbf{x}$ has the same meaning as above, $\mathbf{p} = (p_1^{\pm}, \dots, p_n^{\pm})$, $p_i^{\pm} \in \mathcal{P}$ is a $2n$-tuple of coefficients, B is a totally sign-skew-symmetric exchange matrix. Exchange relations are defined by (3.2). The adjacent seed in a given direction i consists of the new cluster $\mathbf{x}'$ obtained via the corresponding exchange relation, new matrix $B' = \mu_i(B)$ and the new tuple of coefficients $\mathbf{p}'$; the new coefficients are *required* to satisfy relations (3.4) given in Proposition 3.7. Observe that these relations do not define the coefficients uniquely. The way to overcome this problem consists in turning $\mathcal{P}$ into a semifield by introducing an additional operation $\oplus$, and in requiring the *normalization condition* $p_j^+ \oplus p_j^- = 1$ for all j. In case of cluster algebras of geometric type the operation $\oplus$ is given by $\prod_j g_j^{a_j} \oplus \prod_j g_j^{b_j} = \prod_j g_j^{\min\{a_j,b_j\}}$.

3.2. Laurent phenomenon and upper cluster algebras

3.2.1. Laurent phenomenon. Let us look once again at the Example 3.10. We have listed there the cluster variables in all the five distinct clusters expressed via the stable variables and the initial cluster variables. The obtained rational functions possess a remarkable feature: all the denominators are monomials in the initial cluster variables. It turns out that that this feature is not specific for the cluster algebra from Example 3.10, but rather is characteristic for all cluster algebras. More precisely, the following statement, known as the *Laurent phenomenon* holds true.

THEOREM 3.14. *Any cluster variable is expressed via the cluster variables from the initial (or any other) cluster as a Laurent polynomial with coefficients in the group ring* $\mathbb{Z}\mathcal{P}$.

PROOF. Since the nature of coefficients is irrelevant in the proof, we assume that our cluster algebra is general, as defined in Remark 3.13.

Given a seed $\Sigma = (\mathbf{x}, \mathbf{p}, B)$, consider the ring $\mathbb{Z}\mathcal{P}[\mathbf{x}^{\pm 1}]$ of Laurent polynomials in variables $x_1, \dots, x_n$. In fact, Theorem 3.14 claims that the whole cluster algebra is contained in the intersection of such rings over all seeds mutation equivalent to Σ. We will prove that under certain mild restrictions the cluster algebra is contained already in the polynomial ring

$$\mathcal{U}(\mathbf{x}) = \mathbb{Z}\mathcal{P}[\mathbf{x}^{\pm 1}] \cap \mathbb{Z}\mathcal{P}[\mathbf{x}_1^{\pm 1}] \cap \dots \cap \mathbb{Z}\mathcal{P}[\mathbf{x}_n^{\pm 1}],$$

where $\mathbf{x}_1, \dots, \mathbf{x}_n$ are all the clusters adjacent to $\mathbf{x}$. The latter polynomial ring is called the *upper bound* associated with the cluster $\mathbf{x}$.

The main idea of the proof is as follows. It is easy to see that the cluster algebra is contained in the subalgebra of $\mathcal{F}$ generated by the union of upper bounds over all seeds. Therefore, it suffices to prove that under certain mild conditions all upper bounds coincide. It is worth to note that all the computations are done for cluster algebras of ranks 1 and 2.

We start from analyzing the following simple situation. Consider an arbitrary cluster algebra of rank 1. Let $\mathbf{x} = (x_1)$ be the initial cluster and $\mathbf{x}_1 = (x_1')$ be the adjacent cluster. The exchange relation is $x_1 x_1' = P_1$, where $P_1 \in \mathbb{Z}\mathcal{P}$. Observe that even in this simple case P_1 is a sum of two elements of $\mathcal{P}$, and hence itself does not belong to $\mathcal{P}$.

LEMMA 3.15. $\mathbb{Z}\mathcal{P}[x_1^{\pm 1}] \cap \mathbb{Z}\mathcal{P}[(x_1')^{\pm 1}] = \mathbb{Z}\mathcal{P}[x_1, x_1']$.

PROOF. The "$\supseteq$" inclusion is obvious. To prove the opposite inclusion, consider an arbitrary $y \in \mathbb{Z}\mathcal{P}[x_1^{\pm 1}]$ and observe that each monomial in y can be written as $c_k x_1^k = c_k'/(x_1')^k$ with $c_k \in \mathbb{Z}\mathcal{P}$ and $c_k' = c_k P_1^k$. Additional condition $y \in \mathbb{Z}\mathcal{P}[(x_1')^{\pm 1}]$ implies $c_k' \in \mathbb{Z}\mathcal{P}$, which means that c_{-k} is divisible by P_1^k if $k \geq 0$ (since $P_1 \notin \mathcal{P}$). Evidently, for any $k \geq 0$, both $c_k x_1^k$ and $c_{-k}' P_1^k / x_1^k = c_{-k}'(x_1')^k$ belong to $\mathbb{Z}\mathcal{P}[x_1, x_1']$, and hence the same holds for y. □

REMARK 3.16. Observe that we can add an arbitrary tuple $\mathbf{z}$ of independent variables to both sides of the above relation; in other words, the following is true:

$$\mathbb{Z}\mathcal{P}[x_1^{\pm 1}, \mathbf{z}] \cap \mathbb{Z}\mathcal{P}[(x_1')^{\pm 1}, \mathbf{z}] = \mathbb{Z}\mathcal{P}[x_1, x_1', \mathbf{z}].$$

The proof follows immediately from Lemma 3.15 by changing the ground ring from $\mathbb{Z}\mathcal{P}$ to $\mathbb{Z}\mathcal{P}[\mathbf{z}]$.

As a next step, consider an arbitrary cluster algebra of rank 2. Let $\mathbf{x} = (x_1, x_2)$ be the initial cluster, $\mathbf{x}_1 = (x_1', x_2)$ and $\mathbf{x}_2 = (x_1, x_2')$ be the two adjacent clusters (in directions 1 and 2, respectively), and $\mathbf{x}_{1,2} = (x_1', x_2'')$ be the cluster adjacent to $\mathbf{x}_1$ in direction 2.

LEMMA 3.17. $\mathbb{Z}\mathcal{P}[x_1, x_1', x_2, x_2'] = \mathbb{Z}\mathcal{P}[x_1, x_1', x_2, x_2'']$.

PROOF. It is enough to show that $x_2'' \in \mathbb{Z}\mathcal{P}[x_1, x_1', x_2, x_2']$. Consider the 2×2 exchange matrix B corresponding to $\mathbf{x}$.

Assume first that $b_{12} = b_{21} = 0$. Then the exchange relations in direction 2 for the clusters $\mathbf{x}$ and $\mathbf{x}_1$ are $x_2 x_2' = p_2^+ + p_2^-$ and $x_2 x_2'' = (p_2^+)' + (p_2^-)'$, with $(p_2^+)'/(p_2^-)' = p_2^+/p_2^-$. Therefore, $x_2'' = p x_2'$ for some $p \in \mathcal{P}$, and we are done.

Assume now that $b_{12} = b$ and $b_{21} = -c$, so that $bc > 0$. Without loss of generality, we may assume that both b and c are positive (the opposite case is symmetric to this one). The exchange relations for the cluster $\mathbf{x}$ are $x_1 x_1' = p_1^+ x_2^b + p_1^-$ and $x_2 x_2' = p_2^- x_1^c + p_2^+$ for some $p_1^\pm, p_2^\pm \in \mathcal{P}$. Similarly, the exchange relations for cluster $\mathbf{x}_1$ are $x_1' x_1 = p_1^+ x_2^b + p_1^-$ and $x_2 x_2'' = (p_2^+)'(x_1')^c + (p_2^-)'$, where by Proposition 3.7 $(p_2^+)'/(p_2^-)' = p_2^+/p_2^-(p_1^-)^c$. Therefore,

$$\begin{aligned} x_2'' &= \frac{(p_2^+)'(x_1')^c + (p_2^-)'}{x_2} = \frac{(p_2^+)'(x_1')^c (x_2 x_2' - p_2^- x_1^c)}{p_2^+ x_2} + \frac{(p_2^-)'}{x_2} \\ &= \frac{(p_2^+)'(x_1')^c x_2'}{p_2^+} - \left(\frac{(p_2^+)' p_2^-}{p_2^+} \cdot \frac{(x_1 x_1')^c}{x_2} - \frac{(p_2^-)'}{x_2} \right) \\ &= \frac{(p_2^+)'(x_1')^c x_2'}{p_2^+} - \frac{(p_2^-)'}{(p_1^-)^c} \cdot \frac{(p_1^+ x_2^b + p_1^-)^c - (p_1^-)^c}{x_2}. \end{aligned}$$

Evidently, both terms in the last expression belong to $\mathbb{Z}\mathcal{P}[x_1, x_1', x_2, x_2']$, and the Lemma is proved. □

In fact, Lemma 3.17 completes the proof of the Theorem for cluster algebras of rank 2. Indeed, an immediate corollary of Lemma 3.17 is that the whole cluster algebra is generated by x_1, x_1', x_2, x_2'. However, to get the proof for cluster algebras of an arbitrary rank, we have to further study cluster algebras of rank 2.

In the same situation as above, let P_1 and P_2 denote the right hand sides of the exchange relations corresponding to the initial cluster $\mathbf{x}$.

LEMMA 3.18. *Assume that P_1 and P_2 are coprime in $\mathbb{Z}\mathcal{P}[x_1, x_2]$. Then*

$$\mathbb{Z}\mathcal{P}[x_1, x_1', x_2^{\pm 1}] \cap \mathbb{Z}\mathcal{P}[x_1^{\pm 1}, x_2, x_2'] = \mathbb{Z}\mathcal{P}[x_1, x_1', x_2, x_2'].$$

PROOF. Consider the same two cases as in the proof of Lemma 3.17.

In the first case, both P_1 and P_2 lie in $\mathbb{Z}\mathcal{P}$. Therefore,

$$\mathbb{Z}\mathcal{P}[x_1, x_1', x_2, x_2'] = \mathbb{Z}\mathcal{P}[x_1, x_2] + \mathbb{Z}\mathcal{P}[x_1, x_2'] + \mathbb{Z}\mathcal{P}[x_1', x_2] + \mathbb{Z}\mathcal{P}[x_1', x_2'].$$

Applying the same reasoning as in the proof of Lemma 3.15, we see that any $y \in \mathbb{Z}\mathcal{P}[x_1, x_1', x_2, x_2']$ is the sum of monomials of the four following types:

(i) $c_{k,l} x_1^k x_2^l$;

(ii) $c_{k,-l} x_1^k x_2^{-l}$ with $c_{k,-l}$ divisible by P_2^l;

(iii) $c_{-k,l} x_1^{-k} x_2^l$ with $c_{-k,l}$ divisible by P_1^k;

(iv) $c_{-k,-l} x_1^{-k} x_2^{-l}$ with $c_{-k,-l}$ divisible by $P_1^k P_2^l$.

In all the listed cases $k, l \geq 0$.

On the other hand, the proof of Lemma 3.15 together with Remark 3.16 provides a similar description for the monomials of any element in $\mathbb{Z}\mathcal{P}[x_1, x_1', x_2^{\pm 1}] \cap \mathbb{Z}\mathcal{P}[x_1^{\pm 1}, x_2, x_2']$. The only difference is that the coefficient of $x_1^{-k} x_2^{-l}$ should be divisible by P_1^k and P_2^l. However, this condition is equivalent to divisibility by $P_1^k P_2^l$, since P_1 and P_2 are coprime.

In the second case, let us start with proving that

$$\mathbb{Z}\mathcal{P}[x_1, x_1', x_2^{\pm 1}] = \mathbb{Z}\mathcal{P}[x_1, x_1', x_2, x_2'] + \mathbb{Z}\mathcal{P}[x_1, x_2^{\pm 1}]. \tag{3.7}$$

As before, the "$\supseteq$" inclusion is obvious. To prove the opposite inclusion, observe that each monomial $y \in \mathbb{Z}\mathcal{P}[x_1, x_1', x_2^{\pm 1}]$ can be written as

$$y = c_{k_1,k_2,k_3}(p_1^+ x_2^b + p_1^-)^{k_1} x_1^{k_2} x_2^{k_3},$$

where $k_1 \geq 0$, $k_1 + k_2 \geq 0$, while k_2, k_3 have arbitrary signs. If $k_3 \geq 0$ then $y \in \mathbb{Z}\mathcal{P}[x_1, x_1', x_2, x_2']$. If $k_3 < 0$ and $k_2 \geq 0$ then $y \in \mathbb{Z}\mathcal{P}[x_1, x_2^{\pm 1}]$. In the remaining case $k_2 < 0$, $k_3 < 0$, and y can be rewritten as

$$y = c_{k_1,k_2,k_3}(p_1^+ x_2^b + p_1^-)^{k_1+k_2}(x_1')^{-k_2} x_2^{k_3}.$$

It is therefore enough to prove that

$$(x_1')^{-k_2} x_2^{k_3} \in \mathbb{Z}\mathcal{P}[x_1, x_1', x_2, x_2'] + \mathbb{Z}\mathcal{P}[x_1, x_2^{\pm 1}]. \tag{3.8}$$

Indeed, the exchange relation for x_2 implies $(1 - t)/x_2 \in \mathbb{Z}\mathcal{P}[x_1, x_2']$, where $t = -(p_2^-/p_2^+)x_1^c \in \mathbb{Z}\mathcal{P}[x_1, x_2']$. Multiplying $(1 - t)/x_2$ by t^i and summing the obtained products for $i = 0, \dots, -k_2 - 1$ we get $(1 - t^{-k_2})/x_2 \in \mathbb{Z}\mathcal{P}[x_1, x_2']$. Raising the left hand side of this inclusion to power $-k_3$ we get $x_2^{k_3} - x_1^{-k_2} w \in \mathbb{Z}\mathcal{P}[x_1, x_2']$, where $w \in \mathbb{Z}\mathcal{P}[x_1, x_2^{-1}]$. It remains to multiply the left hand side of the latter inclusion by $(x_1')^{-k_2}$ and to use the exchange relation for x_1 to get (3.8), and hence (3.7).

It is now easy to complete the proof of the lemma. Indeed, by (3.7) and an obvious inclusion $\mathbb{Z}\mathcal{P}[x_1, x_1', x_2, x_2'] \subset \mathbb{Z}\mathcal{P}[x_1^{\pm 1}, x_2, x_2']$ we get

$$\mathbb{Z}\mathcal{P}[x_1, x_1', x_2^{\pm 1}] \cap \mathbb{Z}\mathcal{P}[x_1^{\pm 1}, x_2, x_2'] = \mathbb{Z}\mathcal{P}[x_1, x_1', x_2, x_2'] + \mathbb{Z}\mathcal{P}[x_1, x_2^{\pm 1}] \cap \mathbb{Z}\mathcal{P}[x_1^{\pm 1}, x_2, x_2'].$$

It is easy to see that the second term in the above sum equals $\mathbb{Z}\mathcal{P}[x_1, x_2, x_2'] \subset \mathbb{Z}\mathcal{P}[x_1, x_1', x_2, x_2']$, and the result follows. □

The last statement concerning cluster algebras of rank 2 claims that under the coprimality condition, the intersection of three Laurent polynomial rings defined by a cluster and its two neighbors does not depend on the cluster. More precisely, recall that the upper bound associated with $\mathbf{x}$ is

$$\mathcal{U}(\mathbf{x}) = \mathbb{Z}\mathcal{P}[\mathbf{x}^{\pm 1}] \cap \mathbb{Z}\mathcal{P}[\mathbf{x}_1^{\pm 1}] \cap \mathbb{Z}\mathcal{P}[\mathbf{x}_2^{\pm 1}],$$

where $\mathbf{x}_1$ and $\mathbf{x}_2$ are the clusters adjacent to $\mathbf{x}$. Besides, let P_1, P_2 be the exchange polynomials for $\mathbf{x}$, and P_1', P_2' be the exchange polynomials for $\mathbf{x}_1$.

LEMMA 3.19. *Assume that P_1 and P_2 are coprime in $\mathbb{Z}\mathcal{P}[x_1, x_2]$ and P_1' and P_2' are coprime in $\mathbb{Z}\mathcal{P}[x_1', x_2]$. Then $\mathcal{U}(\mathbf{x}) = \mathcal{U}(\mathbf{x}_1)$.*

PROOF. First of all, using Lemma 3.15 and Remark 3.16 we get

$$\mathcal{U}(\mathbf{x}) = \mathbb{Z}\mathcal{P}[x_1, x_1', x_2^{\pm 1}] \cap \mathbb{Z}\mathcal{P}[x_1^{\pm 1}, x_2, x_2'],$$

which by Lemma 3.18 is equal to $\mathbb{Z}\mathcal{P}[x_1, x_1', x_2, x_2']$. Similarly,

$$\mathcal{U}(\mathbf{x}_1) = \mathbb{Z}\mathcal{P}[x_1, x_1', x_2, x_2''].$$

The latter two expressions are equal by Lemma 3.17, which completes the proof. □

The statement of Lemma 3.19 can be immediately extended to cluster algebras of an arbitrary rank by applying the argument of Remark 3.16 to Lemmas 3.17

and 3.18. Assuming that all the exchange polynomials $P_1, \dots, P_n$ are pairwise coprime in $\mathbb{Z}\mathcal{P}[\mathbf{x}]$, and the same is true for $P_1', \dots, P_n'$ in $\mathbb{Z}\mathcal{P}[\mathbf{x}_1]$, we get

$$\begin{aligned}
\mathcal{U}(\mathbf{x}) &= \bigcap_{i=1}^{n} (\mathbb{Z}\mathcal{P}[\mathbf{x}^{\pm 1}] \cap \mathbb{Z}\mathcal{P}[x_1^{\pm 1}, \dots, x_{i-1}^{\pm 1}, (x_i')^{\pm 1}, x_{i+1}^{\pm 1}, \dots, x_n^{\pm 1}]) \\
&= \bigcap_{i=1}^{n} \mathbb{Z}\mathcal{P}[x_1^{\pm 1}, \dots, x_{i-1}^{\pm 1}, x_i, x_i', x_{i+1}^{\pm 1}, \dots, x_n^{\pm 1}] \\
&= \bigcap_{i=2}^{n} (\mathbb{Z}\mathcal{P}[x_1, x_1', x_2^{\pm 1}, \dots, x_n^{\pm 1}] \cap \mathbb{Z}\mathcal{P}[x_1^{\pm 1}, \dots, x_{i-1}^{\pm 1}, x_i, x_i', x_{i+1}^{\pm 1}, \dots, x_n^{\pm 1}]) \\
&= \bigcap_{i=2}^{n} \mathbb{Z}\mathcal{P}[x_1, x_1', x_2^{\pm 1}, \dots, x_{i-1}^{\pm 1}, x_i, x_i', x_{i+1}^{\pm 1}, \dots, x_n^{\pm 1}] \\
&= \bigcap_{i=2}^{n} \mathbb{Z}\mathcal{P}[x_1, x_1', x_2^{\pm 1}, \dots, x_{i-1}^{\pm 1}, x_i, x_i'', x_{i+1}^{\pm 1}, \dots, x_n^{\pm 1}] = \mathcal{U}(\mathbf{x}_1).
\end{aligned}$$

As it is explained at the beginning of the proof, the claim of the theorem follows immediately from this equality.

Finally, to lift the coprimality assumption we use the following argument: we regard the coefficients $p_i^{\pm}$ of the exchange polynomials P_i as indeterminates. Then all the coefficients in all exchange polynomials become "canonical" (i.e. independent of the choice of $\mathcal{P}$) polynomials in these indeterminates, with positive integer coefficients. □

For the cluster algebras of geometric type one can sharpen Theorem 3.14 as follows.

PROPOSITION 3.20. *Any cluster variable in a cluster algebra of geometric type is expressed via the cluster variables from the initial (or any other) cluster as a Laurent polynomial whose coefficients are polynomials in stable variables.*

PROOF. Fix an arbitrary stable variable x. Let $\bar{x}$ be a cluster variable belonging to a seed $\bar{\Sigma}$. We will use induction on $d(\bar{\Sigma}, \Sigma)$ to show that $\bar{x}$, as a function of x, is a polynomial whose constant term is a subtraction-free rational expression in $\mathbf{x}$ and all stable variables except for x.

This statement is trivial when $d(\bar{\Sigma}, \Sigma) = 0$. If $d(\bar{\Sigma}, \Sigma) > 0$, then, by the definition of the distance, $\bar{x}$ can be expressed via the cluster variables belonging to a seed whose distance to Σ is smaller than $d(\bar{\Sigma}, \Sigma)$. Therefore, by the inductive assumption, $\bar{x}$ is a ratio of two polynomials in x. Moreover, since x enters at most one of the two monomials in the exchange relation (3.1), both these polynomials have nonzero constant term. It remains to notice that if a ratio of two polynomials with nonzero constant terms a and b is a Laurent polynomial, then it is, in fact, a polynomial, and its constant term equals a/b. □

3.2.2. Upper cluster algebras. Let $\mathcal{A} = \mathcal{A}(\widetilde{B})$ be a cluster algebra of geometric type. The *upper cluster algebra* $\mathcal{U} = \mathcal{U}(\widetilde{B})$ is defined as the intersection of the upper bounds $\mathcal{U}(\mathbf{x})$ over all clusters $\mathbf{x}$ in $\mathcal{A}$. Upper cluster algebras are especially easy to describe in the case of extended exchange matrices of the maximal rank.

THEOREM 3.21. *Let $\widetilde{B}$ be an $n \times (n+m)$ skew-symmetrizable matrix of rank n. Then the upper bounds do not depend on the choice of* $\mathbf{x}$, *and hence coincide with the upper cluster algebra* $\mathcal{U}(\widetilde{B})$.

PROOF. For cluster algebras of geometric type, the extension of Lemma 3.19 to cluster algebras of an arbitrary rank discussed in Section 3.2.1 can be summarized as follows.

COROLLARY 3.22. *If the exchange polynomials $P_1, \dots, P_n$ are coprime in the ring $\mathbb{Z}[x_1, \dots, x_{n+m}]$ for any cluster $\mathbf{x}$ in $\mathcal{A}(\widetilde{B})$ then the upper bounds do not depend on the choice of* $\mathbf{x}$, *and hence coincide with the upper cluster algebra* $\mathcal{U}(\widetilde{B})$.

Therefore, to prove the theorem it suffices to check the coprimality condition of Corollary 3.22. We start from the following observation.

LEMMA 3.23. *If* $\operatorname{rank} \widetilde{B} = n$, *and $\widetilde{B}'$ is obtained from $\widetilde{B}$ by mutation in direction i then* $\operatorname{rank} \widetilde{B}' = n$.

PROOF. Indeed, apply the following sequence of row and column operations to the matrix $\widetilde{B}$. For any l such that $b_{il} < 0$, subtract the ith column multiplied by b_{il} from the lth column. For any k such that $b_{ki} > 0$, add the ith row multiplied by b_{ki} to the kth row. Finally, multiply the ith row and column by -1. It is easy to see that the result of these operations is exactly $\widetilde{B}'$, and the lemma follows since the elementary row and column operations do not change the rank of a matrix. □

It remains to check that the maximal rank condition implies the coprimality condition.

LEMMA 3.24. *Exchange polynomials $P_1, \dots, P_n$ are coprime in $\mathbb{Z}[x_1, \dots, x_{n+m}]$ if and only if no two rows of $\widetilde{B}$ are proportional to each other with the proportionality coefficient being a ratio of two odd integers.*

PROOF. Let $\widetilde{B}_i$ and $\widetilde{B}_j$ be two rows of $\widetilde{B}$ such that $\widetilde{B}_i = \pm\frac{b}{a}\widetilde{B}_j$ with a and b odd coprime positive integers. Then $P_i = M_1^a + M_2^a$, $P_j = M_1^b + M_2^b$ for some monomials $M_1, M_2 \in \mathbb{Z}[x_1, \dots, x_{n+m}]$, and so $M_1 + M_2$ is a common factor of P_i and P_j.

Assume now that P_i and P_j are not coprime. Let $P_i = M_1 + M_2$. Since M_1 and M_2 do not share common factors, we can rename the variables entering M_1 as $y_1, \dots, y_p$, and those entering M_2 as $z_1, \dots, z_q$, so that $M_1 = \prod_{s=1}^p y_s^{d_s}$ and $M_2 = \prod_{r=1}^q z_r^{\delta_r}$. Clearly, d_s and δ_r constitute positive and negative elements of B_i. Assume that P_i can be factored as $P_i = P'P''$, then each of P' and P'' contain exactly one monomial in y-variables and exactly one monomial in z-variables. Consider P' and denote these monomials M_1' and M_2', and the corresponding exponents d_s' and δ_r', respectively. We want to prove the following fact:

$$d_s'/d_s = \delta_r'/\delta_r = c(P_i) \qquad \text{for any } 1 \le s \le p,\ 1 \le r \le q. \tag{3.9}$$

Let us assign to the variables $y_1, \dots, y_p$ and $z_1, \dots, z_q$ nonnegative weights $w_1, \dots, w_p$ and $\omega_1, \dots, \omega_q$, respectively. The weight of a monomial is defined as the sum of the weights of the variables; for example, the weight of M_1 is equal to $\sum_{s=1}^p w_s d_s$. A polynomial is called *quasihomogeneous* with respect to the weights as above if the weights of all monomials are equal. It is easy to see that if a polynomial is quasihomogeneous with respect to certain weights then all its factors are quasihomogeneous with respect to the same weights as well. To prove (3.9) assign weights $w_s = 1/d_s$, $\omega_r = 1/\delta_r$, $w_{\bar{s}} = \omega_{\bar{r}} = 0$ for any $\bar{s} \neq s$ and $\bar{r} \neq r$. This makes P_i into a quasihomogeneous polynomial, hence P' is quasihomogeneous as well. The weights of the monomials M_1' and M_2' are d_s'/d_s and δ_r'/δ_r, respectively, and (3.9) follows with $c(P_i) \neq 0$. Consequently, the row B_i of $\widetilde{B}$ can be restored up to the sign from the vector of exponents of the variables in P' by dividing it by $c(P_i)$.

Assume now that P' is a common factor of P_i and P_j. Then the row B_j can be restored up to the sign from the same vector by dividing it by $c(P_j)$. Therefore, $c(P_i)B_i = \pm c(P_j)B_j$. Represent the rational number $c(P_j)/c(P_i)$ as b/a, where a and b are coprime positive integers. It remains to prove that a and b are both odd.

Indeed, $P_i = M_3^a + M_4^a$ and $P_j = M_3^b + M_4^b$ for some monomials M_3 and M_4. Consider polynomials $t^a + 1$ and $t^b + 1$ in one variable. If either a or b is even then $t^a + 1$ and $t^b + 1$ are coprime in $\mathbb{Z}[t]$, and hence there eixst $f, g \in \mathbb{Z}[t]$ such that $f(t)(t^a + 1) + g(t)(t^b + 1) = 1$. Substituting $t = M_4/M_3$ we obtain $F(M_3^a + M_4^a) + G(M_3^b + M_4^b) = M_3^c$ in $\mathbb{Z}[x_1, \dots, x_{n+m]}$ for some nonnegative integer c. Therefore the common factor of P_i and P_j is a monomial, a contradiction. □

□

It follows from the discussion in the previous section that $\mathcal{A}(\widetilde{B}) \subseteq \mathcal{U}(\widetilde{B})$. The following example shows that in general these two algebras are different.

EXAMPLE 3.25. Consider the coefficient-free cluster algebra $\mathcal{A}$ introduced in Example 3.12 and the corresponding upper cluster algebra $\mathcal{U}$. This time we make use of valuation $\nu_{\alpha_1\alpha_2\alpha_3}(f)$ for $\alpha_1 = \alpha_2 = \alpha_3 = 1$. Similarly to (3.6) we can prove that

$$\nu_{1,1,1}(\bar{x}_i') = 2\min\{\nu_{1,1,1}(\bar{x}_j), \nu_{1,1,1}(\bar{x}_k)\} - \nu_{1,1,1}(\bar{x}_i) = \nu_{1,1,1}(\bar{x}_i),$$

and hence $\nu_{1,1,1}(x) = 1$ for any cluster variable x. This defines a grading on $\mathcal{A}$, and the zero-degree component in this grading consists of the elements of the ground ring.

On the other hand, consider the element

$$y = \frac{x_1^2 + x_2^2 + x_3^2}{x_1 x_2} \in \mathcal{F}.$$

Using the exchange relations, one can rewrite it as

$$y = \frac{x_1 + x_1'}{x_2} = \frac{x_2 + x_2'}{x_1} = \frac{(x_1^2 + x_2^2)(x_1^2 + x_2^2 + (x_3')^2)}{x_1 x_2 (x_3')^2},$$

so $y \in \mathcal{U}$. Clearly, $\nu_{1,1,1}(y) = 0$ and y does not lie in the ground ring, so $y \notin \mathcal{A}$, and hence $\mathcal{A} \neq \mathcal{U}$.

3.3. Cluster algebras of finite type

We say that a cluster algebra $\mathcal{A}$ is of *finite type* if the number of nonequivalent clusters in $\mathcal{A}$ is finite, or, which is the same, if the exchange graph of $\mathcal{A}$ is finite. Surprisingly, the classification of these algebras is very similar to the Cartan–Killing classification of semisimple Lie algebras. To formulate a precise statement we need to introduce several definitions.

Let A be an integer symmetrizable square matrix. We say that A is *quasi-Cartan* if all the diagonal entries of A are equal to 2, and *Cartan* if, in addition, all off-diagonal entries of A are non-positive. A (quasi)-Cartan matrix is called *positive* if all its principal minors are positive.

Given an integer skew-symmetrizable square matrix B, its *quasi-Cartan companion* is a quasi-Cartan matrix A such that $|a_{i,j}| = |b_{i,j}|$. Observe that B has 2^N quasi-Cartan companions, where N is the number of nonzero entries in B lying above the main diagonal. Exactly one of these quasi-Cartan companions is a Cartan matrix, called the *Cartan companion* of B.

Let B be an integer skew-symmetrizable square matrix. We say that B is *2-finite* if for any matrix B' mutation equivalent to B one has

$$|b'_{ij}b'_{ji}| \leq 3 \tag{3.10}$$

for all pairs of indices i, j.

THEOREM 3.26. *Let $\widetilde{B}$ be a sign-skew-symmetric matrix and B be its principal part, then the following statements are equivalent.*

(1) *The cluster algebra $\mathcal{A}(\widetilde{B})$ is of finite type.*

(2) *B is 2-finite.*

(3) *There exists a matrix B' mutation equivalent to B such that the Cartan companion of B' is positive.*

Furthermore, the Cartan–Killing type of the Cartan matrix in (3) is uniquely determined by B.

REMARK 3.27. As follows from the definitions above, it is implicit in (2) and (3) that B is skew-symmetrizable.

3.3.1. Proof of the implication (1) $\Longrightarrow$ (2). Assume to the contrary that there exist B' mutation equivalent to B and indices i, j such that $|b'_{ij}b'_{ji}| \geq 4$. Without loss of generality, we may restrict ourselves to the case $B' = B$ and $i = 1$, $j = 2$. We want to prove that already the rank 2 subalgebra of $\mathcal{A}(\widetilde{B})$ obtained by the restriction to the first two cluster variables possesses an infinite number of clusters.

Let $b = |b_{12}|$, $c = |b_{21}|$.

The case $b, c \geq 2$, $bc \neq 4$ is quite similar to Example 3.11 above. Indeed, let (y_1, y_2) be a cluster obtained from the initial one by applying mutations in directions 2 and 1 alternatively, the total number of mutations being even. Write y_1, y_2 as rational functions in the initial cluster variables x_1, x_2 and define $\delta(y_1), \delta(y_2)$ as in Example 3.11. Assume that $\delta(y_1) \geq \delta(y_2) > 0$; after applying the mutation in direction 2 followed by the mutation in direction 1 we get a cluster (z_1, z_2) such that

$$\delta(z_1) = bc\delta(y_1) - b\delta(y_2) - \delta(y_1), \qquad \delta(z_2) = c\delta(y_1) - \delta(y_2).$$

It is easy to see that $\delta(z_1) > \delta(z_2)$ provided $bc > b + c$, and $\delta(z_1) > \delta(y_1)$ provided $bc > b + 2$. In our case both inequalities are satisfied, so we get at infinite number of clusters.

The case $b = c = 2$ follows immediately from the Example 3.12 above. Indeed, it suffice to notice that that all the Markov triples obtained from $(1, 1, 1)$ by the process described in Example 3.12 are of the form $(t_1, t_2, 1)$.

The remaining case $b = 1$, $c \geq 4$ (or $b \geq 4$, $c = 1$) is similar to the first one, though slightly more subtle. Consider (y_1, y_2) and (z_1, z_2) as above, and, in addition, the cluster (w_1, w_2) obtained from (z_1, z_2) by the mutation in direction 2 followed by the mutation in direction 1. It is easy to see that

$$\delta(z_1) = c\delta(y_1) - \delta(y_1) - \delta(y_2), \qquad \delta(z_2) = c\delta(y_1) - \delta(y_2);$$

assume that

$$\delta(z_1) > \delta(y_1) > 0, \qquad \delta(z_2) > \delta(y_2) > 0. \tag{3.11}$$

The first inequality in (3.11) is equivalent to $(c-2)\delta(y_1) > \delta(y_2)$, the second one, to

$$c\delta(y_1) > 2\delta(y_2). \tag{3.12}$$

Consequently, the first inequality in (3.11) follows from (3.12), provided $c-2 \geq c/2$, which is equivalent to $c \geq 4$.

Let us check that $c\delta(z_1) > 2\delta(z_2)$, which by the above is equivalent to

$$c(c-3)\delta(y_1) > (c-2)\delta(y_2).$$

Indeed, by (3.12),

$$c(c-3)\delta(y_1) > 2(c-3)\delta(y_2) \geq (c-2)\delta(y_2),$$

since $c \geq 4$. We therefore see that (3.11) implies $\delta(w_1) > \delta(z_1) > 0$, $\delta(w_2) > \delta(z_2) > 0$. To get an infinite number of distinct clusters, it suffices to check (3.11) for $(y_1, y_2) = (x_1, x_2)$. In this case we have $\delta(y_1) = \delta(y_2) = 1$, $\delta(z_1) = c-2$, $\delta(z_2) = c-1$, and the result follows from $c \geq 4$. In the case $b \geq 4$, $c = 1$ it suffices to interchange the order of mutations.

Now, when the inequality (3.10) in the definition of 2-finiteness has been already proved, it remains to check that B is skew-symmetrizable. It is known that a sign-skew-symmetric matrix B is skew-symmetrizable if

$$a_1 a_2 \cdots a_k = (-1)^k b_1 b_2 \cdots b_k \tag{3.13}$$

for any $k \times k$ principal submatrix

$$\widehat{B} = \begin{pmatrix} 0 & a_1 & * & \dots & * & b_k \\ b_1 & 0 & a_2 & \dots & * & * \\ * & b_2 & 0 & \dots & * & * \\ \vdots & \vdots & \vdots & \ddots & \vdots & \vdots \\ * & * & * & \dots & * & a_{k-1} \\ a_k & * & * & \dots & b_{k-1} & * \end{pmatrix}$$

of B (recall that we do not distinguish between matrices obtained by a simultaneous permutation of rows and columns).

For $k = 3$ we have, in fact, to check a statement concerning 3×3 integer matrices. By (3.10), the number of such matrices is finite (and does not exceed 140), so this can be done by direct computer-aided inspection. This inspection reveals that the only cases when a 3×3 sign-skew-symmetric integer matrix together with three adjacent matrices satisfy (3.10) are

$$\pm \begin{pmatrix} 0 & 1 & -1 \\ -1 & 0 & 1 \\ 1 & -1 & 0 \end{pmatrix} \quad \text{and} \quad \pm \begin{pmatrix} 0 & 2 & -1 \\ -1 & 0 & 1 \\ 1 & -2 & 0 \end{pmatrix}. \tag{3.14}$$

Clearly, all these matrices satisfy (3.13) with $k = 3$.

Assume now that (3.13) is already established for $k = 3, \dots, l-1$, and we want to check it for $k = l$. We may assume that all a_i's and b_i's are nonzero, since otherwise (3.13) holds trivially. Suppose that $\hat{b}_{ij} = c \neq 0$, where $1 < j - i < l - 1$, and let $d = \hat{b}_{ji}$. Consider two submatrices of $\widehat{B}$, one formed by rows and columns $1, \dots, i, j, \dots, l$, and the other by rows and columns $i, \dots, j$. Then by induction

$$a_1 \cdots a_{i-1} c a_j \cdots a_l = (-1)^{l+i-j+1} b_1 \cdots b_{i-1} d b_j \cdots b_l,$$
$$a_j \cdots a_{j-1} d = (-1)^{j-i+1} b_j \cdots b_{j-1} c,$$

and (3.13) follows since $cd \neq 0$.

Finally, if all the asterisques in $\widehat{B}$ are equal to zero, we apply the following procedure called *shrinking of cycles*. Consider the mutation in direction j, $1 < j < l$. It is easy to see that all the entries off the principal 3×3 submatrix in the rows and columns $j-1, j, j+1$ remain intact, while this submatrix is transformed as follows:

$$\begin{pmatrix} 0 & a_{j-1} & 0 \\ b_{j-1} & 0 & a_j \\ 0 & b_j & 0 \end{pmatrix} \mapsto \begin{pmatrix} 0 & -a_{j-1} & (|a_{j-1}|a_j + a_{j-1}|a_j|)/2 \\ -b_{j-1} & 0 & -a_j \\ (|b_{j-1}|b_j + b_{j-1}|b_j|)/2 & -b_j & 0 \end{pmatrix}.$$

It follows that if a_{j-1} and a_j have the same sign then (3.13) for $\widehat{B}$ follows from (3.13) for the $(l-1) \times (l-1)$ submatrix of $\mu_j(\widehat{B})$ obtained by deleting the jth row and column. Finally, if the signs of a_j's alternate, it suffices to apply first μ_{j-1} (or μ_{j+1}) followed by μ_j.

3.3.2. Proof of the implication (2) $\Longrightarrow$ (3). The main tool we are going to use in this section is the weighted directed graph $\Gamma(B)$ called the *diagram* of B. It is defined for a skew-symmetrizable $n \times n$ matrix B as follows: $\Gamma(B)$ has n vertices, and an edge goes from i to j if and only if $b_{ij} > 0$; the weight w_{ij} of this edge is set to be $\sqrt{|b_{ij} b_{ji}|}$. The weights thus defined are strictly positive. Slightly abusing notation, we will say that the weight w_{ij} of an edge from i to j is negative (and equals $-w_{ji}$) whenever the edge is directed from j to i, and equals zero whenever there are no edges between i and j in either direction.

The following proposition describes how $\Gamma(B)$ changes with matrix mutations.

LEMMA 3.28. *Let B be skew-symmetrizable, $\Gamma = \Gamma(B)$ and $\Gamma' = \Gamma(\mu_k(B))$. Then Γ' is obtained from Γ as follows:*

(1) *the orientations of all edges incident to k are reversed, their weights remain intact;*

(2) *for any vertices i and j such that both w_{ik} and w_{kj} are positive, the direction of the edge between i and j in Γ' and its weight are uniquely determined by the rule*

$$w_{ij} + w'_{ij} = w_{ik} w_{kj};$$

(3) *the rest of the edges and their weights remain intact.*

PROOF. Let D be a skew-symmetrizer for B, then $S(B) = D^{1/2} B D^{-1/2}$ is evidently skew-symmetric. Moreover, it is easy to check that its entries can be written as $s_{ij} = \operatorname{sgn} b_{ij} \sqrt{|b_{ij} b_{ji}|} = \operatorname{sgn}(b_{ij}) |w_{ij}|$. Therefore the diagram $\Gamma(S(B))$ coincides with $\Gamma(B)$, including the weights of the edges. Besides, $S(\mu_k(B)) = \mu_k(S(B))$ for any k, since mutation rules are invariant under conjugation by a diagonal matrix with positive entries. Translating this statement into the language of diagrams we obtain all three assertions of the Lemma. $\square$

Let us consider abstract diagrams (weighted directed graphs), without relating them to skew-symmetrizable matrices. In view of Lemma 3.28, it is natural to define the *diagram mutation* in direction k by $\mu_k(\Gamma) = \Gamma(\mu_k)$. The notions of mutation equivalence and 2-finiteness are readily carried to diagrams.

To proceed with the proof, recall that indecomposable positive Cartan matrices are classified via their Dynkin diagrams (see Section1.2.2). Given a Dynkin diagram Π, the corresponding *Dynkin graph* $G(\Pi)$ is an undirected weighted graph on the same set of vertices. Two vertices i and j of the graph $G(\Pi)$ are connected by an edge if and only if (i, j) or (j, i) is an edge in Π. The weight of an edge in $G(\Pi)$ equals the square root of the number of edges between the corresponding vertices of Π. Figure 3.1 below presents the list of Dynkin diagrams for indecomposable positive Cartan matrices together with the corresponding Dynkin graphs; all unspecified weights are equal to 1. Observe that the same Dynkin graph corresponds to Dynkin diagrams B_n and C_n.

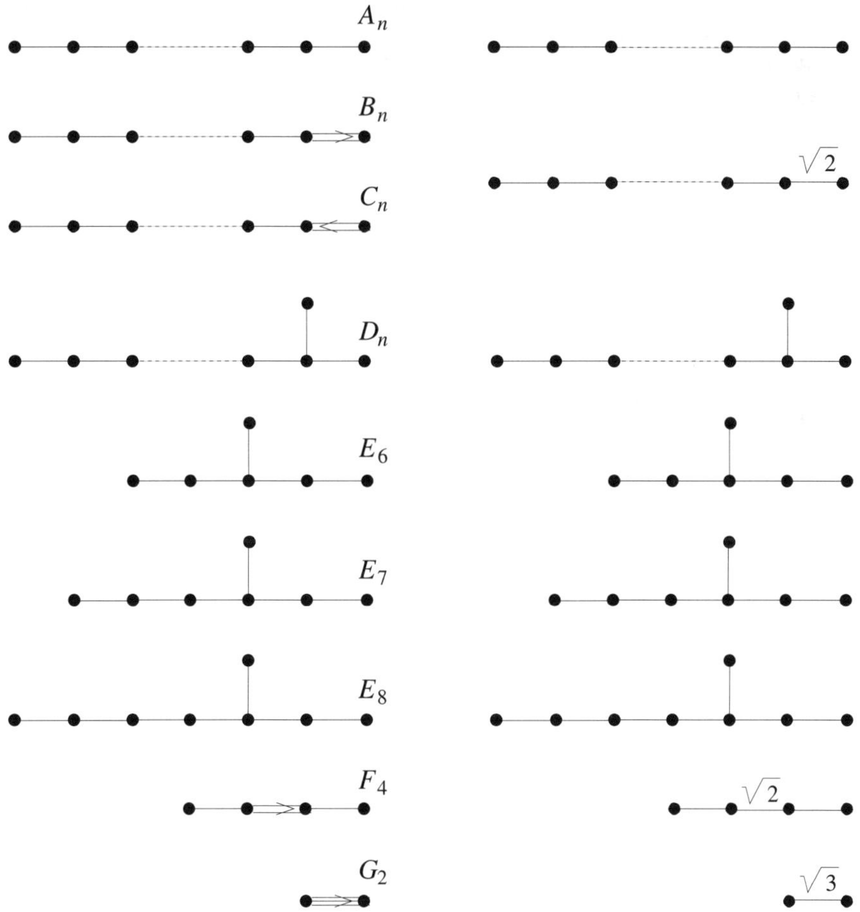

FIGURE 3.1. Dynkin diagrams and Dynkin graphs

By Lemma 3.28, all we need to prove is the following result.

THEOREM 3.29. *Any connected 2-finite diagram is mutation equivalent to an orientation of a Dynkin graph. Furthermore, all orientations of the same Dynkin graph are mutation equivalent to each other.*

PROOF. Let us introduce several useful notions. Recall that a subgraph is called *induced* if it is obtained from the original graph by deleting an arbitrary subset of vertices together with all the edges incident to these vertices. When referring to diagrams, we will write *subdiagram* instead of induced subdiagram. It is clear that any subdiagram of a 2-finite diagram is 2-finite as well.

We say that a diagram is *tree-like* if it is an orientation of a tree. An edge e of a connected diagram Γ is called a *tree edge* if the diagram $\Gamma \setminus \{e\}$ has two connected components and at least one of them is tree-like.

LEMMA 3.30. *Let Γ be a connected diagram, e be a tree edge, and Γ' be the diagram obtained from Γ by reversing the direction of e. Then Γ' and Γ are mutation equivalent.*

PROOF. Let Γ_1 be a tree-like component of $\Gamma \setminus \{e\}$, and $i \notin \Gamma_1$, $j \in \Gamma_1$ be the endpoints of e. Without loss of generality assume that e is directed from i to j. Denote by V_+ (resp., V_-) the set of vertices $k \in \Gamma_1$ such that the last edge in the unique path from j to k is directed from k (resp., to k). To reverse the direction of e it suffices to apply first mutations at the vertices of V_- in the decreasing order of the distance to j, then the mutation at j, and then mutations at the vertices of V_+ in the increasing order of the distance to j (in both cases ties are resolved arbitrarily). Indeed, to see that no new edges arise and that the existing edges do may only change direction, it is enough to notice that each mutation is applied at a vertex that currently has no outcoming edges, and to use Lemma 3.28. Moreover, each edge except for e retains its direction, since we apply mutation once at each vertex of Γ_1. □

Observe that Lemma 3.30 immediately implies the second assertion of Theorem 3.29.

We say that a diagram is *cycle-like* if it is an orientation of a cycle.

LEMMA 3.31. (i) *Any 2-finite cycle-like diagram is an oriented cycle.*

(ii) *The only 2-finite cycle-like diagrams with weights distinct from* 1 *are the three-cycle with weights* $(\sqrt{2}, \sqrt{2}, 1)$ *and the four-cycle with weights* $(\sqrt{2}, 1, \sqrt{2}, 1)$.

PROOF. The proof of this Lemma is basically a translation of the proof of (3.13) to the language of diagrams. The cycles of length 3 correspond to the matrices listed in (3.14), and the validity of the assertions for these matrices follows from inspection. The shrinking of cycles described in the proof of (3.13) immediately implies the first assertion of the Lemma. Next, the product of the edge weights along the cycle, called the *total weight* of the cycle, is preserved under shrinking, so (3.14) imply that this product equals either 1 or 2. In the latter case the initial cycle-like diagram contains two edges of weight $\sqrt{2}$. If the number of edges in the cycle is at least 5, the shrinking process leads to the four-cycle $(\sqrt{2}, \sqrt{2}, 1, 1)$, which is discarded by applying mutation at the vertex incident to both edges of weight $\sqrt{2}$. If there are four edges, the only other possibility is the four-cycle described in Lemma. □

Observe that the property of a diagram to be tree-like or cycle-like is not preserved under diagram mutations.

EXAMPLE 3.32. For any positive integer p, s and any tree-like diagrams T_1, T_2 denote by C_{s,p,T_1,T_2} the diagram on Figure 3.2,a. Applying mutations at vertices $a_1, a_2, \dots, a_s, b_1, b_2, \dots, b_p$ we arrive at the tree-like diagram T_{s,p,T_1,T_2}, see Figure 3.2,b. Moreover, if an additional leave a_0 is connected by a directed edge to a_1, then the same sequence of mutations results in T_{s,p,T_1,T_2} with a_0 connected to b_p.

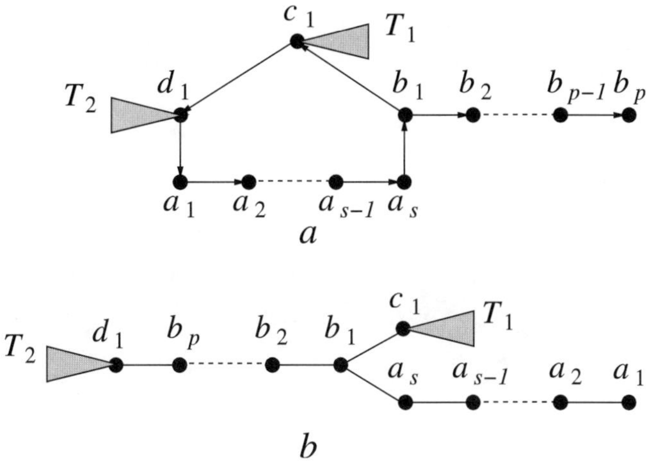

FIGURE 3.2. Transforming a cycle of total weight 1

EXAMPLE 3.33. For any tree-like diagrams T_1, T_2 denote by C_{p,T_1,T_2} the diagram on Figure 3.3,a. Applying mutations at vertices $b_1, b_2, \dots, b_p$ we arrive at the tree-like diagram T_{p,T_1,T_2}, see Figure 3.3,b.

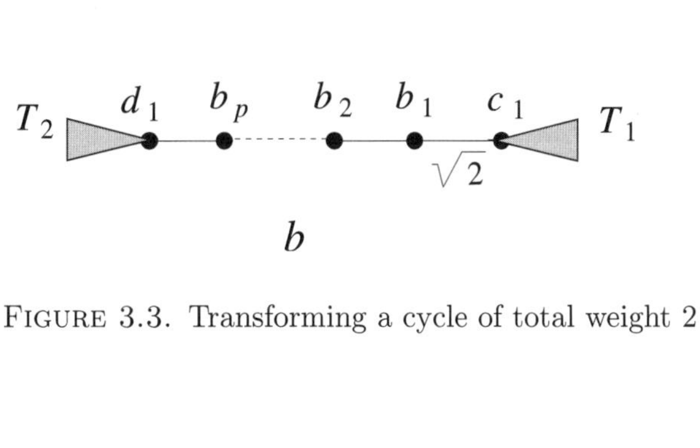

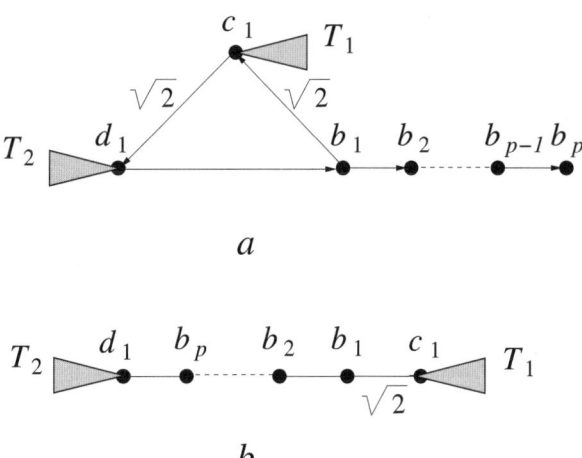

FIGURE 3.3. Transforming a cycle of total weight 2

We can now prove the following weaker version of Theorem 3.29.

LEMMA 3.34. *Any 2-finite tree-like diagram is mutation equivalent to an orientation of a Dynkin graph.*

PROOF. A weighted tree is called *critical* if it is not a Dynkin graph itself, and each of its proper induced subgraphs is a disjoint union of Dynkin graphs. The complete list of critical trees is given on Figure 3.4; here w can take any value in $\{1, \sqrt{2}, \sqrt{3}\}$.

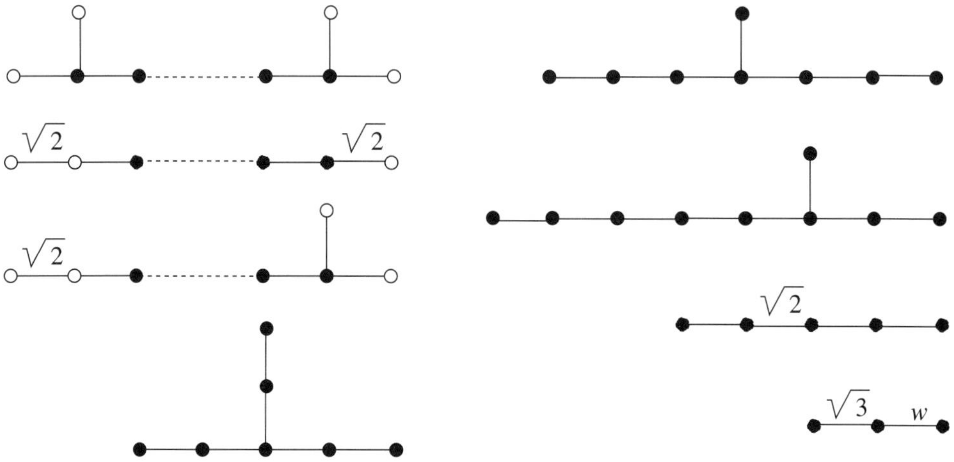

FIGURE 3.4. Critical trees

To prove the Lemma, it is enough to check that any orientation of a critical tree is 2-infinite. By Lemma 3.30, it suffices to consider one particular orientation. For the first three critical trees, direct the edges top-down or leftwards, and apply mutations consecutively at all non-leaves from right to left.

For the first tree, the diagram induced by the white vertices is a non-oriented cycle, in a contradiction with Lemma 3.31. For the second tree, the diagram induced by the white vertices is the three-cycle $(\sqrt{2}, \sqrt{2}, 2)$, in a contradiction with Lemma 3.31. For the third tree, uppend the above sequence by the mutation at the rightmost vertex; the diagram induced by the white vertices is the four-cycle $(\sqrt{2}, \sqrt{2}, 1, 1)$, in a contradiction with Lemma 3.31.

The forth tree is $T_{2,2,P_2,P_1}$, where P_i is a path on i vertices. By Example 3.32, any its orientation is mutation equivalent to $C_{2,2,P_2,P_1}$, which contains an instance of the first critical tree as an induced subgraph (for a reference see vertices a_s, b_1, b_2, c_1, d_1 and a neighbour of c_1 in T_1 on Figure 3.2).

The fifth tree is $T_{1,2,P_3,P_2}$. By the same example, any its orientation is mutation equivalent to $C_{1,2,P_3,P_2}$, which contains an instance of the fourth critical tree as an induced subgraph (for a reference see vertices b_1, b_2, c_1, d_1, d_2 and a path of length 2 in T_1 starting at c_1 on Figure 3.2).

The sixth tree is $T_{1,3,P_2,P_3}$. By the same example, its every orientation is mutation equivalent to $C_{1,3,P_2,P_3}$, which contains an instance of the fourth critical tree as an induced subgraph (for a reference see vertices b_1, b_2, b_3, c_1, d_1, a neighbor of c_1 in T_1 and a path of length 2 in T_2 starting at d_1 on Figure 3.2).

For the remaining two trees the claim follows easily from inspection. □

To complete the proof of Theorem 3.29 we need the following statement.

LEMMA 3.35. *Any connected 2-finite diagram is mutation equivalent to a tree-like diagram.*

PROOF. Assume that assertion of the Lemma is wrong, and let Γ be a counterexample with the minimal number of vertices. This number is at least 4, since for diagrams on three vertices the assertion can be checked easily. Let v be a vertex in Γ such that $\Gamma' = \Gamma \setminus \{v\}$ is connected. We may assume that Γ' is tree-like, and hence, by Lemma 3.34, is an orientation of a Dynkin graph. In particular, Γ' contains at most one edge of weight $\sqrt{2}$, which we denote by e.

If the degree of v in Γ equals 1, then Γ is already tree-like.

If the degree of v in Γ is at least 3, direct all the edges of the Dynkin graph corresponding to Γ' leftwards and top-down (cp. Fig. 3.1). It is easy to see that we immediately get a non-oriented cycle, in a contradiction with Lemma 3.31.

Let the degree of v in Γ be 2, so that Γ contains a unique cycle C passing through v. By Lemma 3.31, the total weight of C is either 1, or 2.

Assume first that the total weight of C equals 1. If C is a three-cycle, the mutation at v turns Γ into a tree-like diagram. Otherwise, Γ is an instance of C_{s,p,T_1,T_2} (possibly, with an additional edge incident to a_1) for some suitable choice of the parameters. By Example 3.32, any such diagram is mutation equivalent to a tree-like diagram.

Assume now that the total weight of C equals 2. By Lemma 3.31, C is either the three-cycle with weights $(\sqrt{2}, \sqrt{2}, 1)$, or the four-cycle with weights $(\sqrt{2}, 1, \sqrt{2}, 1)$. In the former case, Γ is an instance of C_{p,T_1,T_2} for some suitable choice of the parameters; by Example 3.33, any such diagram is mutation equivalent to a tree-like diagram. To discard the remaining case, it suffices to notice that e belongs to C, and that connecting a new vertex by an edge of weight 1 with a vertex of the four-cycle $(\sqrt{2}, 1, \sqrt{2}, 1)$ immediately produces an instance of the first critical tree (cp. Figure 3.4). □

Now the proof of Theorem 3.29 is complete. □

3.3.3. Proof of the implication (3) $\Longrightarrow$ (1). We start from the following general construction. Let $\mathfrak{P}$ be a simple convex polytope in $\mathbb{R}^n$. Assume that the vertices of $\mathfrak{P}$ are labeled by n-tuples of elements of some finite ground set in such a way that i-dimensional faces of $\mathfrak{P}$ correspond bijectively to maximal sets of n-tuples having exactly $n - i$ elements in common. Besides, an integer $n \times n$ sign-skew-symmetric matrix B_v is attached at each vertex v of $\mathfrak{P}$. The rows and columns of B_v are labeled by the elements of the n-tuple at v. Moreover, for every edge $(v, \bar{v})$ of $\mathfrak{P}$, the matrices B_v and $B_{\bar{v}}$ satisfy the following condition: let γ and $\bar{\gamma}$ be the unique elements of the labels at v and $\bar{v}$, respectively that are not common to these two labels, the $B_{\bar{v}}$ is obtained from B_v by a matrix mutation in direction γ followed by replacing the label γ by $\bar{\gamma}$.

Next, fix an arbitrary 2-dimensional face F of $\mathfrak{P}$. The label at any vertex in F contains exactly two elements that are not common to the labels of all other vertices in F; denote these element α and β. It is trivial to check that the integer $|b_{\alpha\beta} b_{\beta\alpha}|$ does not depend on the choice of the vertex in F; we call this number the *type* of F.

Proposition 3.36. *Let $\mathfrak{P}$ be a simple convex polytope and $\{B\}$ be a family of matrices as above. Assume that 2-dimensional faces of $\mathfrak{P}$ are quadrilaterals, pentagons, hexagons and octagons of types 0, 1, 2 and 3, respectively. Then the cluster algebra $\mathcal{A} = \mathcal{A}(\widetilde{B})$ is of finite type for any vertex v of $\mathfrak{P}$ and any $\widetilde{B}$ such that its principal part coincides with B_v.*

Proof. We will prove that the 1-skeleton of $\mathfrak{P}$ covers the exchange graph of $\mathcal{A}$. Let $\Sigma = (\mathbf{x}, \widetilde{B})$ be a seed of $\mathcal{A}$ and let B be the principal part of $\widetilde{B}$. It is convenient to regard the rows and columns of B as being labeled by the entries of $\mathbf{x}$. An *attachment* of Σ at v is a bijection between the labels at v and the cluster variables entering $\mathbf{x}$; such an attachment provides an identification of B and B_v. The *transport* of a seed attachment along and edge $(v, \bar{v})$ is defined as follows: the seed $\bar{\Sigma}$ attached at $\bar{v}$ is obtained from Σ by the mutation in direction $x(\gamma)$, and the new bijection at $\bar{v}$ is uniquely determined by $\bar{x}(\alpha) = x(\alpha)$ for all $\alpha \neq \bar{\gamma}$. Clearly, the transport of $\bar{\Sigma}$ attached at $\bar{v}$ backwards to v recovers the original seed attachment.

Let us fix an arbitrary vertex v of $\mathfrak{P}$ and attach a seed of $\mathcal{A}$ at v. Clearly, we can transport the initial attachment to any other vertex v' along a path from v to v'. To show that the resulting attachment does not depend on the choice of a path, it suffice to prove that transporting the attachment along a loop brings it back unchanged. Since all loops in $\mathfrak{P}$ are generated by 2-dimensional faces, it is enough to prove this fact for quadrilaterals, pentagons, hexagons and octagons of types 0, -1, -2 and -3, respectively. This is a simple exercise on cluster algebras of rank 2, which we leave to the interested reader.

Take an arbitrary seed of $\mathcal{A}$. It is obtained from the seed attached at v by a sequence of mutations. This sequence is uniquely lifted to a path on $\mathfrak{P}$ such that transporting the initial seed attachment at v along the edges of this path produces the chosen sequence of mutations. Therefore, the vertices of $\mathfrak{P}$ are mapped surjectively onto the set of all seeds of $\mathcal{A}$.

Let now v' and v'' be two vertices of $\mathfrak{P}$ such that their labels contain a common element α. Clearly, they can be connected by a path such that each vertex on this path possesses the same property. Therefore, $x'(\alpha) = x''(\alpha)$, and hence the attachment of cluster variables to the elements of the ground set does not depend on the choice of the vertex. It follows from above that this attachment is a surjection. Thus, $\mathcal{A}$ is of finite type, since the ground set is finite. □

The proof of the implication (3) $\Longrightarrow$ (1) consists of the construction of the polytopes $\mathfrak{P}$ for all each of the Cartan–Killing types. In the case of A_n we are in fact done for the following reasons. Let $\mathfrak{P}$ be the associahedron described in Section 2.1.4. It was already explained there that $\mathfrak{P}$ is a simple convex polytope, and that its 2-dimensional faces are quadrilaterals and pentagons. Moreover, let B' be the matrix mentioned in statement (3), then $|b'_{ij}b'_{ji}|$ is either 0, or 1. Recall that the rows and columns of B' are labeled by the cluster variables, that is, by the diagonals of R_m. It follows immediately from the structure of the Plücker relations that $b'_{ij} = 0$ if the diagonals corresponding to i and j are disjoint, and $b'_{ij} = \pm 1$ otherwise. Therefore, quadrilateral 2-faces of $\mathfrak{P}$ are of type 0, while pentagonal 2-faces are of type 1. So, the associahedron satisfies all the assumptions in Proposition 3.36, and hence the implication (3) $\Longrightarrow$ (1) holds true for Cartan–Killing type A_n.

The above construction for type A_n can be described in the following terms. Let Φ be the corresponding root system, Δ be the set of simple roots, Φ_+ be the set of all positive roots (see Section 1.2.2). Define the set of *almost positive roots* $\Phi_{\geq -1}$ as the union of $-\Delta$ and Φ_+. The elements of $\Phi_{\geq -1}$ correspond to the proper diagonals of R_{n+3}, and hence to the cluster variables of the corresponding cluster algebra. Indeed, consider the snake-like triangulation of R_{n+3} formed by the diagonals $(1, n+2), (2, n+2), (2, n+1), (3, n+1), \dots$, see Fig. 2.4 in Section 2.1.4. This diagonals are identified with $-\alpha_1, -\alpha_2, \dots$ for $\alpha_i \in \Pi$. The diagonal intersecting initial diagonals labeled by $-\alpha_j, -\alpha_{j+1}, \dots, -\alpha_k$ is identified with the positive root $\alpha_j + \alpha_{j+1} + \dots + \alpha_k$. A collection of almost positive roots is called *compatible* if the corresponding collection of diagonals is non-crossing. So, the faces of the associahedron are labeled by collections of compatible almost positive roots. In particular, the facets are labeled by the almost positive roots themselves, and the vertices are labeled by maximal compatible collections.

The same construction goes through for other Cartan–Killing types as well. The two main problems are to define the notion of compatibility and to prove that the resulting abstract polytope can be realized as a simple convex polytope satisfying the assumptions of Proposition 3.36. Consequently, the cluster variables of the cluster algebra are identified with the almost positive roots of the corresponding root system. The concrete details of the construction rely on intricate combinatorial properties of finite root systems. Since we will not need these results in the following chapters, they are not going to be reproduced here.

3.3.4. Proof of the uniqueness. Let B'' be another matrix mutation equivalent to B such that its Cartan companion is positive. Denote the Cartan companions of B' and B'' and the corresponding root systems by A', A'' and Φ', Φ'', respectively. It follows from the explanations in the previous section that the sets $\Phi'_{\geq 0}$ and $\Phi''_{\geq 0}$ are isomorphic, hence Φ' and Φ'' have the same rank and the same cardinality. A direct check shows that the only different Cartan–Killing types with this property are B_n and C_n for all $n \geq 3$, and also E_6, which has the same data as B_6 and C_6. To distinguish between these types, note that mutation-equivalent matrices share the same skew-symmetrizer D. Furthermore, D is a skew-symmetrizer for a skew-symmetrizable matrix, it is also a symmetrizer for its Cartan companion. Therefore, the diagonal entries of D are given by $d_i = (\alpha_i, \alpha_i)$, see Section 1.2.2. Since the root system of type B_n has one short simple root and $n-1$ long ones, while that of type C_n has one long root and $n-1$ short ones, the corresponding matrices cannot be mutation equivalent. The same is true for E_6 and B_6 or C_6 since all simple roots for E_6 are of the same length. Therefore, A' and A'' have the same Cartan–Killing type.

Let now B' and B'' be two sign-skew-symmetric matrices such that their Cartan companions A' and A'' are of the same Cartan–Killing type. Clearly, the corresponding diagrams $\Gamma(B')$ and $\Gamma(B'')$ are tree-like. By Lemma 3.30, we may assume that $\Gamma(B') = \Gamma(B'') = \Gamma$. It is easy to see that any D that serves a symmetrizer for A' (and hence for A'') is also a skew-symmetrizer for B' and B''. By the proof of Lemma 3.28, the skew-symmetric matrices $D^{1/2}B'D^{-1/2}$ and $D^{1/2}B''D^{-1/2}$ have the same diagram Γ, hence they coincide. Therefore, B' and B'' coincide as well.

3.4. Cluster algebras and rings of regular functions

As we have seen in Chapter 2, one can find cluster-like structures in the rings of functions related to Schubert varieties, such as the ring of regular functions on a double Bruhat cell, or a homogeneous coordinate ring of a Grassmannian. The goal of this section is to establish a corresponding phenomenon in a context of general Zariski open subsets of the affine space.

Let V be a Zariski open subset in $\mathbb{C}^{n+m}$, $\mathcal{A}_{\mathbb{C}}$ be a cluster algebra of geometric type tensored with $\mathbb{C}$. We assume that the variables in some extended cluster are identified with a set of algebraically independent rational functions on V. This allows us to identify cluster variables in any cluster with rational functions on V as well, and thus to consider $\mathcal{A}_{\mathbb{C}}$ as a subalgebra of $\mathbb{C}(V)$. Finally, we denote by $\mathcal{A}_{\mathbb{C}}^V$ the localization of $\mathcal{A}_C$ with respect to the stable variables that do not vanish on V; assuming that the latter variables are $x_{n+1}, \dots, x_{n+k}$, taking the localization is equivalent to changing the ground ring $\mathbb{A}$ from $\mathbb{Z}[x_{n+1}, \dots, x_{n+m}]$ to $\mathbb{Z}[x_{n+1}^{\pm 1}, \dots, x_{n+k}^{\pm 1}, x_{n+k+1}, \dots, x_{n+m}]$.

Proposition 3.37. *Let V and $\mathcal{A}$ as above satisfy the following conditions:*
(i) *each regular function on V belongs to $\mathcal{A}_{\mathbb{C}}^V$;*
(ii) *there exists an extended cluster $\widetilde{\mathbf{x}} = (x_1, \dots, x_{n+m})$ in $\mathcal{A}_{\mathbb{C}}$ consisting of algebraically independent functions regular on V;*
(iii) *any cluster variable x'_k, $k \in [1, n]$, obtained by the cluster transformation (3.1) applied to $\widetilde{\mathbf{x}}$ is regular on V.*
Then $\mathcal{A}_{\mathbb{C}}^V$ coincides with the ring $\mathcal{O}(V)$ of regular functions on V.

Proof. All we have to prove is that any element in $\mathcal{A}_{\mathbb{C}}^V$ is a regular function on V. The proof consists of three steps.

Lemma 3.38. *Let $\widetilde{\mathbf{z}} = (z_1, \dots, z_{n+m})$ be an arbitrary extended cluster in $\mathcal{A}_{\mathbb{C}}$. If a Laurent monomial $M = z_1^{d_1} \cdots z_{n+m}^{d_{n+m}}$ is regular on V then $d_i \geq 0$ for $i \in [1, n]$.*

Proof. Indeed, assume that $d_k < 0$ and consider the cluster $\widetilde{\mathbf{z}}_k$. By (3.1), M can be rewritten as $M = M'(z'_k)^{-d_k}/B^{-d_k}$, where M' is a Laurent monomial in common variables of $\widetilde{\mathbf{z}}$ and $\widetilde{\mathbf{z}}_k$, and B is the binomial (in the same variables) that appears in the right hand side of (3.1). By condition (i) and Theorem 3.14, M can be written as a Laurent polynomial in the variables of $\widetilde{\mathbf{z}}_k$. Equating two expressions for M, we see that B^{-d_k} times a polynomial in variables of $\widetilde{\mathbf{z}}_k$ equals a Laurent monomial in the same variables. This contradicts the algebraic independence of variables in $\widetilde{\mathbf{z}}_k$, which follows from the algebraic independence of variables in $\widetilde{\mathbf{x}}$. □

Lemma 3.39. *Let z be a cluster variable in an extended cluster $\widetilde{\mathbf{z}}$, and assume that z is a regular function on V. Then z is irreducible in the ring of regular functions on V.*

Proof. Without loss of generality, assume that $z_{n+1} = x_{n+1}, \dots, z_{n+m'} = x_{n+m'}$ do not vanish on V, and $z_{n+m'+1} = x_{n+m'+1}, \dots, z_{n+m} = x_{n+m}$ may vanish on V. Moreover, assume to the contrary that $z = fg$, where f and g are non-invertible regular functions on V. By condition (i) and Proposition 3.20, both f and g are Laurent polynomials in $z_1, \dots, z_{n+m'}$ whose coefficients are polynomials in $z_{n+m'+1}, \dots, z_{n+m}$. Applying the same argument as in the proof of Lemma 3.38, we see that both f and g are, in fact, Laurent monomials in $z_1, \dots, z_{n+m}$ and that $z_{n+1}, \dots, z_{n+m'}$ enter both f and g with a non-negative degree. Moreover, by Lemma 3.38, each cluster variable $z_1, \dots, z_n$ enters both f and g with a non-negative degree. This can only happen if one of f and g is invertible in $\mathcal{O}(V)$, a contradiction. □

Denote by $U_0 \subset V$ the locus of all $t \in V$ such that $x_i(t) \neq 0$ for all $i \in [1, n]$. Besides, denote by $U_k \subset V$ the locus of all $t \in V$ such that $x_i(t) \neq 0$ for all $i \in [1, n] \setminus k$ and $x'_k(t) \neq 0$. Note a similarity between sets U_0, U_i and sets $U_{\mathbf{i}}, U_{n,\mathbf{i}}$ defined in Lemma 2.11. The following statement is a straightforward generalization of the codimension 2 result implicitly comprising the first part of Theorem 2.16.

LEMMA 3.40. *Let* $U = \cup_{i=0}^n U_i$*, then* $\operatorname{codim} V \setminus U \geq 2$.

PROOF. Follows immediately from Lemma 3.39 and conditions (ii) and (iii). □

We can now complete the proof of Proposition 3.37. Inclusion $\mathcal{A}_{\mathbb{C}}^V \subseteq \mathcal{O}(U)$ is an immediate corollary of Proposition 3.20. The rest of the proof follows literally the proof of Corollary 2.18. □

In some situations condition (i) of Proposition 3.37 is difficult to check. In this case one may attempt prove a weaker result, with cluster algebras replaced by upper cluster algebras, by imposing additional conditions on the variables x_i, $i \in [1, n+m]$ and x'_k, $k \in [1, n]$. As an example, consider reduced double Bruhat cells $L^{u,v}$ studied in Section 2.2. In Section 2.2.3 we associated to each reduced word $\mathbf{i}$ a symmetric $(l(u)+l(v)) \times (l(u)+l(v))$ matrix $C = C_{\mathbf{i}}$ and a directed graph $\Sigma_{\mathbf{i}}$. Based on $C_{\mathbf{i}}$ and $\Sigma_{\mathbf{i}}$, we define a matrix $\widetilde{B} = \widetilde{B}_{\mathbf{i}}$ as follows. The rows of $\widetilde{B}$ correspond to $\mathbf{i}$-bounded indices, and the columns of $\widetilde{B}$ correspond to all indices. For an $\mathbf{i}$-bounded index k and an arbitrary index l, $b_{kl} = c_{kl}$ if $(k \to l) \in \Sigma_{\mathbf{i}}$ and $b_{kl} = -c_{kl}$ if $(l \to k) \in \Sigma_{\mathbf{i}}$; all the other entries of $\widetilde{B}$ are equal to zero. So, $\widetilde{B}$ can be considered as a submatrix of the incidence matrix of $\Sigma_{\mathbf{i}}$ weighted by the corresponding elements of $C_{\mathbf{i}}$.

PROPOSITION 3.41. *The matrix* $\widetilde{B}_{\mathbf{i}}$ *satisfies the assumptions of Theorem 3.21.*

PROOF. Clearly, $\widetilde{B}_{\mathbf{i}}$ is skew-symmetric. To see that it has a full rank, consider the square submatrix B of $\widetilde{B}_{\mathbf{i}}$ formed by the columns indexed by all p such that $p = q^-$ for some $\mathbf{i}$-bounded index q. Let us rearrange the rows of B in such a way that its diagonal represents the one-to-one correspondance $q \mapsto q^-$. It follows immediately from Definition 2.7 that the rearranged B is upper-triangular and all its diagonal entries equal ± 1. □

Choose functions M_i, $1 \leq i \leq l(u) + l(v)$, discussed in Lemma 2.11 (and described explicitly in the SL_n case immediately after that Lemma) as an initial cluster and denote by $\mathcal{U}_{\mathbf{i}}$ the corresponding upper cluster algebra $\mathcal{U}(\widetilde{B}_{\mathbf{i}})$. In these terms, Corollary 2.18 can be restated as follows.

THEOREM 3.42. *For any reduced word* $\mathbf{i}$ *for the pair* (u, v)*, the ring* $\mathcal{O}(L^{u,v})$ *of regular functions on the reduced double Bruhat cell* $L^{u,v}$ *coincides with the upper cluster algebra* $\mathcal{U}_{\mathbf{i}}$ *tensored with* $\mathbb{C}$.

PROOF. Indeed, by Corollary 2.18, $\mathcal{O}(L^{u,v})$ coincides with the ring of regular functions on $U = U_{\mathbf{i}} \cup U_{1,\mathbf{i}} \cup \dots \cup U_{n,\mathbf{i}}$. The latter is, by definition, the upper bound corresponding to the initial cluster, which in view of Theorem 3.21 coincides with the upper cluster algebra $\mathcal{U}_{\mathbf{i}}$. □

The case of double Bruhat cells $\mathcal{G}^{u,v}$ can be treated in a similar way. More exactly, given a reduced word $\mathbf{i}$ for (u,v), define $\bar{\mathbf{i}}$ as $\mathbf{i}$ written in the reverse direction appended with the word $-1\ \ -2\ \cdots -r$, where r is the rank of $\mathcal{G}$. Define the $(l(u)+l(v)+r)\times(l(u)+l(v)+r)$ matrix $C_{\bar{\mathbf{i}}}$ and the directed graph $\Sigma_{\bar{\mathbf{i}}}$ on $l(u)+l(v)+r$ vertices similarly to $C_{\mathbf{i}}$ and $\Sigma_{\mathbf{i}}$. The graph $\Sigma'_{\mathbf{i}}$ is obtained from $\Sigma_{\bar{\mathbf{i}}}$ via the following modifications:

(i) erase all inclined edges with both endpoints lying in $[l(u)+l(v)+1, l(u)+l(v)+r]$;

(ii) reverse the direction of all remaining edges;

(iii) reverse the ordering of the vertices as folows: $k\mapsto \bar{k}$ with

$$\bar{k}=\begin{cases} l(u)+l(v)+1-k & \text{if } k\in[1,l(u)+l(v)],\\ l(u)+l(v)-k & \text{if } k\in[l(u)+l(v)+1, l(u)+l(v)+r]. \end{cases}$$

As a result, the vertex set of $\Sigma'_{\mathbf{i}}$ is $[-r,-1]\cup[1,l(u)+l(v)]$.

EXAMPLE 3.43. Let $\mathbf{i}=(1\ 2\ 1\ -1\ -2\ -1)$. Then $\bar{\mathbf{i}}=(-1\ -2\ -1\ 1\ 2\ 1\ -1\ -2)$, and the corresponding graph $\Sigma_{\bar{\mathbf{i}}}$ is shown on Fig. 3.5a. To get the graph $\Sigma'_{\mathbf{i}}$ we delete the edge between the vertices 7 and 8, reverse the directions of eges and reorder the vertices. Since the vertices in $\Sigma_{\bar{\mathbf{i}}}$ are ordered from left to right, the latter operation can be represented as the symmetry with respect to the vertical axes. The resulting graph $\Sigma'_{\mathbf{i}}$ is shown on Fig. 3.5b.

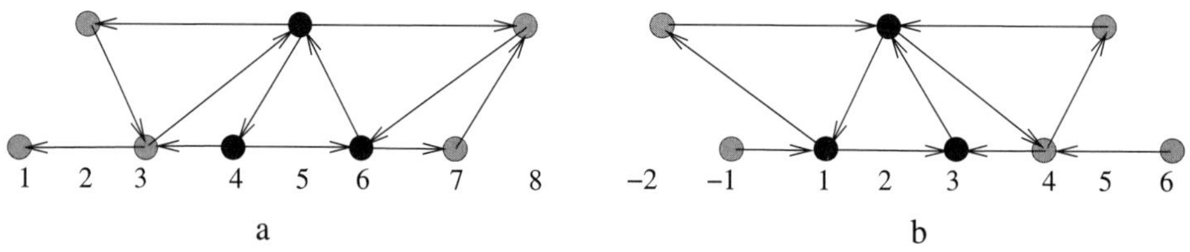

FIGURE 3.5. Transforming $\Sigma_{\bar{\mathbf{i}}}$ to $\Sigma'_{\mathbf{i}}$

An index $\bar{k}$ is called $\mathbf{i}$-*exchangeable* if the corresponding index k is $\mathbf{i}$-bounded. Based on this combinatorial data, define the matrix $\widetilde{B}'_{\mathbf{i}}$ similarly to $\widetilde{B}_{\mathbf{i}}$; its rows correspond to $\mathbf{i}$-exchangeable indices among in $[1,l(u)+l(v)]$, and its columns correspond to all indices. It follows immediately from Proposition 3.41 that $\widetilde{B}'_{\mathbf{i}}$ satisfies the assumptions of Theorem 3.21.

For $k\in[1,l(u)+l(v)]$ denote

$$u_{\le k}=u_{\le k}(\mathbf{i})=\prod_{l=1,\dots,k} s_{|i_l|}^{\frac{1-\theta(l)}{2}},\qquad v_{>k}=v_{>k}(\mathbf{i})=\prod_{l=l(u)+l(v),\dots,k+1} s_{|i_l|}^{\frac{1+\theta(l)}{2}} \tag{3.15}$$

(see Remark 1.1 for notation). Besides, for $k\in-[1,r]$ set $u_{\le k}$ to be the identity and $v_{>k}$ to be equal to v^{-1}. For $k\in[-r,-1]\cup[1,l(u)+l(v)]$ put

$$\Delta(k;\mathbf{i})=\Delta_{u_{\le k}\omega_{|i_k|},v_{>k}\omega_{|i_k|}}, \tag{3.16}$$

where the right hand side is a generalized minor defined by (1.11). Take the functions $\Delta(k;\mathbf{i})$ as the initial cluster and define the corresponding upper cluster algebra $\mathcal{U}'_{\mathbf{i}}=\mathcal{U}(\widetilde{B}'_{\mathbf{i}})$. An analog of Theorem 3.42 can be stated as follows.

THEOREM 3.44. *For any reduced word* $\mathbf{i}$ *for the pair* (u, v), *the ring* $\mathcal{O}(\mathcal{G}^{u,v})$ *of regular functions on the double Bruhat cell* $\mathcal{G}^{u,v}$ *coincides with the upper cluster algebra* $\mathcal{U}'_{\mathbf{i}}$ *tensored by* $\mathbb{C}$.

3.5. Conjectures on cluster algebras

Besides the general results on cluster algebras presented in the previous two sections, there are also many general conjectures. One can divide the conjectures into several groups.

The conjectures in the first group deal with the structure of the exchange graph of a cluster algebra.

CONJECTURE 3.45. *The exchange graph of a cluster algebra depends only on the initial exchange matrix* B.

The vertices and the edges of the exchange graph are conjecturally described as follows.

CONJECTURE 3.46. (i) *Every seed is uniquely defined by its cluster; thus, the vertices of the exchange graph can be identified with the clusters, up to a permutation of cluster variables.*

(ii) *Two clusters are adjacent in the exchange graph if and only if they have exactly* $n-1$ *common cluster variables.*

The last conjecture in the first group says that if two clusters contain the same variable then they can be reached one from the other without changing this variable.

CONJECTURE 3.47. *For any cluster variable* x, *the seeds whose clusters contain* x *form a connected subgraph of the exchange graph.*

The second group of conjectures treats in more detail the Laurent phenomenon described in Section 3.2. Recall that by Theorem 3.14, any cluster variable x can be uniquely expressed via the initial cluster variables $x_1, \dots, x_n$ as

$$x = \frac{P(x_1, \dots, x_n)}{x_1^{d_1} \dots x_n^{d_n}}, \tag{3.17}$$

where P is a polynomial not divisible by any of x_i's. The exponent d_i as defined above may depend on the choice of the cluster containing x and on the choice of the initial cluster. The first conjecture in the second group says that there is no such dependence.

CONJECTURE 3.48. *The exponent* d_i *depends only on* x *and* x_i.

The following conjecture deals with the properties of the exponents d_i.

CONJECTURE 3.49. (i) *The exponents* d_i, $i \in [1, n]$, *are nonnegative for any cluster variable* $x \neq x_i$.

(ii) *The exponent* d_i *vanishes if and only if there exists a cluster containing both* x *and* x_i.

The third, and the most interesting group of conjectures deal with cluster monomials. A *cluster monomial* is a monomial in cluster variables all of which belong to the same cluster; in other words, if $\bar{\mathbf{x}} = (\bar{x}_1, \dots, \bar{x}_n)$ is a cluster then $\prod_{i=1}^n \bar{x}_i^{a_i}$ is a cluster monomial for any choice of nonnegative exponents a_i, $i \in [1, n]$.

CONJECTURE 3.50. *Cluster monomials in any cluster algebra are linearly independent over the ground ring.*

Writing down each variable in a cluster monomial in form (3.17), one can extend the definition of exponents d_i to cluster monomials.

CONJECTURE 3.51. *If the exponents d_i for two cluster monomials coincide for all $i \in [1, n]$ then the monomials themselves coincide.*

We say that a nonzero element in a cluster algebra $\mathcal{A}$ is *positive* if its Laurent expansion in terms of the variables from any seed in $\mathcal{A}$ has nonnegative coefficients. Further, a positive element is called *indecomposable* if it cannot be written as a sum of two positive elements.

CONJECTURE 3.52. *Every cluster monomial is an indecomposable positive element.*

Conjectures 3.45–3.50 are known to be true for the algebras of finite type. In what follows we exhibit several other classes of cluster algebras for which some of these conjectures hold. Conjectures 3.51 and 3.52 are even less accessible. They are known to be true for cluster algebras of classical type, that is, for the cluster algebras of finite type corresponding to classical semisimple Lie algebras.

3.6. Summary

- A seed of geometric type is a pair $(\mathbf{x}, \widetilde{B})$, where $\mathbf{x}$ is an $(n+m)$-tuple of variables called an extended cluster and $\widetilde{B}$ is an $n \times (n+m)$ integer matrix such that its principal part $\widetilde{B}[n;n]$ is skew-symmetrizable. The first n entries of $\widetilde{\mathbf{x}}$ are called cluster variables, and the last m entries are called stable variables.
- Two seeds $(\mathbf{x}, \widetilde{B})$ and $(\mathbf{x}', \widetilde{B}')$ are adjacent in direction k if x_k and x'_k are related by exchange relation (3.1), $x_i = x'_i$ for $i \neq k$, and $\widetilde{B}'$ is obtained from $\widetilde{B}$ by the matrix mutation in direction k. By iterating this procedure we obtain an n-regular tree $\mathbb{T}_n$ whose vertices correspond to seeds and edges connect adjacent clusters.
- The cluster algebra of geometric type $\mathcal{A}(\widetilde{B})$ over a ground ring $\mathbb{A}$ is the $\mathbb{A}$-subalgebra of the ambient field generated by all cluster variables at all the vertices of $\mathbb{T}_n$.
- Two seeds are equivalent if they are obtained one from another by a permutation of cluster variables and the corresponding permutation of rows and columns of the exchange matrix. The exchange graph of a cluster algebra is the quotient of $\mathbb{T}_n$ by this equivalence relation.
- Any cluster variable is a Laurent polynomial in the initial cluster variables (the Laurent phenomenon, Theorem 3.14).
- The ring of functions that can be written as Laurent polynomials in the variables of a given cluster $\mathbf{x}$, as well as in the variables of any of the adjacent clusters, is called the upper bound associated with $\mathbf{x}$. The upper cluster algebra $\mathcal{U}$ is the intersection of all upper bounds. If the exchange matrix $\widetilde{B}$ has a full rank then the upper bounds do not depend on the choice of $\mathbf{x}$, and hence coincide with $\mathcal{U}(\widetilde{B})$ (Theorem 3.21). The upper cluster algebra $\mathcal{U}(\widetilde{B})$ contains $\mathcal{A}(\widetilde{B})$, and in general these two algebras do not coincide (Example 3.25).

- The classification of cluster algebras having a finite exchange graph (called algebras of finite type) coincides with the Cartan-Killing classification of semisimple Lie algebras (Theorem 3.26).

Bibliographical notes

Our exposition in this chapter is based on the three consecutive papers [**FZ2, FZ4, BFZ2**], which laid the foundation of the theory of cluster algebras.

3.1. For the complete classification of cluster algebras of rank 2, see [**FZ2, ShZ**]. Example 3.12 involving Markov triples is borrowed from [**BFZ2**]. For relations of Markov numbers to various mathematical problems see [**CuFl**] and references therein. The definition of general cluster algebras in Remark 3.13 follows [**FZ2**]. The approach based on normalization condition is suggested in [**FZ4**].

3.2. Theorem 3.14 was proved in [**FZ2**] (see also [**FZ3**] for an extended discussion of this subject, going beyond the context of cluster algebras). Our proof here follows, with minor modifications, that of [**BFZ2**]. To lift the coprimality assumption at the end of the proof, we use the argument suggested in [**FZ2**], p.507. Proposition 3.20 can be found in [**FZ4**]. The treatment of upper cluster algebras follows [**BFZ2**]. Lemma 3.23 is borrowed from [**GSV2**].

3.3. The classification theorem for cluster algebras of finite type was obtained in [**FZ4**].

Exposition in Section 3.3.1 follows [**BaGZ**] with some deviations. For condition (3.13) see, e.g., Theorem 4 in [**PY**] or Theorem 4 in [**Ma**]. Section 3.3.2 follows [**BaGZ**] with certain modifications. For the complete proof of the implication (3)$\Longrightarrow$(1) in Theorem 3.26 based on the notion of generalized associahedra, see [**FZ4**]. Theory of generalized associahedra was developed in [**FZ5, CFZ**]. For further generalizations of associahedra see [**FRe, HLT**] and references therein.

3.4. Proposition 3.37 can be found in [**GSV7**]. Theorem 3.42, apparently, has not appeared in literature in this form. Theorem 3.44 is proved in [**BFZ2**] based on an analog of Lemma 2.11.

3.5. The list of conjectures is borrowed from [**FZ6**].

CHAPTER 4

Poisson structures compatible with the cluster algebra structure

The first goal of this chapter is to give a geometric explanation for cluster algebra matrix mutation rules from the Poisson point of view. Namely, we introduce the notion of a Poisson structure compatible with the cluster algebra structure. Compatibility means that in any set of cluster variables, the Poisson structure is given by the simplest possible kind of a homogeneously quadratic bracket. Then exchange relation are related to simple transvections with respect to the Poisson structure. In particular, matrix mutations (see Definition 3.5) can be simply explained as transformations of the coefficients of a compatible Poisson structure under a transvection. In the first section we give a complete description of all Poisson brackets compatible with a given cluster algebra $\mathcal{A} = \mathcal{A}(\widetilde{B})$, provided $\widetilde{B}$ is of full rank.

The second goal of this chapter is to restore the cluster algebra structure on a Grassmannian starting from the standard Poisson structure. This is done in the second section.

Finally, in the third section, we show that the restriction of the standard Poisson-Lie structure on a simple Lie group to a double Bruhat cell is compatible with the cluster algebra structure on this cell defined at the end of Section 3.4.

4.1. Cluster algebras of geometric type and Poisson brackets

In this Section we assume that the basic field $\mathbb{F}$ is $\mathbb{R}$ or $\mathbb{C}$, however the results are true for any field of characteristic 0.

4.1.1. Compatible Poisson brackets. Let $\{\cdot,\cdot\}$ be a Poisson bracket on the ambient field $\mathcal{F}$ considered as the field of rational functions in $n+m$ independent variables with rational coefficients. We say that functions $f_1,\dots,f_{n+m}$ are *log-canonical* with respect to $\{\cdot,\cdot\}$ if $\{f_i,f_j\} = \omega_{ij} f_i f_j$, where $\omega_{ij} \in \mathbb{Z}$ are constants for all $i,j \in [1,n+m]$. The matrix $\Omega^f = (\omega_{ij})$ is called the *coefficient matrix* of $\{\cdot,\cdot\}$ (in the basis f); evidently, Ω^f is skew-symmetric.

Given a cluster algebra of geometric type $\mathcal{A} = \mathcal{A}(\widetilde{B})$, we consider the following question: *is there a Poisson structure on $\mathcal{F}$ such that all extended clusters in $\mathcal{A}$ are log-canonical with respect to it?* Such a Poisson structure (if any) is called *compatible* with $\mathcal{A}$. The importance of this definition for the cluster algebra theory follows from Example 4.1 and Theorem 4.5 below.

EXAMPLE 4.1. Consider the cluster algebra of Example 3.10 that corresponds to the homogeneous coordinate ring of the Grassmannian $G_2(5)$. Define a Poisson bracket by the coefficient matrix

$$\Omega^x = \begin{pmatrix} 0 & 1 & -1 & -1 & 1 & 1 & 0 \\ -1 & 0 & -1 & -1 & 0 & 1 & -1 \\ 1 & 1 & 0 & 1 & 2 & 2 & 1 \\ 1 & 1 & -1 & 0 & 1 & 2 & 0 \\ -1 & 0 & -2 & -1 & 0 & 1 & 0 \\ -1 & -1 & -2 & -2 & -1 & 0 & -1 \\ 0 & 1 & -1 & 0 & 0 & 1 & 0 \end{pmatrix}$$

in the initial variables $x_1, \dots, x_7$. Let us check that both adjacent clusters are log-canonical with respect to the above bracket. Indeed, the adjacent cluster in direction 1 contains a new cluster variable $x_1' = (x_2x_5 + x_4x_6)/x_1$. Therefore, to check that $\{x_1', x_i\}$ is proportional to $x_1'x_i$ for $i \in [2, 7]$, it suffices to verify that the sum of the second and fifth rows of Ω^x differs from the sum of the fourth and the sixth rows only in the first column, which corresponds to the initial cluster variable x_1. To compute the actual value of $\{x_1', x_i\}/x_1'x_i$ one has to subtract the first row of Ω^x from the above sum. We thus get the coefficient matrix of our bracket in the new basis:

$$\Omega^{x'} = \begin{pmatrix} 0 & -1 & -2 & -1 & -1 & 1 & -1 \\ 1 & 0 & -1 & -1 & 0 & 1 & -1 \\ 2 & 1 & 0 & 1 & 2 & 2 & 1 \\ 1 & 1 & -1 & 0 & 1 & 2 & 0 \\ 1 & 0 & -2 & -1 & 0 & 1 & 0 \\ -1 & -1 & -2 & -2 & -1 & 0 & -1 \\ 1 & 1 & -1 & 0 & 0 & 1 & 0 \end{pmatrix}.$$

Similar calculations for all the five clusters convince us that the bracket in question is compatible with the cluster algebra structure. We will see later that this bracket is, in fact, a Poisson homogeneous bracket on $G_2(5)$.

Note that a compatible Poisson structure does not necessarily exist if $\operatorname{rank}(\widetilde{B}) < n$.

EXAMPLE 4.2. Consider a seed consisting of a cluster $\mathbf{x} = (x_1, x_2, x_3)$ and a 3×3 skew-symmetric matrix $\widetilde{B} = B = \begin{pmatrix} 0 & 1 & -1 \\ -1 & 0 & 1 \\ 1 & -1 & 0 \end{pmatrix}$. Note that B is degenerate.

The exchange relations imply $x_1' = \frac{x_2+x_3}{x_1}$, $x_2' = x_2$, $x_3' = x_3$.

Let $\{\cdot,\cdot\}$ be a Poisson structure compatible with the coefficient free cluster algebra $\mathcal{A}(B)$. Then, by the definition of compatibility, $\{x_1, x_2\} = \lambda x_1x_2$, $\{x_2, x_3\} = \mu x_2x_3$, $\{x_1, x_3\} = \nu x_1x_3$. Besides, $\{x_1', x_2'\} = -\lambda x_2^2/x_1 - \mu x_2x_3/x_1 - \lambda x_2x_3/x_1$. On the other hand, the compatibility yields $\{x_1', x_2'\} = \alpha x_1'x_2' = \alpha x_2^2/x_1 + \alpha x_2x_3/x_1$. Comparing these two expressions we immediately get $\mu = 0$, which means that $\{x_2, x_3\} = 0$. Similarly, $\{x_1, x_2\} = \{x_1, x_3\} = 0$. Hence we see that the only Poisson structure compatible with this cluster algebra is the trivial one.

4.1.2. τ-coordinates. Keeping in mind Example 4.2 we assume for the rest of this chapter that matrix $\widetilde{B}$ has the maximal rank n. Such an assumption is justified by Lemma 3.23.

For our further purposes it is convenient to consider, along with cluster and stable variables $\widetilde{\mathbf{x}}$, another $(n+m)$-tuple of rational functions. In what follows this

$(n+m)$-tuple is denoted $\tau = (\tau_1, \dots, \tau_{n+m})$ and called *τ-coordinates*. It is related to the initial $(n+m)$-tuple $(x_1, \dots, x_n, x_{n+1}, \dots, x_{n+m})$ as follows.

Let $D = \operatorname{diag}(d_1, \dots, d_n)$ be a skew-symmetrizer of $\widetilde{B}$. Define an $(n+m) \times (n+m)$ diagonal matrix $\widehat{D} = \operatorname{diag}(d_1, \dots, d_n, 1, \dots, 1)$ and fix an arbitrary $\widehat{D}$-skew-symmetrizable matrix $\widehat{B}$ such that $\widehat{B}[n; n+m] = \widetilde{B}$ (here and in what follows $A[p; q]$ stands for the $p \times q$ submatrix of A whose entries lie in the first p rows and the first q columns). As before, we slightly abuse notation and denote the entries of $\widehat{B}$ by b_{ij}, $1 \le i, j \le n+m$. Finally, put

$$\tau_j = x_j^{\varkappa_j} \prod_{k=1}^{n+m} x_k^{b_{jk}}, \tag{4.1}$$

where $\varkappa_j$ is an integer, $\varkappa_j = 0$ for $1 \le j \le n$. We say that the entries τ_i, $i \in [1, n+m]$, form a *τ-cluster*.

The transformation $\widetilde{\mathbf{x}} \mapsto \tau$ is *nondegenerate* if

$$\det(\widehat{B} + K) \ne 0, \tag{4.2}$$

where $K = \operatorname{diag}(\varkappa_1, \dots, \varkappa_{n+m})$. It is easy to see that if the transformation $\widetilde{\mathbf{x}} \mapsto \tau$ is nondegenerate and the entries of the extended cluster $\widetilde{\mathbf{x}}$ are functionally independent, then so are the entries of the τ-cluster.

LEMMA 4.3. *An $(n+m)$-tuple of $\varkappa_i$ such that transformation $\widetilde{\mathbf{x}} \mapsto \tau$ is nondegenerate exists if and only if* $\operatorname{rank} \widetilde{B} = n$.

PROOF. The "only if" part is trivial. To prove the "if" part, assume that $\operatorname{rank} \widetilde{B} = n$ and $\operatorname{rank} B = k \le n$, where $B = \widetilde{B}[n; n]$ as before. Then there exists a nonzero $n \times n$ minor Q of $\widetilde{B}$ consisting of the columns $j_1 < j_2 < \dots < j_n$ such that $j_k \le n$, $j_{k+1} > n$ (here $j_0 = 0$, $j_{n+1} = n+m+1$). Without loss of generality assume that $j_i = i$ for $i \in [1, k]$. Define

$$\varkappa_j = \begin{cases} 0 & \text{for } j = 1, \dots, n, \\ 1 & \text{for } j = j_{k+1}, \dots, j_n, \\ \varkappa & \text{otherwise.} \end{cases}$$

Let us prove that there exists an integer $\varkappa$ such that $\det(\widehat{B} + K) \ne 0$. Indeed, the leading coefficient of Q (regarded as a polynomial in $\varkappa$) is equal to the $(2n-k) \times (2n-k)$ minor consisting of the rows and columns with numbers $1, 2, \dots, n, j_{k+1}, \dots, j_n$. It is easy to see that applying the same sequence of elementary operations to rows and columns of $\widehat{B}$ one can make it into a matrix

$$M = \begin{pmatrix} B_1 & 0 & 0 \\ 0 & B_2 & B_3 \\ 0 & B_4 & B_5 \end{pmatrix},$$

where B_1 is just $\widetilde{B}[k; k]$, B_2 is an $(n-k) \times (n-k)$ matrix depending only on the entries of B, B_3, B_4, B_5 are $(n-k) \times (n-k)$ matrices. Moreover, $B_2 = 0$, since otherwise $\operatorname{rank} B$ would exceed k, and hence $\det M = \det B_1 \det B_3 \det B_4$. On the other hand, condition $\operatorname{rank} B = n$ implies $\det B_1 \det B_3 \ne 0$, while the skew-symmetrizability of $\widetilde{B}$ implies $\det B_4 \ne 0$. Therefore, the leading coefficient of $\det(\widehat{B} + K)$ is not zero, and we are done. □

Exchange relation in direction i for $\widetilde{\mathbf{x}}$ induces a transformation $T_i : \tau \mapsto \tau'$. This transformation has a rather simple form. First, we extend the mutation rules 3.5 to all entries of $\widehat{B}$. Observe that for any $\widehat{B}$ as above, one has $\widehat{B}'[n; n+m] = \widetilde{B}'$. Moreover, the coordinate change $\widetilde{\mathbf{x}}' \mapsto \tau'$ remains nondegenerate, due to Lemma 4.3 and Lemma 3.23.

Exchange relations for coordinates τ are described by the following statement.

LEMMA 4.4. *Let $i \in [1, n]$ and let $\tau'_j = T_i(\tau_j)$ for $j \in [1, n+m]$. Then $\tau'_i = 1/\tau_i$ and $\tau'_j = \tau_j \psi_{ji}(\tau_i)$, where*

$$\psi_{ji}(\xi) = \begin{cases} \left(\dfrac{1}{\xi} + 1\right)^{-b_{ji}} & \text{for } b_{ji} > 0, \\ (\xi + 1)^{-b_{ji}} & \text{for } b_{ji} < 0, \\ 1 & \text{for } b_{ji} = 0 \text{ and } j \neq i. \end{cases}$$

PROOF. Let us start from the case $j = i$. Since $i \in [1, n]$ and $b_{ii} = 0$, we have

$$\tau'_i = \prod_{k \neq i} (x'_k)^{b'_{ik}} = \prod_{k \neq i} x_k^{-b'_{ik}} = \frac{1}{\tau_i},$$

as required.

Now, let $j \neq i$. Then

$$\begin{aligned} \tau'_j &= (x'_j)^{\varkappa_j} (x'_i)^{b'_{ji}} \prod_{k \neq i} (x'_k)^{b'_{jk}} \\ &= x_j^{\varkappa_j} \left(\prod_{b_{ik} > 0} x_k^{b_{ik}} + \prod_{b_{ik} < 0} x_k^{-b_{ik}} \right)^{-b_{ji}} x_i^{b_{ji}} \prod_{k \neq i} x_k^{b_{jk}} \prod_{k \neq i} x_k^{(|b_{ji}| b_{ik} + b_{ji} |b_{ik}|)/2} \\ &= \tau_j \left(\prod_{b_{ik} > 0} x_k^{b_{ik}} + \prod_{b_{ik} < 0} x_k^{-b_{ik}} \right)^{-b_{ji}} \prod_{k \neq i} x_k^{(|b_{ji}| b_{ik} + b_{ji} |b_{ik}|)/2}. \end{aligned}$$

If $b_{ji} = 0$, then evidently $\tau'_j = \tau_j$.
If $b_{ji} > 0$, then

$$\begin{aligned} \bar{\tau}_j &= \tau_j \left(\prod_{b_{ik} > 0} x_k^{b_{ik}} + \prod_{b_{ik} < 0} x_k^{-b_{ik}} \right)^{-b_{ji}} \prod_{b_{ik} > 0} x_k^{b_{ji} b_{ik}} \\ &= \tau_j \left(\prod_{b_{ik} \neq 0} x_k^{-b_{ik}} + 1 \right)^{-b_{ji}} = \tau_j (1/\tau_i + 1)^{-b_{ji}}. \end{aligned}$$

If $b_{ji} < 0$, then

$$\begin{aligned} \tau'_j &= \tau_j \left(\prod_{b_{ik} > 0} x_k^{b_{ik}} + \prod_{b_{ik} < 0} x_k^{-b_{ik}} \right)^{-b_{ji}} \prod_{b_{ik} < 0} x_k^{-b_{ji} b_{ik}} \\ &= \tau_j \left(\prod_{b_{ik} \neq 0} x_k^{b_{ik}} + 1 \right)^{-b_{ji}} \prod_{b_{ik} \neq 0} x_k^{-b_{ij} b_{ik}} = \tau_j (\tau_i + 1)^{-b_{ji}}, \end{aligned}$$

as required. □

4.1.3. Characterization of compatible Poisson brackets. We say that a square matrix A is *reducible* if there exists a permutation matrix P such that PAP^T is a block-diagonal matrix, and *irreducible* otherwise; $r(A)$ is defined as the maximal number of diagonal blocks in PAP^T. The partition into blocks defines an obvious equivalence relation $\sim$ on the rows (or columns) of A.

Now we are ready to formulate the existence result for compatible Poisson brackets.

THEOREM 4.5. *Assume that $\widetilde{B}$ is D-skew-symmetrizable and $\operatorname{rank}\widetilde{B} = n$. Then all compatible with $\mathcal{A}(\widetilde{B})$ Poisson brackets form a vector space of dimension $r(B)+\binom{m}{2}$, where $B = \widetilde{B}[n;n]$. Moreover, the coefficient matrices of these Poisson brackets in the basis τ are characterized by the equation $\Omega^\tau[n;n+m] = \Lambda\widetilde{B}\widehat{D}^{-1}$, where $\Lambda = \operatorname{diag}(\lambda_1,\dots,\lambda_n)$ with $\lambda_i = \lambda_j$ whenever $i \sim j$. In particular, if B is irreducible, then $\Omega^\tau[n;n+m] = \lambda\widetilde{B}\widehat{D}^{-1}$.*

PROOF. Let us note first that τ-coordinates are expressed in a monomial way in terms of initial extended cluster $\widetilde{\mathbf{x}}$, and by Lemma 4.3 this transformation is invertible. It is easy to check that all extended clusters in $\mathcal{A}(\widetilde{B})$ are log-canonical w.r.t. some bracket $\{\cdot,\cdot\}$ if and only if so are all τ-clusters. Simple computation shows that $\Omega^\tau = (\widehat{B}+K)\Omega^{\widetilde{\mathbf{x}}}(\widehat{B}+K)^T$. By definition, the exchange relation in direction i preserves the log-canonicity if and only if for any $j \neq i$,

$$\{x_i', x_j'\} = \omega_{ij}' x_i' x_j' \tag{4.3}$$

for some constant ω_{ij}' provided $\{x_i, x_j\} = \omega_{ij} x_i x_j$. Using (3.1) we get

$$\begin{aligned}\{x_i', x_j'\} &= \left\{\frac{1}{x_i}\left(\prod_{b_{ik}>0} x_k^{b_{ik}} + \prod_{b_{ik}<0} x_k^{-b_{ik}}\right), x_j\right\}\\ &= \frac{x_j}{x_i}\prod_{b_{ik}>0} x_k^{b_{ik}}\left(\sum_{b_{ik}>0} b_{ik}\omega_{kj} - \omega_{ij}\right) + \frac{x_j}{x_i}\prod_{b_{ik}<0} x_k^{b_{ik}}\left(-\sum_{b_{ik}<0} b_{ik}\omega_{kj} - \omega_{ij}\right),\end{aligned}$$

and hence conditions (4.3) are satisfied if and only if

$$\sum_{b_{ik}>0} b_{ik}\omega_{kj} - \omega_{ij} = -\sum_{b_{ik}<0} b_{ik}\omega_{kj} - \omega_{ij}$$

for $j \neq i$. This means that

$$(\widehat{B}+K)\Omega^{\widetilde{\mathbf{x}}}[n;n+m] = \widetilde{B}\Omega^{\widetilde{\mathbf{x}}} = \begin{pmatrix}\Delta & 0\end{pmatrix}, \tag{4.4}$$

where Δ is a diagonal matrix. Consequently, we get $\Omega^\tau[n;n+m] = \Delta\widehat{B}[n+m;n]^T$, and hence $\Delta B^T = \Omega^\tau[n;n]$ is skew-symmetric. Therefore, $\Delta = -\Lambda D^{-1}$ where $\Lambda = \operatorname{diag}(\lambda_1,\dots,\lambda_n)$ with $\lambda_i = \lambda_j$ whenever $i \sim j$. It remains to notice that $\widehat{B} = -\widehat{D}^{-1}\widehat{B}^T\widehat{D}$, and therefore $\widehat{B}[n+m;n] = -\widehat{D}^{-1}\widetilde{B}^T D$, and the equation $\Omega^\tau[n;n+m] = \Lambda\widetilde{B}\widehat{D}^{-1}$ follows. The entries $\omega_{ij}^{\widetilde{\mathbf{x}}}$ for $n+1 \le i < j \le n+m$ are free parameters. □

EXAMPLE 4.6. Let us proceed with the cluster algebra considered above in Examples 3.10 and 4.1. Define $b_{ij} = 0$ for $i, j \in [3, 7]$, then

$$\widehat{B} + K = \begin{pmatrix} 0 & 1 & 0 & -1 & 1 & -1 & 0 \\ -1 & 0 & 1 & 0 & 0 & 1 & -1 \\ 0 & -1 & \varkappa & 0 & 0 & 0 & 0 \\ 1 & 0 & 0 & \varkappa & 0 & 0 & 0 \\ -1 & 0 & 0 & 0 & \varkappa & 0 & 0 \\ 1 & -1 & 0 & 0 & 0 & \varkappa & 0 \\ 0 & 1 & 0 & 0 & 0 & 0 & \varkappa \end{pmatrix}.$$

It is easy to see that $\varkappa = 1$ implies $\operatorname{rank}(\widehat{B} + K) = 7$. We therefore can define

$$\tau_1 = \frac{x_2 x_5}{x_4 x_6}, \quad \tau_2 = \frac{x_3 x_6}{x_1 x_7}, \quad \tau_3 = \frac{x_3}{x_2}, \quad \tau_4 = x_1 x_4,$$
$$\tau_5 = \frac{x_5}{x_1}, \quad \tau_6 = \frac{x_1 x_6}{x_2}, \quad \tau_7 = x_2 x_7.$$

Evidently, coordinates $\tau_1, \dots, \tau_7$ are log-canonical for the compatible bracket defined above in Example 4.1. A simple computation shows that the coefficient matrix of this bracket in coordinates $\tau_1, \dots, \tau_7$ equals

$$\Omega^\tau = \begin{pmatrix} 0 & 2 & 0 & -2 & 2 & -2 & 0 \\ -2 & 0 & 2 & 0 & 0 & 2 & -2 \\ 0 & -2 & 0 & 4 & 0 & 2 & 3 \\ 2 & 0 & -4 & 0 & 1 & 2 & 2 \\ -2 & 0 & 0 & -1 & 0 & 0 & -1 \\ 2 & -2 & -2 & -2 & 0 & 0 & 0 \\ 0 & 2 & -3 & -2 & 1 & 0 & 0 \end{pmatrix}.$$

We thus see that $\Omega^\tau[2;7] = 2\widehat{B}[2;7] = \widetilde{B}$, as predicted by Theorem 4.5.

In conclusion, we want to point out that mutations of $\widehat{B}$ can be viewed as linear transformations of a skew-symmetric bilinear form on an $(n+m)$-dimensional vector space under a certain basis change. Since mutations commute with a skew-symmetrization, it is sufficient to consider only mutations of skew-symmetric matrices. Consider an $(n+m)$-dimensional vector space L spanned by $\log \tau_i$. We treat log as a formal expression subject to the relations $\log(fg) = \log f + \log g$ and $\partial \log f / \partial f = 1/f$.

A compatible Poisson bracket $\{\cdot,\cdot\}$ can be identified with a skew-symmetric form ω on L defined by $\omega(\log \tau_i, \log \tau_j) = \{\log \tau_i, \log \tau_j\} = \omega_{ij} \in \mathbb{Z}$. For any element $v \in L$ define a *symplectic transvection* $\sigma_v : L \to L$ as the linear transformation that maps x to $x - \omega(x, v) \cdot v$. Given a basis $\{e_i = \log \tau_i\}$ in L, we define its mutation in direction j as a set of n vectors e_i',

$$e_i' = \begin{cases} -e_j & \text{if } i = j, \\ e_i & \text{if } \omega(e_j, e_i) \geq 0 \text{ and } i \neq j, \\ \sigma_{e_j}(e_i) & \text{if } \omega(e_j, e_i) < 0. \end{cases}$$

Note that e_i' also form a basis in L. It is easy to check that the matrix obtained from $\widehat{B}$ by mutation in direction j coincides with the coefficient matrix of the form ω with respect to the basis e_i'.

4.2. Poisson and cluster algebra structures on Grassmannians

In this section we consider examples of varieties that admit a maximal family of regular functions log-canonical with respect to a given (quadratic) Poisson structure. These examples are provided by the theory of Poisson-Lie groups and Poisson homogeneous spaces and include our running example, real Grassmannians. We will use the latter to show how a structure of a Poisson homogeneous space leads to a construction of a cluster algebra. Slightly abusing terminology, we will say that this cluster algebra is *compatible* with the corresponding Poisson structure.

4.2.1. Poisson structure and log-canonical coordinates. Let $\mathcal{P}$ be a Lie subgroup of a Poisson-Lie group $\mathcal{G}$. A Poisson structure on the homogeneous space $\mathcal{P}\backslash\mathcal{G}$ is called *Poisson homogeneous* if the action map $\mathcal{P}\backslash\mathcal{G}\times\mathcal{G}\to\mathcal{P}\backslash\mathcal{G}$ is Poisson. In particular, if $\mathcal{P}$ is a parabolic subgroup of a simple Lie group $\mathcal{G}$ equipped with the standard Poisson-Lie structure, then $\mathcal{P}\backslash\mathcal{G}$ is a Poisson homogeneous space. We will be interested in the case when $\mathcal{G}=SL_n$ equipped with the bracket (1.25) and

$$\mathcal{P}=\mathcal{P}_k=\left\{\begin{pmatrix} A & 0\\ B & C\end{pmatrix}\ : A\in SL_k, C\in SL_{n-k}\right\}.$$

The resulting homogeneous space is the Grassmannian $G_k(n)$.

We will need an explicit expression of the Poisson homogeneous brackets on an open cell $G_k^0(n)$ in $G_k(n)$ characterized by non-vanishing of the dense Plücker coordinate $x_{[1,k]}$. Elements of $G_k^0(n)$ are parametrized by $k\times(n-k)$ matrices in the following way: if $W\in SL_n$ admits a factorization into block-triangular matrices

$$W=\begin{pmatrix} W_1 & 0\\ Y' & W_2\end{pmatrix}\begin{pmatrix} \mathbf{1}_k & Y\\ 0 & \mathbf{1}_{n-k}\end{pmatrix}=VU, \tag{4.5}$$

then $Y=Y(W)$ represents an element of the cell $G_k^0(n)$.

It is easy to check that Plücker coordinates x_I, $I=\{i_1,\dots,i_k\ :1\le i_1<\dots<i_k\le n\}$, and minors $Y_{\alpha_1\dots\alpha_l}^{\beta_1\dots\beta_l}=\det(y_{\alpha_i,\beta_j})_{i,j=1}^{l}$ of Y are related via

$$Y_{\alpha_1\dots\alpha_l}^{\beta_1\dots\beta_l}=(-1)^{kl-l(l-1)/2-(\alpha_1+\dots+\alpha_l)}\frac{x_{([1,k]\setminus\{\alpha_1\dots\alpha_l\})\cup\{\beta_1+k\dots\beta_l+k\}}}{x_{[1,k]}}.$$

Note that, if the row index set $\{\alpha_1\dots\alpha_l\}$ in the above formula is contiguous then the sign in the right hand side can be expressed as $(-1)^{(k-\alpha_l)l}$.

It will be convenient to extend the Sklyanin bracket (1.25) from SL_n to the Poisson bracket on the associative algebra Mat_n of $n\times n$ matrices given by the same formula:

$$\begin{aligned}&\{f_1,f_2\}_{\mathrm{Mat}_n}(W)\\ &=\frac12(R(\mathrm{grad}f_1(W)\ W),\mathrm{grad}f_2(W)\ W)-\frac12(R(W\ \mathrm{grad}f_1(W)),W\ \mathrm{grad}f_2(W)),\end{aligned}\tag{4.6}$$

where R is the standard R-matrix (1.23).

For any function f on Mat_n, denote by Δf its variation in the direction ΔW:

$$\Delta f(W)=\frac{d}{dt}f(W+t\Delta W)|_{t=0}\ .$$

We can view Y in (4.5) as a function of W. Then it is not hard to see that the variation of $Y=Y(W)$ is given by

$$\Delta Y=\begin{pmatrix}\mathbf{1}_k & 0\end{pmatrix}V^{-1}\Delta W U^{-1}\begin{pmatrix}0\\ \mathbf{1}_{n-k}\end{pmatrix},$$

and, therefore, for any function f on $G_k^0(n)$ (or, equivalently, on the space of all $k \times (n-k)$ matrices), the gradient of a superposition $f \circ Y$ at W can be computed as

$$\operatorname{grad}(f \circ Y) = U^{-1} \begin{pmatrix} 0 \\ \mathbf{1}_{n-k} \end{pmatrix} \operatorname{grad} f \circ Y \begin{pmatrix} \mathbf{1}_k & 0 \end{pmatrix} V^{-1}$$

and

$$\operatorname{grad}(f \circ Y)W = \operatorname{Ad}_{U^{-1}} \left(\begin{pmatrix} 0 \\ \mathbf{1}_{n-k} \end{pmatrix} \operatorname{grad} f \circ Y \begin{pmatrix} \mathbf{1}_k & 0 \end{pmatrix} \right) = \begin{pmatrix} -Y \\ \mathbf{1}_{n-k} \end{pmatrix} \operatorname{grad} f \circ Y \begin{pmatrix} \mathbf{1}_k & Y \end{pmatrix},$$

$$W \operatorname{grad}(f \circ Y) = \operatorname{Ad}_V \left(\begin{pmatrix} 0 \\ \mathbf{1}_{n-k} \end{pmatrix} \operatorname{grad} f \circ Y \begin{pmatrix} \mathbf{1}_k & 0 \end{pmatrix} \right).$$

In particular, we see that $W\operatorname{grad}(f \circ Y)$ is strictly lower triangular.

Thus we obtain from (4.6)

$$\begin{aligned} &\{(f_1 \circ Y, f_2 \circ Y\}_{\mathrm{Mat}_n}(W) \\ &= \frac{1}{2} \left(\begin{pmatrix} -R(Y\operatorname{grad} f_1(Y)) & -Y\operatorname{grad} f_1(Y)Y \\ -\operatorname{grad} f_1(Y) & R(\operatorname{grad} f_1(Y)Y) \end{pmatrix}, \begin{pmatrix} -Y\operatorname{grad} f_2(Y) & -Y\operatorname{grad} f_2(Y)Y \\ \operatorname{grad} f_2(Y) & \operatorname{grad} f_2(Y)Y \end{pmatrix} \right) \\ &= \frac{1}{2} \left(R(\operatorname{grad} f_1(Y)Y),\ \operatorname{grad} f_2(Y)Y \right) + \frac{1}{2} \left(R(Y\operatorname{grad} f_1(Y)),\ Y\operatorname{grad} f_2(Y) \right). \end{aligned}$$

It follows that the last expression defines a Poisson bracket on $G_k^0(n)$ (which, to avoid cumbersome notations we will simply denote by $\{\cdot,\cdot\}$) such that

$$\{f_1 \circ Y, f_2 \circ Y\}_{\mathrm{Mat}_n} = \{f_1, f_2\} \circ Y\ .$$

In terms of matrix elements y_{ij} of Y, this bracket looks as follows:

(4.7) $$2\{y_{ij}, y_{\alpha\beta}\} = (\operatorname{sign}(\alpha - i) - \operatorname{sign}(\beta - j)) y_{i\beta} y_{\alpha j}.$$

Next, we will introduce new coordinates on $G_k^0(n)$, log-canonical with respect to $\{\cdot,\cdot\}$. This will require some preparation.

Let $I = \{i_1, \dots, i_r\}$, $J = \{j_1, \dots, j_r\}$ be two multi-indices. As in (1.1), we denote by $I(i_p \to \alpha)$ the result of replacing i_p with α in I Furthermore, $I \setminus i_p$ is the multi-index obtained by deleting i_p from I and αI is the multi-index $I = \{\alpha, i_1, \dots, i_r\}$. Then the Laplace expansion formula implies

(4.8) $$\sum_{p=1}^{r} w_{i_p\beta} W^J_{I(i_p\to\alpha)} = w_{\alpha\beta} W^J_I - W^{\beta J}_{\alpha I}.$$

We will say that $\alpha < I$ (resp. $\alpha > I$), if α is less than the minimal index in I (resp., the maximal index in I is less than α). We define $\operatorname{sign}(\alpha - I) = -\operatorname{sign}(I - \alpha)$ to be $-1, 0$ or 1, if $\alpha < I, \alpha \in I$ or $\alpha > I$, resp. Otherwise, $\operatorname{sign}(\alpha - I)$ is not defined.

LEMMA 4.7. *If* $\operatorname{sign}(\alpha - I)$ *and* $\operatorname{sign}(\beta - J)$ *are defined and*

(4.9) $$|\operatorname{sign}(\alpha - I) - \operatorname{sign}(\beta - J)| \le 1,$$

then

(4.10) $$\{y_{\alpha\beta}, Y^J_I\} = -\left(\operatorname{sign}(\alpha - I) - \operatorname{sign}(\beta - J)\right) y_{\alpha\beta} Y^J_I.$$

PROOF. It is evident from (4.7) that $\{y_{\alpha\beta}, Y^J_I\} = 0$ if $\alpha < I$, $\beta < J$, or $\alpha > I$, $\beta > J$, so in these cases (4.10) holds true.

In general, one obtains from (4.7)

$$\begin{aligned}
\{y_{\alpha\beta}, Y_I^J\} &= \sum_{p,q=1}^{r} (-1)^{p+q}\{y_{\alpha\beta}, y_{i_p}^{j_q}\} Y_{I\setminus i_p}^{J\setminus j_q} \\
&= \sum_{p=1}^{r} \operatorname{sign}(i_p-\alpha) y_{i_p\beta} \sum_{q=1}^{r} (-1)^{p+q} y_{\alpha j_q} Y_{I\setminus i_p}^{J\setminus j_q} \\
&\quad - \sum_{q=1}^{r} \operatorname{sign}(j_q-\beta) y_{\alpha j_q} \sum_{p=1}^{r} (-1)^{p+q} y_{i_p\beta} Y_{I\setminus i_p}^{J\setminus j_q} \\
&= \sum_{p=1}^{r} \operatorname{sign}(i_p-\alpha) y_{i_p\beta} Y_{I(i_p\to\alpha)}^{J} - \sum_{q=1}^{r} \operatorname{sign}(j_q-\beta) y_{\alpha j_q} Y_I^{J(j_q\to\beta)}.
\end{aligned}$$

Assume that $\beta \in J$. Then, in the second sum above, $Y_I^{J(j_q\to\beta)}$ can be nonzero only if $\operatorname{sign}(j_q-\beta) = 0$, which implies that the sum is zero. If in addition $\alpha \in I$, the first sum is equal to zero as well, and thus $\{y_{\alpha\beta}, Y_I^J\} = 0$ if $\alpha \in I, \beta \in J$, which is consistent with (4.10). Otherwise, by our assumptions, $\operatorname{sign}(i_p-\alpha)$ does not depend on p and is equal to $\operatorname{sign}(I-\alpha)$. In this case, (4.8) implies

$$\{y_{\alpha\beta}, Y_I^J\} = \operatorname{sign}(I-\alpha)(y_{\alpha\beta}Y_I^J - Y_{\alpha I}^{\beta J}) = \operatorname{sign}(I-\alpha) y_{\alpha\beta} Y_I^J.$$

This agrees with (4.10). The remaining case $\alpha \in I$, $\operatorname{sign}(J-\beta) = \pm 1$ can be treated in the same way. □

The following Corollary drops out immediately from Lemma 4.7 together with the Leibniz rule for Poisson brackets.

COROLLARY 4.8. *Let $A = \{\alpha_1, \dots, \alpha_l\}, B = \{\beta_1, \dots, \beta_l\}$ be such that for every pair (α_p, β_q), $p, q \in [1, l]$, condition (4.9) is satisfied. Then*

$$\{Y_A^B, Y_I^J\} = -\left(\sum_{p=1}^{l} (\operatorname{sign}(\alpha_p - I) - \operatorname{sign}(\beta_p - J))\right) Y_A^B Y_I^J. \tag{4.11}$$

For every (i, j)-entry of Y define

$$l(i, j) = \min(i-1, n-k-j) \tag{4.12}$$

and

$$F_{ij} = Y_{[i-l(i,j),i]}^{[j,j+l(i,j)]}, \tag{4.13}$$

see Fig. 4.1 where the diagonal line separates two regions in which $l(i, j)$ is calculated via distinct linear expressions.

Note that F_{ij} depends only on $y_{\alpha\beta}$ with $\alpha \le i$, $\beta \ge j$ and dependence of F_{ij} on y_{ij} is linear, which implies that the change of coordinates $(y_{ij}) \mapsto (F_{ij})$ is a birational transformation.

LEMMA 4.9. *Put*

$$t_{ij} = \frac{F_{ij}}{F_{i-1,j+1}}. \tag{4.14}$$

Then

$$\{\ln t_{ij}, \ln t_{\alpha\beta}\} = \operatorname{sign}(j-\beta)\delta_{i\alpha} - \operatorname{sign}(i-\alpha)\delta_{j\beta}. \tag{4.15}$$

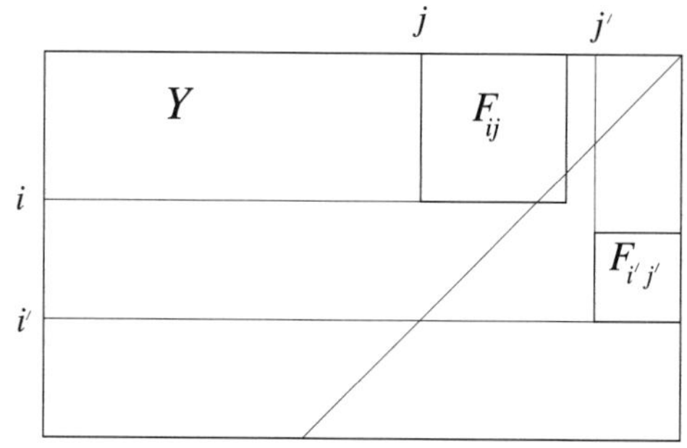

FIGURE 4.1. To the definition of F_{ij}

PROOF. First, we will show that coordinates F_{ij} are log-canonical. For this one needs to check that conditions of Corollary 4.8 are satisfied for every pair $F_{\alpha\beta}, F_{ij}$. One has the following seven cases to consider.

(i) $\alpha \le i$, $\beta \le j$, $i + j \le n - k + 1$;
(ii) $\alpha \le i$, $\beta > j$, $\max(\alpha + \beta, i + j) \le n - k + 1$;
(iii) $\alpha \le i$, $\beta \le j$, $\alpha + \beta > n - k + 1$;
(iv) $\alpha \le i$, $\beta > j$, $\min(\alpha + \beta, i + j) > n - k + 1$;
(v) $\alpha \le i$, $\beta \le j$, $\alpha + \beta \le n - k + 1 < i + j$;
(vi) $\alpha \le i$, $\beta > j$, $\alpha + \beta \le n - k + 1 < i + j$;
(vii) $\alpha \le i$, $\beta > j$, $i + j \le n - k + 1 < \alpha + \beta$.

Direct inspection shows that choosing $Y_A^B = F_{ij}$, $Y_I^J = F_{\alpha\beta}$ in case 3 and $Y_A^B = F_{\alpha\beta}$, $Y_I^J = F_{ij}$ in all the remaining cases ensures that conditions of Corollary 4.8 hold true. Moreover, one can use (4.11) to compute $\{\ln F_{\alpha\beta}, \ln F_{ij}\}$ in each of these cases; the answers are

(i) $-\min(\alpha, j - \beta)$;
(ii) $\max(\alpha + \beta, i + j) - \max(\beta, i + j)$;
(iii) $i - \max(\alpha, i + j + k - n - 1))$;
(iv) $\min(\alpha, i + j + k - n - 1) - \min(\alpha + \beta + k - n - 1, i + j + k - n - 1)$;
(v) $n + 1 - j - k + \max(\alpha, i + j + k - n - 1) - \max(\alpha + \beta, i)$;
(vi) $-(n + 1 - k - \max(\beta, i + j))$;
(vii) $-\min(\alpha, i + j + k - n - 1)$.

Note that formulae above remain valid if we replace all strict inequalities used to describe cases (i) through (vii) by non-strict ones. Now (4.15) can be derived from the formulae above via the case by case verification, simplified by noticing that due to the previous remark, for any (i, j) and (α, β), all the four quadruples $(i, j), (\alpha, \beta)$; $(i - 1, j + 1), (\alpha, \beta)$; $(i, j), (\alpha - 1, \beta + 1)$ and $(i - 1, j + 1), (\alpha - 1, \beta + 1)$ satisfy the same set of conditions out of (i)–(vii) (with inequalities relaxed). □

Denote $n - k$ by m. If we arrange variables $\ln t_{ij}$ into a vector

$$\tilde{t} = (\ln t_{11}, \ldots, \ln t_{1m}, \ldots, \ln t_{k1}, \ldots, \ln t_{km}), \tag{4.16}$$

the coefficient matrix Ω_{km} of Poisson brackets (4.15) will look as follows:

$$\Omega_{km} = \begin{pmatrix} A_m & \mathbf{1}_m & \mathbf{1}_m & \cdots & \mathbf{1}_m \\ -\mathbf{1}_m & A_m & \mathbf{1}_m & \cdots & \mathbf{1}_m \\ -\mathbf{1}_m & -\mathbf{1}_m & A_m & \cdots & \mathbf{1}_m \\ \vdots & \vdots & \vdots & \ddots & \vdots \\ -\mathbf{1}_m & -\mathbf{1}_m & -\mathbf{1}_m & \cdots & A_m \end{pmatrix} = A_m \otimes \mathbf{1}_k - \mathbf{1}_m \otimes A_k, \tag{4.17}$$

where $A_1 = 0$ and

$$A_m = \Omega_{1m}^T = \begin{pmatrix} 0 & -1 & -1 & \cdots & -1 \\ 1 & 0 & -1 & \cdots & -1 \\ 1 & 1 & 0 & \cdots & -1 \\ \vdots & \vdots & \vdots & \ddots & \vdots \\ 1 & 1 & 1 & \cdots & 0 \end{pmatrix}.$$

We now proceed to compute a maximal dimension of a symplectic leaf of the bracket (4.7).

First note that left multiplying $(\lambda \mathbf{1}_m + A_m^T)$ by $C_m = \mathbf{1}_m + e_{1m} - \sum_{i=2}^m e_{i,i-1}$, where e_{ij} is a $(0,1)$-matrix with a unique 1 at position (i,j), results in the following matrix:

$$B_m(\lambda) = \begin{pmatrix} \lambda-1 & 0 & 0 & \cdots & 0 & \lambda+1 \\ -\lambda-1 & \lambda-1 & 0 & \cdots & 0 & 0 \\ 0 & -\lambda-1 & \lambda-1 & \cdots & 0 & 0 \\ \vdots & \vdots & \vdots & \ddots & \vdots & \vdots \\ 0 & 0 & 0 & \cdots & -\lambda-1 & \lambda-1 \end{pmatrix}.$$

Since the determinant of $B_m(\lambda)$ is easily computed to be equal to $(\lambda-1)^m+(\lambda+1)^m$, it follows that the spectrum of A_m is given by

$$\left(\frac{\lambda+1}{\lambda-1}\right)^m = -1. \tag{4.18}$$

Next, we observe that using block row transformations similar to row transformations applied to $(\lambda \mathbf{1}_m + A_m^T)$ above, one can reduce Ω_{km} to a matrix $B_k(A_m)$ obtained from $B_k(\lambda)$ by replacing λ with A_m and 1 with $\mathbf{1}_m$. Since $(A_m - \mathbf{1}_m)$ is invertible by (4.18), we can left multiply $B_k(A_m)$ by $\operatorname{diag}((A_m-\mathbf{1}_m)^{-1}, \dots, (A_m-\mathbf{1}_m)^{-1})$ and conclude that the kernel of Ω_{km} coincides with that of the matrix

$$\begin{pmatrix} \mathbf{1}_m & 0 & 0 & \cdots & Q \\ -Q & \mathbf{1}_m & 0 & \cdots & 0 \\ 0 & -Q & \mathbf{1}_m & \cdots & 0 \\ \vdots & \vdots & \vdots & \ddots & \vdots \\ 0 & 0 & 0 & \cdots & \mathbf{1}_m \end{pmatrix},$$

where $Q = (A_m + \mathbf{1}_m)(A_m - \mathbf{1}_m)^{-1}$. The kernel consists of vectors of the form $\begin{pmatrix} v^T & Qv^T & \dots & Q^{k-1}v^T \end{pmatrix}^T$, where v satisfies condition $(Q^k + \mathbf{1}_m)v^T = 0$. In other words, the kernel of Ω_{km} is parametrized by the (-1)-eigenspace of Q^k. Due to (4.18), the dimension of this eigenspace is equal to

$$d_{km} = \#\left\{\nu \in \mathbb{C} : \ \nu^k = \nu^m = -1\right\}.$$

Moreover, it is not hard to check that

$$Q = (A_m + \mathbf{1}_m)(A_m - \mathbf{1}_m)^{-1} = -e_{m1} + \sum_{i=2}^{m} e_{i-1,i}, \tag{4.19}$$

and therefore $Q^k = \sum_{i=k+1}^{m} e_{i-k,i} - \sum_{i=1}^{k} e_{i+m-k,i}$. A (-1)-eigenspace of Q^k is non-trivial if and only if there exist natural numbers p, q such that $(2p-1)k = (2q-1)m$ (we can assume $(2p-1)$ and $(2q-1)$ are co-prime). Let $l = \gcd(k, m)$, then every non-trivial (-1)-eigenvector of Q^k is a linear combination of vectors $v(i) = (v(i)_j)_{j=1}^m$, $i \in [1, l]$, that can be described as follows: $v(i)_{i+\alpha l} = (-1)^\alpha$ for $\alpha = 0, \dots, \frac{m}{l} - 1$, and all the other entries of $v(i)$ vanish. To analyze the corresponding element of the kernel of P_{km}, we represent it as a $k \times m$ matrix

$$V(i) = \begin{pmatrix} v(i) \\ v(i)Q^T \\ \vdots \\ v(i)(Q^{k-1})^T \end{pmatrix}.$$

From the form of Q one concludes that $V(i)$ is a matrix of 0's and ± 1's that has a Hankel structure, i.e. its entries do not change along anti-diagonals. More precisely,

$$V(i)_{pq} = \begin{cases} (-1)^\alpha & \text{if } p + q = i + \alpha l, \\ 0 & \text{otherwise} \end{cases}$$

Here α changes from 0 to $\frac{n}{l} - 1$. Since to each element V of the kernel of Ω_{km} (represented as a $k \times m$ matrix) there corresponds a Casimir function I_V of $\{\cdot, \cdot\}$ given by $I_V = \prod_{i=1,j=1}^{k,m} t_{ij}^{V_{ij}}$, the observation above together with (4.14) implies that *on an open dense subset of $G_k^0(n)$ the algebra of Casimir functions is generated by monomials in*

$$J_1 = F_{11}, \dots, J_k = F_{k1}, J_{k+1} = F_{k2}, \dots, J_{n-1} = F_{km}. \tag{4.20}$$

In particular,

$$I_{V(i)} = \prod_{\alpha=0}^{\frac{n}{l}-1} J_{i+\alpha l}^{(-1)^\alpha} \tag{4.21}$$

(we assume that $J_n = 1$).

Thus we have proved

THEOREM 4.10. *Let $l = \gcd(k, n)$. The codimension of a maximal symplectic leaf of $G_k^0(n)$ is equal to 0 if $\frac{k}{l}$ is even or $\frac{n-k}{l}$ is even, and is equal to l otherwise. In the latter case, a symplectic leaf via a point in general position is parametrized by values of Casimir functions $I_{V(i)}$, $i \in [1, l]$, defined in* (4.21).

We call a symplectic leaf in $G_k^0(n)$ *generic* if it contains a point Y such that

$$J_1(Y) \cdots J_{n-1}(Y) \neq 0.$$

Recall now that right multiplication of a $k \times n$ matrix $\tilde{X}$ that represents an element $X \in G_k(n)$ by elements of GL_n defines the natural action of GL_n on $G_k(n)$. In particular, the action of the maximal torus $T = \{\text{diag}(t_1, \dots, t_n)\}$ of GL_n on $G_k(n)$ amounts to multiplying the i-th column of $\tilde{X}$ by t_i. This action descends to an action of the torus $(\mathbb{C}^*)^{n-1}$ in $G_k^0(n)$: for $t = (t_1, \ldots, t_n)$ such that $t_1 \times \cdots \times t_n = 1$ and $Y \in G_k^0(n)$

$$(tY)_{ij} = t_i^{-1} y_{ij} t_{k+j} \quad 1 \le i \le k,\ 1 \le j \le m\ . \tag{4.22}$$

Then (i) the action (4.22) restricts to the action of $(\mathbb{R}^*)^{n-1} = \{\operatorname{diag}(t_1, \ldots, t_n) : t_1 \times \cdots \times t_n = \pm 1\}$ of $G_k^0(n)0$; (ii) the latter action is Poisson with respect to the bracket (4.7). Indeed,

$$\begin{aligned}\{t_i^{-1} y_{ij} t_{k+j}, t_\alpha^{-1} y_{\alpha\beta} t_{k+\beta}\} &= t_i^{-1} t_\alpha^{-1} \{y_{ij}, y_{\alpha\beta}\} t_{k+j} t_{k+\beta} \\ &= \frac{1}{2}(\operatorname{sign}(\alpha - i) - \operatorname{sign}(\beta - j))(t_\alpha^{-1} y_{\alpha j} t_{k+j})(t_i^{-1} y_{i\beta} t_{k+\beta}).\end{aligned}$$

PROPOSITION 4.11. *The locus* $S_k(n) = \{Y \in G_k^0(n)\ : J_1(Y) \cdots J_{n-1}(Y) \neq 0\}$ *is the union of generic symplectic leaves and the action* (4.22) *is locally transitive on* $S_k(n)$.

PROOF. Recall that by (4.13) and (4.20),

$$J_i = \begin{cases} Y_{[1,i]}^{[1,i]} & i \le k, \\ Y_{[1,k]}^{[i-k+1,i]} & k < i \le m, \\ Y_{[i-m+1,k]}^{[i-k+1,m]} & m < i \le n. \end{cases}$$

Thus, for all i, α, β conditions of the Lemma 4.7 are satisfied for J_i and $y_{\alpha\beta}$. As a result, the locus $\{J_i = 0\}$ is a Poisson submanifold in $G_k^0(n)$. Therefore, any generic symplectic leaf lies in $S_k(n)$.

Clearly, the toric action (4.22) preserves $S_k(n)$. By Theorem 4.10, if the codimension of a generic symplectic leaf in $G_k^0(n)$ is l, then $k = (2r+1)l$, $n = 2pl$ for some integers r, l such that $\gcd(2r+1, p) = 1$. We can rewrite Casimir functions in (4.21) as

$$I_{V(i)} = \prod_{\alpha=0}^{\mathrm{p}-1} \frac{J_{i+2\alpha l}}{J_{i+(2\alpha+1)l}}\ .$$

Using relations between minors of Y and Plücker coordinates of the element $X \in G_k^0(n)$ that is represented by Y, on can see that, up to a sign, $I_{V(i)}$ coincides with

$$I_i = \prod_{j=i \bmod 2l} \frac{x_{[j,j+k-1]}}{x_{[j+l,j+l+k-1]}}\ ,$$

where, as usual, indices defining Plücker coordinates are considered mod n. Then, since under the toric action on $G_k(n)$ every $x_{i_1 \cdots i_k}$ changes to $x_{i_1, \cdots, i_k} t_{i_1} \times \cdots \times t_{i_k}$, I_i are transformed under the action (4.22) as follows:

$$I_i(tY) = I_i(Y) \prod_{\beta=0}^{l-1} \prod_{j=i+\beta \bmod 2l} \frac{t_{j+l}}{t_j}\ .$$

Denote $\kappa_\beta = \prod_{j=\beta \bmod 2l} t_{j+l}/t_j$. Then $\kappa_{\beta+l} = 1/\kappa_\beta$, and we have for $i \in [1, l]$

$$I_i(tY) = I_i(Y) \prod_{\beta=1}^{i-1} \kappa_\beta^{-1} \prod_{\beta=i}^{l} \kappa_\beta\ .$$

Define

$$C_i(Y) = \begin{cases} I_i(Y)/I_{i+1}(Y) & i \in [1, l-1], \\ I_l(Y) I_1(Y) & i = l. \end{cases}$$

Then

$$C_i(tY) = \kappa_i^2 C_i(tY) \tag{4.23}$$

for $i \in [1,l]$. Note that the map $(\mathbb{R}^*)^{n-1} \ni t \mapsto \kappa = (\kappa_1, \dots, \kappa_l) \in (\mathbb{R}^*)^l$ is surjective: the preimage of κ contains, e.g., an element t with $t_i = \sqrt{|\kappa_i|}$ for $i \in [1,l]$, $t_i = \kappa_{i-l}/\sqrt{|\kappa_{i-l}|}$ for $i \in [l+1, 2l]$ and $t_i = 1$ for $i > 2l$. This means that any point Y in $S_k(n)$ can be moved by the the action (4.22) into a point tY such that $C_i(tY) = \pm 1$, $i \in [1,l]$. □

4.2.2. Cluster algebra structures on Grassmannians compatible with the standard Poisson-homogeneous structure. Our next goal is to build a cluster algebra $\mathcal{A}^{G_k^0(n)}$ associated with the Poisson bracket (4.7). The initial extended cluster consists of functions

$$f_{ij} = (-1)^{(k-i)(l(i,j)-1)} F_{ij} = \frac{x_{([1,k]\setminus[i-l(i,j),\ i])\cup[j+k, j+l(i,j)+k]}}{x_{[1,k]}}, \tag{4.24}$$

$$i \in [1,k], \quad j \in [1,m];$$

(here $m = n-k$ as above, and $l(i,j)$ is defined by (4.12)). We designate functions $f_{11}, f_{21}, \dots, f_{k1}, f_{k2}, \dots, f_{km}$ (cf. (4.20)) to serve as stable coordinates. This choice is motivated by the last statement of Theorem 4.10 and by the following observation: let I, J be the row and column sets of the minor that represents one of the functions (4.20), then for any pair (α, β), $\alpha \in [1,k]$, $\beta \in [1,m]$, condition (4.9) is satisfied and thus, functions (4.20) have log-canonical brackets with all coordinate functions $y_{\alpha\beta}$.

Now we need to define the matrix $\widetilde{B}$ that gives rise to cluster transformations compatible with the Poisson structure. We want to choose $\widetilde{B}$ in such a way that the submatrix of $\widetilde{B}$ corresponding to cluster coordinates will be skew-symmetric and irreducible. According to (4.4) and to our choice of stable coordinates, this means that $\widetilde{B}$ must satisfy

$$\widetilde{B}\Omega^f = const \cdot \begin{pmatrix} \operatorname{diag}(P, \dots, P) & 0 \end{pmatrix}$$

where $P = \sum_{i=1}^{m-1} e_{i,i+1}$ is a $(m-1) \times m$ matrix and Ω^f is the coefficient matrix of Poisson brackets $\{\cdot,\cdot\}$ in the basis f_{ij}.

Let $\tilde{t}$ be defined as in (4.16), and let

$$\tilde{f} = (\ln f_{11}, \dots, \ln f_{1m}, \dots, \ln f_{k1}, \dots, \ln f_{km}).$$

Then $\tilde{t} = M\tilde{f}^T$, where

$$M = \begin{pmatrix} \mathbf{1}_m & 0 & \cdots & 0 & 0 \\ -S & \mathbf{1}_m & \cdots & 0 & 0 \\ \vdots & \vdots & \ddots & \vdots & \vdots \\ 0 & 0 & \cdots & -S & \mathbf{1}_m \end{pmatrix}$$

and $S = \sum_{i=2}^{m} e_{i-1,i}$. Then $\Omega^f = M\Omega_{km}M^T$.

Define a $(k-1)m \times km$ block bidiagonal matrix

$$V = \begin{pmatrix} \mathbf{1}_m - Q^{-1} & Q^{-1} - \mathbf{1}_m & \cdots & 0 & 0 \\ 0 & \mathbf{1}_m - Q^{-1} & \cdots & 0 & 0 \\ \vdots & \vdots & \ddots & \vdots & \vdots \\ 0 & 0 & \cdots & \mathbf{1}_m - Q^{-1} & Q^{-1} - \mathbf{1}_m \end{pmatrix}.$$

Observe that P is the upper $(m-1)\times m$ submatrix of S, and $PQ^{-1} = PS^T$.

Since by (4.17), (4.19),

$$V\Omega_{km} = 2\begin{pmatrix} \mathbf{1}_m & -Q^{-1} & 0 & \dots & 0 & 0 \\ 0 & \mathbf{1}_m & -Q^{-1} & \dots & 0 & 0 \\ \vdots & \vdots & \vdots & \ddots & \vdots & \vdots \\ 0 & 0 & 0 & \dots & \mathbf{1}_m & -Q^{-1} \end{pmatrix},$$

we obtain

$$\frac{1}{2}\operatorname{diag}(P,\dots,P)V\Omega_{km} = \big(\operatorname{diag}(P,\dots,P)\quad 0\big)\, M^T.$$

Define

$$\widetilde{B} = \operatorname{diag}(P,\dots,P)VM = \begin{pmatrix} B_0 & B_1 & 0 & \dots & 0 \\ B_{-1} & B_0 & B_1 & \dots & 0 \\ 0 & B_{-1} & B_0 & \dots & 0 \\ \vdots & \vdots & \vdots & \ddots & \vdots \\ 0 & 0 & 0 & \dots & B_1 \end{pmatrix}, \tag{4.25}$$

where $B_0 = P(\mathbf{1}_m - Q^{-1})(\mathbf{1}_m + S)$, $B_1 = -P(\mathbf{1}_m - Q^{-1})$, and $B_{-1} = -P(\mathbf{1}_m - Q^{-1})S$. Then

$$\widetilde{B}\Omega^f = \operatorname{diag}(P,\dots,P)V\Omega_{km}(M^T)^{-1} = 2\,\big(\operatorname{diag}(P,\dots,P)\quad 0\big)\,,$$

as needed. Note that for $z = (z_{11},\dots,z_{1,m},\dots,z_{k1},\dots,z_{k,m})$ one has

$$(\widetilde{B}z)_{ij} = z_{i+1,j} + z_{i,j-1} + z_{i-1,j+1} - z_{i+1,j-1} - z_{i,j+1} - z_{i-1,j}.$$

It is easy to see that the submatrix B of $\widetilde{B}$ corresponding to the non-stable coordinates is indeed skew-symmetric and irreducible.

The matrix $\widetilde{B}$ thus obtained can be conveniently represented by its diagram $\Gamma(\widetilde{B})$. The latter is defined similarly to the definition of the diagram of a square skew-symmetrizable matrix (see Section 3.3.2). The only difference is that the vertices of $\Gamma(\widetilde{B})$ correspond to all columns of $\widetilde{B}$, and hence there are no edges between the vertices corresponding to stable variables. In our case, $\Gamma(\widetilde{B})$ is a directed graph with vertices forming a rectangular $k\times m$ array and labeled by pairs of integers (i,j), and edges $(i,j)\to(i,j+1)$, $(i+1,j)\to(i,j)$ and $(i,j)\to(i+1,j-1)$ (cf. Fig. 4.2).

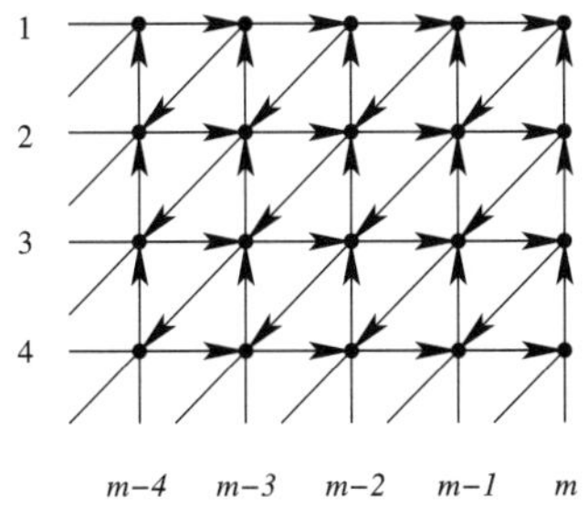

FIGURE 4.2. Diagram corresponding to $G_k^0(n)$

Cluster transformations can now be "read off" the diagram:

$$(4.26)\qquad f'_{ij}f_{ij} = \begin{cases} f_{i+1,j-1}f_{i,j+1}f_{i-1,j} + f_{i+1,j}f_{i,j-1}f_{i-1,j+1}, & i>1,\ j<m,\\ f_{1,i-1}f_{2j} + f_{1,j+1}f_{2,j-1}, & i=1,\ j=m,\\ f_{i,m-1}f_{i+1,m} + f_{i+1,m-1}f_{i-1,m}, & i>1,\ j=m,\\ f_{1,m-1}f_{2m} + f_{2.m-1}, & i=1,\ j=m. \end{cases}$$

The following proposition can be considered as an analog of the third claim in Lemma 2.11.

LEMMA 4.12. *For every $i \in [1, k-1]$ and every $j \in [2, n-k]$, a cluster variable f'_{ij} obtained via the cluster transformation from the initial cluster (4.24) is a regular function on $G^0_k(n)$.*

PROOF. The proof is based on Jacobi's determinantal identity

$$(4.27)\qquad A^{\gamma\delta J}_{\alpha\beta I}A^J_I = A^{\gamma J}_{\alpha I}A^{\delta J}_{\beta I} - A^{\delta J}_{\alpha I}A^{\gamma J}_{\beta I}\ .$$

We will consider the following cases:

(i) $1 < i < m+1-j$. In this case $f_{ij} = (-1)^{(k-i)i}Y^{[j,j+i-1]}_{[1,i]}$, and (4.27), (4.24) imply

$$\begin{aligned}(-1)^{i(k+i)}f_{i+1,j-1}f_{i,j+1}f_{i-1,j} &= Y^{[j-1,j+i-1]}_{[1,i+1]}Y^{[j,j+i-2]}_{[1,i-1]}Y^{[j+1,j+i]}_{[1,i]}\\ &= \left(Y^{[j,j+i-1]}_{[1,i+1]\setminus i}Y^{[j-1,j+i-2]}_{[1,i]} - Y^{[j-1,j+i-2]}_{[1,i+1]\setminus i}Y^{[j,j+i-1]}_{[1,i]}\right)Y^{[j+1,j+i]}_{[1,i]}\end{aligned}$$

and

$$\begin{aligned}(-1)^{i(k+i)}f_{i+1,j}f_{i,j-1}f_{i-1,j+1} &= Y^{[j,j+i]}_{[1,i+1]}Y^{[j+1,j+i-1]}_{[1,i-1]}Y^{[j-1,j+i-2]}_{[1,i]}\\ &= \left(Y^{[j+1,j+i]}_{[1,i+1]\setminus i}Y^{[j,j+i-1]}_{[1,i]} - Y^{[j,j+i-1]}_{[1,i+1]\setminus i}Y^{[j+1,j+i]}_{[1,i]}\right)Y^{[j+1,j+i]}_{[1,i]}.\end{aligned}$$

Therefore, by (4.26),

$$\begin{aligned}f'_{ij} &= \frac{f_{i+1,j-1}f_{i,j+1}f_{i-1,j} + f_{i+1,j}f_{i,j-1}f_{i-1,j+1}}{f_{ij}}\\ &= Y^{[j+1,j+i]}_{[1,i+1]\setminus i}Y^{[j-1,j+i-2]}_{[1,i]} - Y^{[j-1,j+i-2]}_{[1,i+1]\setminus i}Y^{[j+1,j+i]}_{[1,i]}.\end{aligned}$$

Other cases can be treated similarly. Below, we present corresponding expressions for f'_{ij}.

(ii) $m+1-i < j < m$. Then

$$f'_{ij} = Y^{[j-1,m]\setminus j}_{[\alpha+1,i+1]}Y^{[j,m]}_{[\alpha-1,i-1]} - Y^{[j-1,m]\setminus j}_{[\alpha-1,i-1]}Y^{[j,m]}_{[\alpha+1,i+1]},$$

where $\alpha = i + j - m$.

(iii) $m+1-i = j < m$. Then

$$f'_{ij} = (-1)^{k-i}\left(Y^{[j-1,m]}_{[1,i+1]}Y^{[j+1,m-1]}_{[2,i-1]} - Y^{[j-1,m-1]}_{[2,i+1]}Y^{[j+1,m]}_{[1,i-1]}\right).$$

(iv) $i = 1$, $j < m$. Then $f'_{1j} = Y^{j-1,j+1}_{12}$.

(v) $i > 1$, $j = m$. Then $f'_{1j} = -Y^{m-1,m}_{i-1,i+1}$.

(vi) $i = 1$, $j = m$. Then $f'_{1,m} = (-1)^k y_{2,m-1}$.

In all six cases, f'_{ij} is a polynomial in variables y_{pq}, which proves the assertion. □

LEMMA 4.13. *For every $i \in [1, k-1]$ and every $j \in [2, n-k]$, the coordinate function y_{ij} belongs to some cluster obtained from the initial one.*

PROOF. Let $\widehat{Y}$ be the matrix obtained from Y by deleting the first row and the last column. Denote by $\widehat{F}_{ij}$, $\hat{f}_{ij}$, $i \in [2, k]$, $j \in [1, m-1]$, the functions defined by (4.13), (4.24) with Y replaced by $\widehat{Y}$. Define also $f = (f_{ij})$ and $\hat{f} = (\hat{f}_{ij})$.

Let us consider the following composition of cluster transformations:

$$T = T_{k-1} \circ \cdots \circ T_1, \tag{4.28}$$

where

$$T_\gamma = T_{k-1,m-\gamma+1} \circ \cdots \circ T_{\gamma+1,m-\gamma+1} \circ T_{\gamma\ 2} \circ \cdots \circ T_{\gamma,m-\gamma+1}$$

for $\gamma \in [1, k-1]$. Note that every cluster transformation T_{ij}, $i \in [2, k]$, $j \in [1, m-1]$, features in (4.28) exactly once.

We claim that

$$\begin{aligned} (Tf)_{ij} &= \hat{f}_{ij}, && i \in [1, k-1];\ j = 2, m-1, \\ (T\widetilde{B})_{(ij),(\alpha\beta)} &= b_{(ij),(\alpha\beta)}, && i, \alpha \in [1, k-1];\ j, \beta = 2, m-1; \\ & && j + \beta > 2;\ i + \alpha < 2k. \end{aligned} \tag{4.29}$$

In particular, $(Tf)_{1j} = Y_{2,k-1}$, $j \in [2, m]$, and $(Tf)_{im} = Y_{i+1,m-1}$, $j \in [2, m]$. Applying the same strategy to $\widehat{Y}$ etc., we will eventually recover all matrix entries of Y.

To prove (4.29), it is convenient to work with diagrams $\Gamma(\widetilde{B})$, rather than with matrices themselves. Using the initial graph given on Fig. 4.2 and Lemma 3.28, it is not hard to convince oneself that at the moment when T_{ij} (considered as a part of the composition (4.28)) is applied, the corresponding graph changes according to Fig. 4.3. In the latter figure, the white circle denotes the vertex (i, j) and only the vertices connected with (i, j) are shown. If $i = 1$ (resp., $j = m$) then vertices above (resp. to the right of) (i, j) and edges that connect them to (i, j) should be ignored.

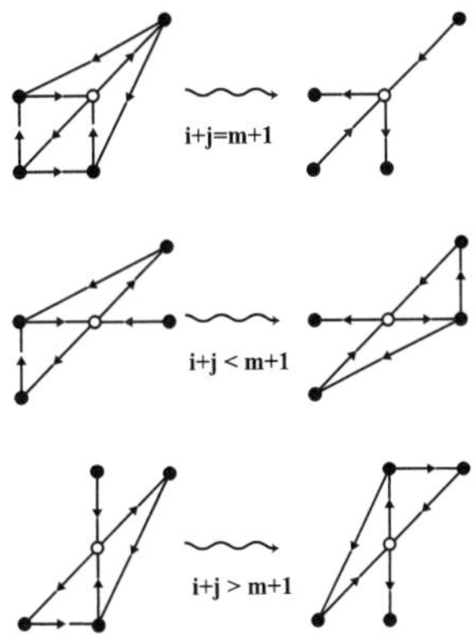

FIGURE 4.3. Mutations

It follows from Fig. 4.3 that

(i) the direction of an edge between (α, β) and $(\alpha+1, \beta-1)$ changes when $T_{\alpha\beta}$ is applied and is restored when $T_{\alpha+1,\beta-1}$ is applied;

(ii) a horizontal edge between (α,β) and $(\alpha,\beta+1)$ is erased when $T_{\alpha-1,\beta-1}$ is applied and is restored with the original direction when $T_{\alpha+1,\beta}$ is applied;
(iii) a vertical edge between (α,β) and $(\alpha+1,\beta)$ is erased when $T_{\alpha,\beta+1}$ is applied and is restored with the original direction when $T_{\alpha-1,\beta+1}$ is applied;
(iv) an edge between $(\alpha,\beta+1)$ and $(\alpha+1,\beta-1)$ is introduced when $T_{\alpha\beta}$ is applied and is erased when $T_{\alpha+1,\beta}$ is applied;
(v) an edge between $(\alpha-1,\beta)$ and $(\alpha+1,\beta-1)$ is introduced when $T_{\alpha\beta}$ is applied and is erased when $T_{\alpha,\beta-1}$ is applied.

Thus, after applying all transformations that constitute T in (4.28), one obtains a directed graph whose upper right $(k-1)\times(m-1)$ part (not taking into account edges between the vertices in the first column and the last row) coincides with that of the initial graph on Fig. 4.2. This proves the second equality in (4.29).

To prove the first equality in (4.29), note that, by the definition of (4.28), a cluster coordinate f_{ij} changes to $(Tf)_{ij}$ at the moment when T_{ij} is applied and stays unchanged afterwards. In particular, on Fig. 4.3, coordinates that correspond to vertices above and to the right of (i,j) are coordinates of Tf, while coordinates that correspond to vertices above and to the right of (i,j) are coordinates of f. Thus, (4.29) will follow if we show that

$$f_{ij}\hat{f}_{ij} = f_{i+1,j-1}\hat{f}_{i-1,j+1} + f_{i,j-1}\hat{f}_{i+1,j}, \quad i+j=m+1,$$
$$f_{ij}\hat{f}_{ij} = f_{i+1,j-1}\hat{f}_{i-1,j+1} + f_{i,j-1}\hat{f}_{i,j+1}, \quad i+j<m+1,$$
$$f_{ij}\hat{f}_{ij} = f_{i+1,j-1}\hat{f}_{i-1,j+1} + f_{i+1,j}\hat{f}_{i-1,j}, \quad i+j>m+1.$$

But referring to definitions of f and $\bar{f}$, one finds that the three equations above are just another instances of Jacobi's identity (4.27). The proof is complete. □

We are now in a position to prove the following

THEOREM 4.14. *$\mathcal{A}^{G_k^0(n)}$ tensored with $\mathbb{C}$ coincides with the ring of regular function on $G_k^0(n)$.*

PROOF. The proof is based on Proposition 3.37. First of all, observe that every stable variable attains zero value at some point in $G_k^0(n)$, hence, there is no need to use localization, and $\mathcal{A}_{\mathbb{C}}^{G_k^0(n)}$ is just $\mathcal{A}^{G_k^0(n)}$ tensored by $\mathbb{C}$. Second, $G_k^0(n)$ is isomorphic to $\mathbb{C}^{k(n-k)}$, so Proposition 3.37 applies with $V=G_k^0(n)$. Condition (i) in Proposition 3.37 follows from Lemma 4.13. Condition (ii) holds by construction of the initial cluster. Finally, condition (iii) is guaranteed by Lemma 4.12. □

Next, we will modify our construction of $\mathcal{A}^{G_k^0(n)}$ by replacing the initial cluster (4.24) with the collection of all Plücker coordinates involved in the definition of f_{ij}, that is with the collection

(4.30)
$$\mathbf{x}=\mathbf{x}(k,n)=\{x_{[1,k]},\ x_{([1,k]\setminus[i-l(i,j),\ i])\cup[j+k,j+l(i,j)+k]},\quad i\in[1,k],\ j\in[1,m]\},$$

where $l(i,j)$ is defined by (4.12). It will be convenient to denote the index set

$$([1,k]\setminus[i-l(i,j),\ i])\cup[j+k,j+l(i,j)+k]$$

by $I_{ij}=I_{ij}(k,m)$. In what follows we usually write x_{ij} instead of $x_{I_{ij}}$. Besides, we write $x_{0,m+1}$ instead of $x_{[1,k]}$, since $l(0,m+1)=-1$, and hence $I_{0,m+1}=[1,k]$.

Now, there is only one natural way to also modify the set of stable coordinates and the integer matrix that defines initial cluster transformations in $\mathcal{A}^{G_k^0(n)}$ (we will denote this matrix $\bar{B}$). Stable coordinates now are $x_{0,m+1}, x_{11}, \dots, x_{k1}, x_{k2}, \dots, x_{km}$. Note that $I_{i1} = [i+1, k+i]$ for $i \in [1,k]$ and

$$I_{kj} = \begin{cases} [k+j, 2k+j-1], & \text{if } j \in [1, m-k+1] \\ [1, k+j-m-1] \cup [k+j, n], & \text{if } j \in [m-k+2, m]; \end{cases}$$

in other words, stable coordinates coincide with cyclically dense Plücker coordinates. To define the $(k-1)(m-1) \times (km+1)$ matrix $\bar{B}$, note that for every initial transformation T_{ij} in $\mathcal{A}^{G_k^0(n)}$, one has $f_{ij}T_{ij}(f_{ij}) = P_{ij}/x_{0,m+1}^r$, where P_{ij} is a binomial in variables $x_{\alpha\beta}$ not divisible by $x_{0,m+1}$, and $r = 2$ if $i = 1$ or $j = m$ and $r = 3$ otherwise. The exponent of every $x_{\alpha\beta}$ in P_{ij} coincides with the exponent of $f_{\alpha\beta}$ in $f_{ij}T_{ij}(f_{ij})$. Moreover, only P_{1m} contains $x_{0,m+1}$: $P_{1m} = x_{1,m-1}x_{2,m} + x_{2,m-1}x_{0,m+1}$. Thus, $\bar{B} = \bar{B}(k,n)$ is obtained from $\widetilde{B}$ by attaching one column that contains a single non-zero entry. Equivalently, the diagram $\Gamma(\bar{B})$ is obtained from $\Gamma(\widetilde{B})$ by attaching a single vertex to the upper-right vertex in Fig. 4.2 with an edge directed towards the added vertex (for convenience, we will think of this additional vertex as also being part of the grid and having coordinates $(0, m+1)$):

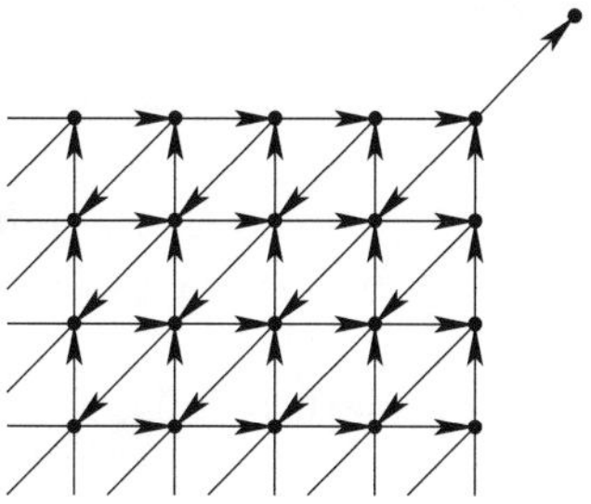

FIGURE 4.4. Diagram corresponding to $G_k(n)$

Denote the cluster algebra with the initial seed $(\mathbf{x}, \bar{B})$ by $\mathcal{A}^{G_k(n)}$. It follows from the above discussion that $\mathcal{A}^{G_k^0(n)}$ is obtained from $\mathcal{A}^{G_k(n)}$ by freezing the stable variable $x_{0,m+1}$.

Note that, similar to the case of $\mathcal{A}^{G_k^0(n)}$, each cluster variable x_{ij} is associated with (placed at) the vertex with coordinates (i,j) of the grid in Fig. 4.4 for $i \in [1,k]$, $j \in [1,m]$, while an additional vertex in Fig. 4.4 corresponds to the cluster variable $x_{0,m+1}$.

PROPOSITION 4.15. *Every Plücker coordinate in* (4.30) *is a cluster variable in* $\mathcal{A}^{G_k(n)}$.

PROOF. We use a strategy similar to one employed in the proof of Lemma 4.13 and look at the result of applying two specially designed sequences, R and S, of cluster transformations to the initial cluster (4.30). As in Lemma 4.13, each cluster transformation T_{ij} appears in both R and S exactly once:

$$R = R_t \circ \dots \circ R_1, \quad t = \min\{k, m\} - 1,$$

where

$$R_i = (T_{k-1,m+1-i} \circ \dots \circ T_{i+1,m+1-i}) \circ (T_{i2} \circ \dots \circ T_{i,m+1-i}),$$

and

$$S = S_3 \circ \cdots \circ S_{k+m-1},$$

where

$$S_i = \begin{cases} T_{i-m,m} \circ \cdots \circ T_{k-1,i-k+1} & \text{if } i > m, \\ T_{1,i-1} \circ \cdots \circ T_{i-2,2} & \text{if } i \le m. \end{cases}$$

In other words, R consists of applying cluster transformations to variables associated with vertices of the grid depicted in Fig. 4.4 going right to left along the first row, then down along the last column starting from the second vertex, then right to left along the second row starting from the second vertex etc., while to obtain S we apply transformations going up along anti-diagonals of the grid starting with the rightmost and ending with the leftmost anti-diagonal.

We are interested in the seeds $R(\mathbf{x}, \bar{B})$ and $S(\mathbf{x}, \bar{B})$, where $(\mathbf{x}, \bar{B})$ is the initial cluster in $\mathcal{A}^{G_k(n)}$ described by (4.30). In order to describe $\Gamma(R(\bar{B})$ we will place, for every $i \in [1, k]$, $j \in [1, m]$, the new cluster variable $R(x_{ij})$ at (i, j)-vertex of the grid and keep $x_{0,m+1}$ at the additional vertex. However, when dealing with the seed $S(\mathbf{x}, \bar{B})$, it will be more convenient to place $S(x_{i,j})$ at (i, j)-vertex of the grid only for $i \in [1, k-1]$, $j \in [2, m]$, while rotating the stable cluster variables counterclockwise, i.e. placing $x_{0,m+1}$ at $(1, 1)$-vertex, x_{11} at $(2, 1)$-vertex, ..., and finally x_{km} at the vertex $(0, m+1)$.

Next, we define , for every $i \in [1, k]$, $j \in [1, m]$, k-element subsets $R(I_{ij})$, $S(I_{ij})$ as follows:

- For $i \in [1, k-1]$, $j \in [2, m]$, $R(I_{ij})$ is the union of $\{1\}$ and the $(k-1)$-element subset of $[2, n-1]$ obtained by shifting every element of $I_{i,j-1}(k-1, m-1)$ up by 1.
- For $i = k$ or $j = 1$, or $(i, j) = (0, m+1)$, $R(I_{ij}) = I_{ij}$.
- For every pair i, j, $S(I_{ij})$ is obtained from I_{ij} by shifting (mod n) every index in I_{ij} down by 1.

Clearly, for any cyclically dense index set I, $S(I)$ is also cyclically dense.

LEMMA 4.16. (i) *All cluster variables in the cluster $R(\mathbf{x})$ are Plücker coordinates, moreover, for every pair i, j*

$$R(x_{I_{ij}}) = x_{R(I_{ij})}.$$

(ii) *The diagram obtained from $\Gamma(R(\bar{B}))$ by deleting vertices (i, j) with $i = k$ or $j = 1$ coincides with $\Gamma(\bar{B}(k-1, m-1))$ (recall that a diagram is defined up to edges between stable variables).*

(iii) *All cluster variables in the cluster $S(\mathbf{x})$ are Plücker coordinates, moreover, for every pair i, j*

$$S(x_{I_{ij}}) = x_{S(I_{ij})}.$$

(iv) *The diagrams $\Gamma(S(\bar{B}))$ and $\Gamma(\bar{B})$ coincide.*

PROOF. We will argue inductively to see that at the moment when we reach a transformation T_{ij} in the sequence of transformations prescribed by the definition of R (resp. S) the following conditions are satisfied:

(i) all of the already transformed variables are Plücker coordinates (i.e. to every vertex of the grid there corresponds a k-element subset of $[1, n]$);

(ii) the vertex v with coordinates (i, j) is connected to precisely four other vertices of the grid, u_1, u_2, u_3, u_4;

(iii) exactly two of these edges are directed towards v (we assume those are the edges connecting v with u_2 and u_4) ;

(iv) an intersection, J, of the index set I_{ij} that corresponds to v with four index sets that correspond to the neighbors of v has $(k-2)$ elements;

(v) if one writes I_{ij} as $J \cup \{\alpha, \beta\}$ for some $\alpha, \beta \in [1, n]$, then there exist such $\gamma, \delta \in [1, n]$ that index sets corresponding to u_1, u_2, u_3, u_4 are $J \cup \{\alpha, \gamma\}$, $J \cup \{\beta, \gamma\}$, $J \cup \{\beta, \delta\}$ and $J \cup \{\alpha, \delta\}$, respectively.

Then it will follow from the short Plücker identity that an application of T_{ij} to x_{ij} will produce the Plücker coordinate $x_{J\cup\{\gamma,\delta\}}$.

Let us first concentrate on R. Note that for every $i \in [1, k-1], j \in [2, m]$, index sets I_{ij} and $R(I_{ij})$ have a $(k-2)$-element intersection

$$J_{ij} = \begin{cases} [i+2, k] \cup [k+j, k+j+i-2] & \text{if } i+j \leq m+1, \\ [1, i+j-m-1] \cup [i+2, k] \cup [k+j, n-1] & \text{if } i+j > m+1. \end{cases}$$

The first transformation to be applied is T_{1m}. It is easy to check that conditions (i)-(v) above are met with $u_1 = (0, m+1)$, $u_2 = (2, m)$, $u_3 = (2, m-1)$, $u_4 = (1, m-1)$, $J = [3, k]$, $\alpha = 2$, $\beta = n$, $\gamma = 1$ and $\delta = n-1$, and so

$$R(x_{I_{1m}}) = T_{1m}(x_{I_{1m}}) = x_{[3,k]\cup\{1,n-1\}} = x_{R(I_{1m})}.$$

The corresponding transformation of the diagram $\Gamma(\bar{B})$ consists in reversing directions of four edges incident to $(1, m)$, erasing edges $(2, m-1) \to (1, m-1)$ and $(2, m-1) \to (2, m)$ and introducing edges $(1, m-1) \to (0, m+1)$ and $(2, m) \to (0, m+1)$.

Proceeding further, we see that prior to applying T_{1j}, $j \in [2, m-1]$, the vertex $(1, j)$ satisfies conditions (i)-(v) with $u_1 = (0, m+1)$, $u_2 = (1, j+1)$, $u_3 = (2, j-1)$, $u_4 = (1, j-1)$, $J = [3, k]$, $\alpha = 2$, $\beta = k+j$, $\gamma = 1$ and $\delta = k+j-1$. Moreover, $R(x_{I_{1j}}) = x_{[3,k]\cup\{1,k+j-1\}} = x_{R(I_{1j})}$, and the corresponding transformations of the diagram are as follows: all edges incident to $(1, j)$ reverse their directions (in particular, the edge joining $(1, j)$ and $(1, j+1)$ is restored to its original direction, which was reversed at the previous step), edges $(0, m+1) \to (1, j+1)$ and $(2, j-1) \to (1, j-1)$ are erased and new edges $(1, j-1) \to (0, m+1)$ and $(1, j+1) \to (2, j-1)$ are introduced. Therefore, after applying transformations $T_{1m}, \dots, T_{12}$ we get the diagram that differs from the original one in that

- the edge $(1, m) \to (0, m+1)$ has been erased and new edges $(1, 2) \to (0, m+1)$ and $(2, m) \to (0, m+1)$ has been added;
- the edge $(2, m) \to (1, m)$ and the edges $(2, j-1) \to (1, j)$, $j \in [2, m]$, have changed their directions;
- every vertical edge $(2, j) \to (1, j)$, $j \in [2, m-1]$, has been erased;
- new edges $(1, j) \to (2, j-2)$, $j \in [3, m]$, has been introduced.

Moving to the last column of the grid, we find that conditions (i)-(v) are now valid for the vertex $(2, m)$ with $u_1 = (0, m+1)$, $u_2 = (3, m)$, $u_3 = (3, m-1)$, $u_4 = (1, m)$, $J = \{1\} \cup [4, k]$, $\alpha = 3$, $\beta = n$, $\gamma = 2$ and $\delta = n-1$. Applying T_{2m}, we obtain $R(x_{I_{1,m-1}}) = x_{[4,k]\cup\{1,2,n-1\}} = x_{R(I_{2m})}$, while the diagram is transformed as follows: all edges incident to $(2, m)$ reverse their directions (in particular, the edge joining $(2, m)$ and $(1, m)$ is restored to its original direction), edges $(2, m-2) \to (2, m-1)$ and $(2, m-2) \to (1, m-2)$ are erased, a new edge $(1, m) \to (3, m-1)$ is introduced and the edge $(1, m) \to (0, m+1)$ is restored (with its original orientation). Similarly to the above, we see that prior to applying T_{im}, $i \in [3, k-1]$, the vertex (i, m)

satisfies conditions (i)-(v) with $u_1 = (0, m+1)$, $u_2 = (i+1, m)$, $u_3 = (i+1, m-1)$, $u_4 = (i-1, m)$, $J = [1, k] \setminus [i, i+1]$, $\alpha = i+1$, $\beta = n$, $\gamma = i$ and $\delta = n-1$. Moreover, $R(x_{I_{im}}) = x_{J\cup\{i,n-1\}} = x_{R(I_{1j})}$, and the corresponding transformations of the diagram are similar to those described above.

Figure 4.5 below presents the above transformations for the Grassmannian $G_4(9)$. Note that the edges between the stable variables are omitted.

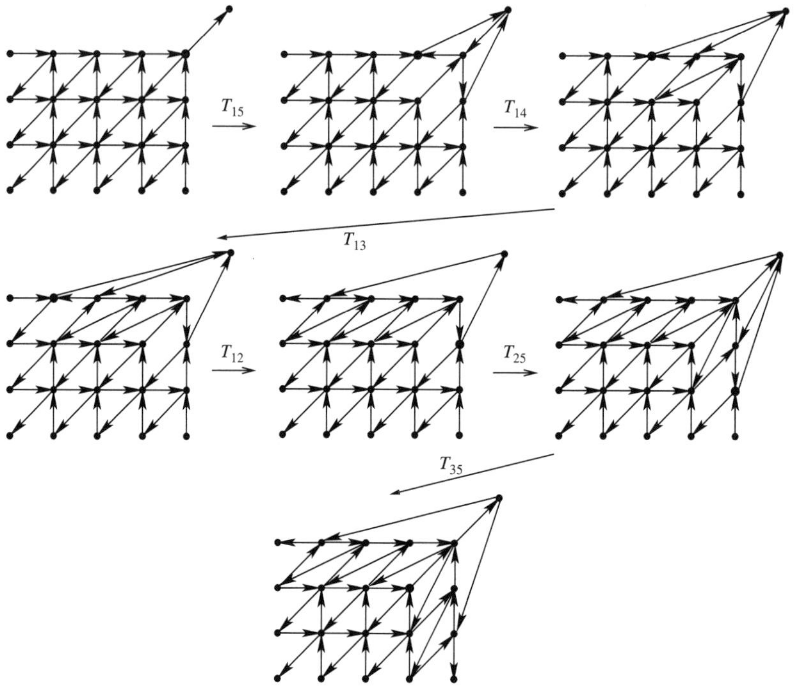

FIGURE 4.5. Transformations of the diagram of $G_4(9)$: acting by R_1

Now the first two claims of the Lemma follow by induction if we assume that at the moment when T_{ij} is applied, the index sets associated with vertices (p, q) located above and to the right from (i, j) are $R(I_{pq})$ and then show that conditions (i)-(v) are satisfied with $J = J_{ij}$ and

- $u_1 = (i-1, j+1)$, $u_2 = (i, j+1)$, $u_3 = (i+1, j-1)$, $u_4 = (i, j-1)$, $\alpha = i+1$, $\beta = k+j+i-1$, $\gamma = 1$, $\delta = k+j-1$ if $i+j \le m+1$, $i > 1$, $j < m$;
- $u_1 = (i-1, j+1)$, $u_2 = (i+1, j)$, $u_3 = (i+1, j-1)$, $u_4 = (i-1, j)$, $\alpha = i+1$, $\beta = n$, $\gamma = i+j-m$, $\delta = k+j-1$ if $i+j > m+1$, $i > 1$, $j < m$.

To do this we discern the pattern in transformations of the diagram. Namely, for i, j such that $i \in [2, k], j \in [1, m-1]$ and $i+j \ne m+1$, let us consider a parallelogram $CDEF$ in in the initial diagram $\Gamma(\bar{B})$ with vertices $C = (i, j)$, $D = (i, j+1)$, $E = (i-1, j+2)$, $F = (i-1, j+1)$ if $i+j < m+1$ and $C = (i, j)$, $D = (i-1, j)$, $E = (i-2, j+1)$, $F = (i-1, j+1)$ if $i+j > m+1$ and edges $C \to D$, $E \to D$, $F \to E$, $F \to C$ and $D \to F$ in both cases. The order in which cluster transformations enter the definition of R guarantees that these edges are affected *only* when transformations that correspond to vertices of $CDEF$ are applied. These transformations enter R in the following order : first, T_E, then T_F, then T_D, then T_C (here, we slightly abuse notation by replacing T_{ij} with $T_{(i,j)}$ etc.) As a result, first the edge $D \to F$ is erased, then the edge $E \to C$ is introduced, then the edge $E \to C$ is erased and, finally, the edge $D \to F$ is restored, while every side of $CDEF$ changes direction twice. Thus, restrictions of $\Gamma(\bar{B})$ and $\Gamma(R(\bar{B}))$ to $CDEF$ coincide (see Fig. 4.6 for details in the case of $G_4(9)$). A straightforward though tedious check shows that these transformations are consistent with induction assumptions above. Furthermore, a direct inspection shows that in $\Gamma(R(\bar{B}))$, the vertex $(0, m+1)$ is connected to, in addition to $(1, m)$, only two other vertices : $(1, 2)$ and $(k-1, m)$. This establishes the claims of the Lemma concerning R. Claim (ii) of the Lemma is illustrated on Fig. 4.6: the part of the last diagram enclosed in the box coincides with the diagram for the Grassmannian $G_3(7)$, up to the edges between the stable variables of the new diagram.

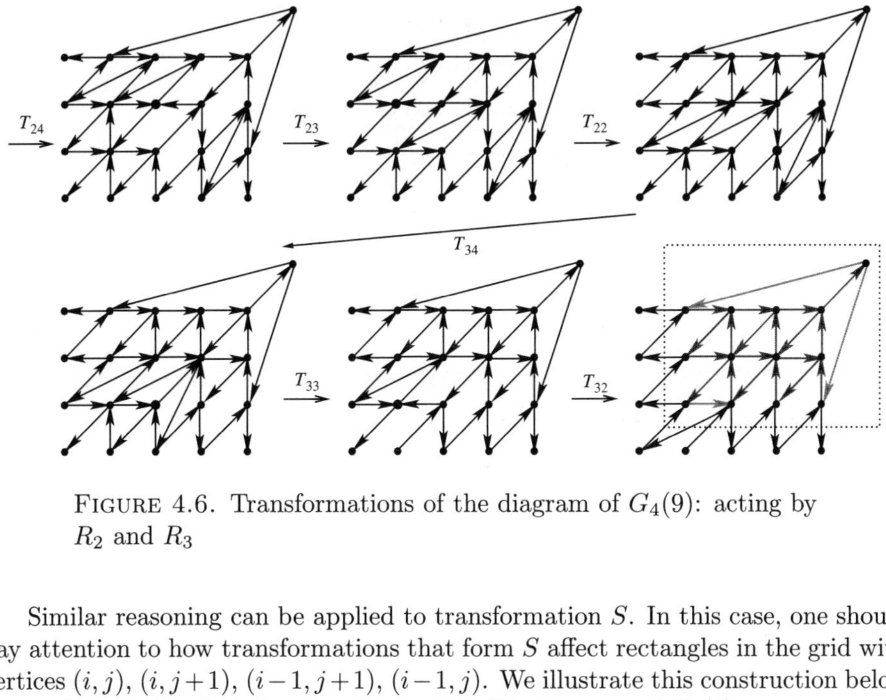

FIGURE 4.6. Transformations of the diagram of $G_4(9)$: acting by R_2 and R_3

Similar reasoning can be applied to transformation S. In this case, one should pay attention to how transformations that form S affect rectangles in the grid with vertices (i, j), $(i, j+1)$, $(i-1, j+1)$, $(i-1, j)$. We illustrate this construction below in two figures that describe transformations S in the case of $G_4(9)$.

□

Now we can finish the proof of the Proposition 4.15 using an induction on n and k. Indeed, for $k = 2$, the statement was shown to be true in Chapter 2. Now, consider a k-element subset I of $[1, n]$. Suppose first that $1 \in I$ and $n \notin I$, that is $I = \{1, j_1 + 1, \dots, j_{k-1} + 1\}$, where $J = \{j_1, \dots, j_{k-1}\} \in [1, n-2]$. For every such I, define y_J to be equal to x_I. Clearly, the family $\{y_J\}$ forms a collection of Plücker coordinates of an element in $G_{k-1}(n-2)$. Consider the cluster $R(\mathbf{x}(k, n)) \in \mathcal{A}^{G_k(n)}$. Due to claim (i) of Lemma 4.16, for $i < k$ or $j > 1$, the variable $R(x_{I_{ij}(k,n)})$ coincides with $y_{I_{i-1,j}(k-1,n-2)}$. Moreover, by claim (2) of

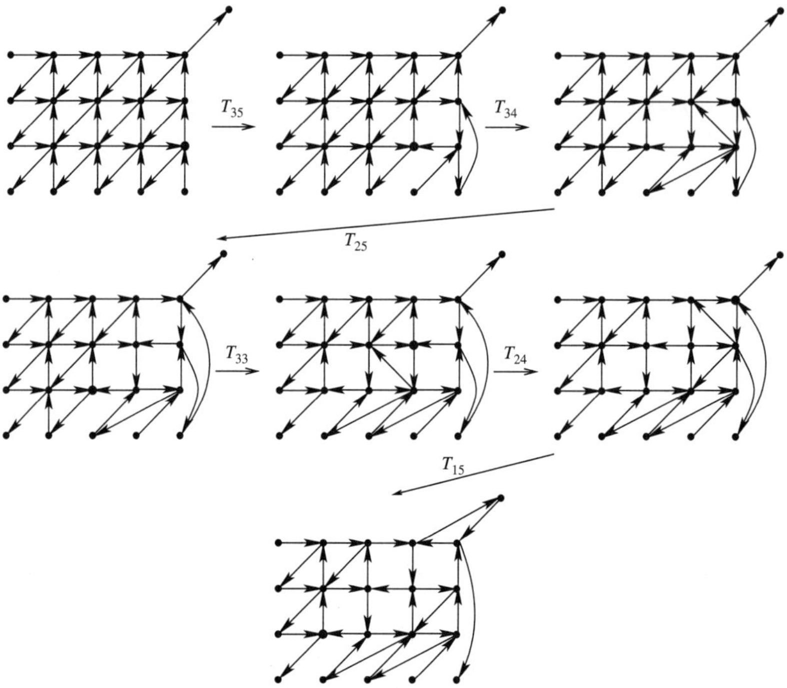

FIGURE 4.7. Transformations of the diagram of $G_4(9)$: acting by S_8, S_7 and S_6

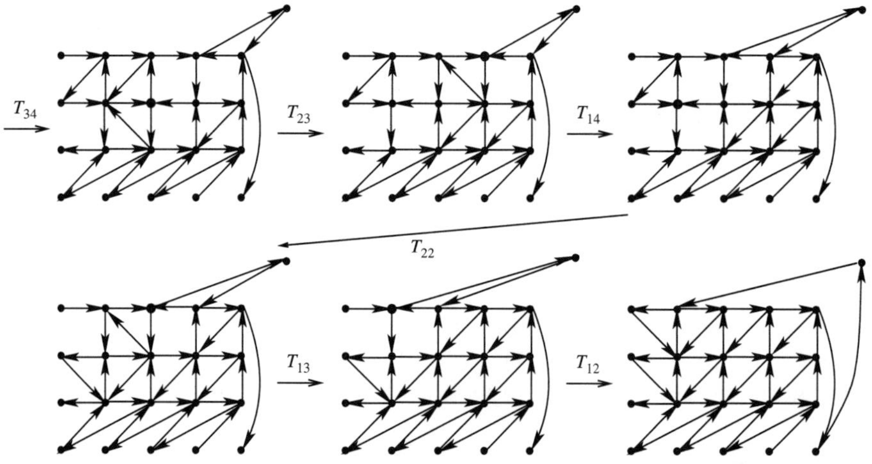

FIGURE 4.8. Transformations of the diagram of $G_4(9)$: acting by S_5, S_4 and S_3

Lemma 4.16 the restriction of $\mathcal{A}^{G_k(n)}$ obtained by fixing, in addition to all cyclically dense Plücker coordinates, cluster variables $R(x_{I_{ij}(k,n)})$ for $i = k-1$ or $j = 2$ has the same initial exchange matrix as $\mathcal{A}^{G_{k-1}(n-2)}$. Then, by the induction hypothesis, for any $J \subset [1, n-2]$, y_J is a cluster variable of $\mathcal{A}^{G_{k-1}(n-2)} \subset \mathcal{A}^{G_k(n)}$. Thus we proved the proposition for any $I \subset [1, n]$ such that $1 \in I$ and $n \notin I$.

If $I = \{i_1, \dots, i_k\}$ is not of this kind, let us fix an index $i \notin I$ such that $(i+1)(\operatorname{mod} n) \notin I$. For any $L = \{l_1, \dots, l_k\} \subset [1, n]$ define $\tilde{L} = \{(l_1 - i)(\operatorname{mod} n), \dots, (l_k - i)(\operatorname{mod} n)\}$. Clearly, $1 \in \tilde{I}$ and $n \notin \tilde{I}$. Define also $z_L = x_{\tilde{L}}$, then the family z_L forms a collection of Plücker coordinates of an element in $G_k(n)$. Claims (iii) and (iv) of Lemma 4.16 imply that

$$S^i((\mathbf{x}(k,n), \bar{B})) = (\{z_{[1,k]}, z_{I(i,j)}, i \in [1,k]\ ,\ j \in [1,m]\}, \bar{B})$$

and so, by the argument above, $z_{\tilde{I}} = x_I$ is a cluster variable in $\mathcal{A}^{G_k(n)}$.

□

We conclude this section with the following statement.

THEOREM 4.17. *$\mathcal{A}^{G_k(n)}$ tensored with $\mathbb{C}$ coincides with the homogeneous coordinate ring of $G_k(n)$.*

PROOF. Slightly abusing notation, denote $\mathcal{A}^{G_k(n)} \otimes_{\mathbb{Q}} \mathbb{C}$ by $\mathcal{A}_{\mathbb{C}}^{G_k(n)}$. Direct application of Proposition 3.37 is impossible, since $G_k(n)$ is not a Zariski open subset of an affine space. To circumvent this problem, consider the standard Plücker embedding $G_k(n) \to \mathbb{C}P^{\binom{n}{k}-1}$, and denote by $\mathcal{C}(G_k(n)) \subset \mathbb{C}^{\binom{n}{k}}$ the affine cone over the image of this embedding. Starting with the initial extended cluster in $\mathcal{A}^{G_k^0(n)}$ given by (4.24) and the corresponding initial extended cluster $\mathbf{x}$ in $\mathcal{A}^{G_k(n)}$ given by (4.30), define the sets $U_0 \subset G_k^0(n)$ and $\widetilde{U}_0 \subset \mathcal{C}(G_k(n))$ as the nonzero loci for all the cluster variables constituting these two clusters, respectively. Similarly, define the corresponding nonzero loci U_i and $\widetilde{U}_i$ for all the neighboring clusters $\mathbf{x}_i$, $i \in [1, (k-1)(m-1)]$. Finally, put

$$U = \bigcup_{i=0}^{(k-1)(m-1)} U_i, \qquad \widetilde{U} = \bigcup_{i=0}^{(k-1)(m-1)} \widetilde{U}_i \tag{4.31}$$

(cp. with Lemma 3.40).

Proposition 4.15 means that the ring $\mathcal{O}(\mathcal{C}(G_k(n)))$ of regular functions on $\mathcal{C}(G_k(n))$ (which is generated by Plücker coordinates) is contained in $\mathcal{A}_{\mathbb{C}}^{G_k(n)}$. Similarly to the proof of Proposition 3.37, we would like to check that $\mathcal{A}_{\mathbb{C}}^{G_k(n)}$ is contained in the ring $\mathcal{O}(\widetilde{U})$ of regular functions on $\widetilde{U}$, and that the latter coincides with $\mathcal{O}(\mathcal{C}(G_k(n)))$.

Indeed, by Proposition 3.20, for any $i \in [0, (k-1)(m-1)]$, any element ψ in $\mathcal{A}_{\mathbb{C}}^{G_k(n)}$ is a Laurent polynomial in the cluster variables of $\mathbf{x}_i$, whose coefficients are polynomials in the stable variables. Therefore, at any point of $\widetilde{U}$ one can find a suitable cluster that guarantees the regularity of ψ at this point, which means that $\mathcal{A}_{\mathbb{C}}^{G_k(n)} \subseteq \mathcal{O}(\widetilde{U})$.

Let us prove, similarly, to Lemma 3.40, that the complement $\widetilde{W}$ to $\widetilde{U}$ in $\mathcal{C}(G_k(n))$ is of codimension at least 2 in $\mathcal{C}(G_k(n))$. Indeed, $\widetilde{W} = \widetilde{W}' \cup \widetilde{W}''$, where $\widetilde{W}'$ is the affine cone over $W' = G_k^0(n) \setminus U$ and $\widetilde{W}'' = \widetilde{W} \cap \{x_{[1,k]} = 0\}$. Clearly, $\operatorname{codim} \widetilde{W}' \geq 2$, since $\operatorname{codim} W' \geq 2$ in $G_k^0(n)$ (the latter fact is true since, as we have checked in the proof of Theorem 4.14, $V = G_k^0(n)$ satisfies all the conditions of Proposition 3.37, and hence Lemma 3.40 applies).

Further, let $\widetilde{W}_0$ be the complement to $\widetilde{U}_0$ in $\mathcal{C}(G_k(n))$, and $\widetilde{W}_0'' = \widetilde{W}_0 \cap \{x_{[1,k]} = 0\}$. It is easy to see that $\widetilde{W}_0''$ is the affine cone over a proper closed subvariety of a hyperplane section of $G_k(n)$. This subvariety is given by vanishing of at least one Plücker coordinate involved in $\mathbf{x}$. Since Plücker coordinates are irreducible, we see that $\operatorname{codim} \widetilde{W}_0'' \geq 2$, and hence $\operatorname{codim} \widetilde{W}'' \geq 2$, since $\widetilde{W}'' \subset \widetilde{W}_0''$. So, $\operatorname{codim} \widetilde{W} \geq 2$, and, hence $\mathcal{A}_{\mathbb{C}}^{G_k(n)}$ coincides with $\mathcal{O}(\mathcal{C}(G_k(n)))$, since the Grassmannian is projectively normal in the Plücker embedding. It remains to recall that the ring of regular functions on $\mathcal{C}(G_k(n))$ is isomorphic to the homogeneous coordinate ring of $G_k(n)$. □

4.3. Poisson and cluster algebra structures on double Bruhat cells

To illustrate the concept of compatible Poisson brackets, let us consider the example that was discussed at the end of Section 3.4. Recall that with a double Bruhat cell $\mathcal{G}^{u,v}$ and a fixed reduced word $\mathbf{i}$ for (u, v) we have associated an $(l(u) + l(v)) \times (l(u)+l(v)+r)$ matrix $\widetilde{B}_{\mathbf{i}}'$ of rank $l(u)+l(v)$ and an initial cluster $x_k = \Delta(k; \mathbf{i})$, $k \in [-r, -1] \cup [1, l(u)+l(v)]$. This data defines a cluster algebra $\mathcal{A}(\widetilde{B}_{\mathbf{i}}')$ (in Section 3.4 we were interested only in the corresponding upper cluster algebra $\mathcal{U}(\widetilde{B}_{\mathbf{i}}')$). We will prove

THEOREM 4.18. *For any reduced word* $\mathbf{i}$ *for a pair* $(u, v) \in W \times W$, *the restriction of the standard Poisson-Lie structure* $\{\cdot,\cdot\}_{\mathcal{G}}$ *to* $\mathcal{G}^{u,v}$ *is compatible with the cluster algebra structure* $\mathcal{A}(\widetilde{B}_{\mathbf{i}}')$.

PROOF. First, we will show that functions x_k, $k \in [-r, -1] \cup [1, l(u) + l(v)]$, are log-canonical with respect to the restriction of the standard Poisson-Lie structure $\{\cdot,\cdot\}_{\mathcal{G}}$ to $\mathcal{G}^{u,v}$. This will require some preparation.

For any $u, v \in W$ and $\xi \in \mathfrak{h}$ define a function on $\mathcal{G}$ by the formula

$$\varphi_{u,v}^{\xi}(x) = \left(\log[u^{-1}xv]_0, \xi\right), \tag{4.32}$$

where $(\cdot,\cdot)$ is the Killing form that identifies $\mathfrak{h}^*$ with $\mathfrak{h}$, and then also serves as a W-invariant inner product on $\mathfrak{h}^*$, see Section 1.2.2. Note that for any generalized minor $\Delta_{u\omega_i, v\omega_i}(x)$ defined in (1.11), one has $\log\left(\Delta_{u\omega_i, v\omega_i}(x)\right) = \varphi_{u,v}^{\omega_i}(x)$.

PROPOSITION 4.19. *For any* $u, u', v, v' \in W$ *such that* $l(uu') = l(u) + l(u')$ *and* $l(vv') = l(v) + l(v')$ *and for any* $\xi, \eta \in \mathfrak{h}$,

$$\left\{\varphi_{u,vv'}^{\eta}, \varphi_{uu',v}^{\xi}\right\}_{\mathcal{G}} = \frac{(\eta, \operatorname{Ad}_{u'}\xi) - (\operatorname{Ad}_{v'}\eta, \xi)}{2}. \tag{4.33}$$

PROOF. We start with the following

LEMMA 4.20. (i) *Left and right gradients of* $\varphi_{u,v}^{\xi}$ *are given by*

$$\nabla'\varphi_{u,v}^{\xi} = \operatorname{Ad}_{v[u^{-1}xv]_+^{-1}}\xi, \qquad \nabla\varphi_{u,v}^{\xi} = \operatorname{Ad}_{u[u^{-1}xv]_-^{-1}}\xi.$$

In particular,

$$\left(\nabla'\varphi_{u,v}^{\xi}\right)_0 = \operatorname{Ad}_v\xi, \qquad \left(\nabla\varphi_{u,v}^{\xi}\right)_0 = \operatorname{Ad}_u\xi. \tag{4.34}$$

(ii) $\nabla'\varphi_{u,v}^{\xi} \perp \operatorname{Ad}_v\mathfrak{n}_+$, $\nabla\varphi_{u,v}^{\xi} \perp \operatorname{Ad}_u\mathfrak{n}_-$.

PROOF. To simplify notation, we will prove part (i) viewing $\mathcal{G}$ as a subgroup of a matrix group with the Killing form proportional to the trace-form. The function $\varphi^\xi_{u,v}$ can be written as a superposition $\varphi^\xi_{u,v} = \ell_\xi \circ \psi_{u,v}$ of a map $\psi_{u,v}(x) = [u^{-1}xv]_0$ from $\mathcal{G}$ to $\mathcal{H}$ and a function $\ell_\xi(h) = (\log h, \xi)$ on $\mathcal{H}$. Let $\delta x \in \mathfrak{g}$. Define

$$\delta_- = \frac{d}{dt}[u^{-1}xe^{t\delta x}v]_-\Big|_{t=0};$$

δ_0 and δ_+ are defined similarly.

It easy to see that $\nabla \ell_\xi(h) = \nabla' \ell_\xi(h) = \xi$. Therefore,

$$\frac{d}{dt}\varphi^\xi_{u,v}(xe^{t\delta x})\Big|_{t=0} = \left(\nabla'\ell_\xi(\psi_{u,v}(x)), \psi_{u,v}(x)^{-1}\frac{d}{dt}\psi_{u,v}(xe^{t\delta x})\Big|_{t=0}\right) = \left(\xi, \psi_{u,v}(x)^{-1}\delta_0\right).$$

Further,

$$\frac{d}{dt}u^{-1}xe^{t\delta x}v\Big|_{t=0} = \delta_-[u^{-1}xv]_0[u^{-1}xv]_+ + [u^{-1}xv]_-\delta_0[u^{-1}xv]_+ + [u^{-1}xv]_-[u^{-1}xv]_0\delta_+.$$

Clearly, the left hand side equals $u^{-1}x\delta xv$, which implies

$$\mathrm{Ad}_{[u^{-1}xv]_+v^{-1}}\delta x = \mathrm{Ad}_{[u^{-1}xv]_0^{-1}}\left([u^{-1}xv]_-^{-1}\delta_-\right) + [u^{-1}xv]_0^{-1}\delta_0 + \delta_+[u^{-1}xv]_+^{-1}.$$

The first, the second and the third terms in the right-hand side of the above equality belong to $\mathfrak{n}_-$, $\mathfrak{h}$ and $\mathfrak{n}_+$, respectively, therefore

$$[u^{-1}xv]_0^{-1}\delta_0 = \left(\mathrm{Ad}_{[u^{-1}xv]_+v^{-1}}\delta x\right)_0.$$

Thus,

$$\frac{d}{dt}\varphi^\xi_{u,v}(xe^{t\delta x})\Big|_{t=0} = \left(\xi, \psi_{u,v}(x)^{-1}\delta_0\right) = \left(\xi, \left(\mathrm{Ad}_{[u^{-1}xv]_+v^{-1}}\delta x\right)_0\right).$$

Since $\xi \in \mathfrak{h}$, the latter expression is equal to $\left(\mathrm{Ad}_{v[u^{-1}xv]_+^{-1}}\xi, \delta x\right)$, and hence $\nabla'\varphi^\xi_{u,v}(x) = \mathrm{Ad}_{v[u^{-1}xv]_+^{-1}}\xi$, as required. Furthermore,

$$\left(\nabla'\varphi^\xi_{u,v}(x)\right)_0 = \left(\mathrm{Ad}_{v[u^{-1}xv]_+^{-1}}\xi\right)_0 = \mathrm{Ad}_v\left(\mathrm{Ad}_{[u^{-1}xv]_+^{-1}}\xi\right)_0 = \mathrm{Ad}_v\xi,$$

which coincides with (4.34). The formulas for $\nabla\varphi^\xi_{u,v}(x)$ can be obtained similarly.

To prove part (ii), observe that for any $y \in v\mathcal{N}_+v^{-1}$, and $z \in u\mathcal{N}_-u^{-1}$,

$$[u^{-1}(zxy)v]_0 = [(u^{-1}zu)u^{-1}xv(v^{-1}yv)]_0 = [u^{-1}xv]_0,$$

hence $\varphi^\xi_{u,v}$ is right-invariant with respect to $v\mathcal{N}_+v^{-1}$ and left-invariant with respect to $u\mathcal{N}_-u^{-1}$. Now the claim follows from the definition (1.21) of the right and left gradients. □

Now we can complete the proof of Proposition 4.19. Denote $\varphi^\xi_{uu',v}$ by g and $\varphi^\eta_{u,vv'}$ by f. By (1.22) we have

$$2\{f,g\}_\mathcal{G} = (R(\nabla'f), \nabla'g) - (R(\nabla f), \nabla g) = -(\nabla'f, R(\nabla'g)) - (R(\nabla f), \nabla g).$$

By (1.23), the first term in the right hand side can be rewritten as $-(\nabla'f, R(\nabla'g)) = -(\nabla'f, (\nabla'g)_+ - (\nabla'g)_-)$. We claim that

$$(\nabla'f, (\nabla'g)_-) = 0. \tag{4.35}$$

For an arbitrary $w \in W$ define the following subgroups of $\mathcal{G}$:

$$\mathcal{N}_w = \mathcal{N}_+ \cap w\mathcal{N}_+ w^{-1}, \qquad \mathcal{N}'_w = \mathcal{N}_+ \cap w\mathcal{N}_- w^{-1},$$
$$\bar{\mathcal{N}}_w = \mathcal{N}_- \cap w\mathcal{N}_- w^{-1}, \qquad \bar{\mathcal{N}}'_w = \mathcal{N}_- \cap w\mathcal{N}_+ w^{-1},$$

and the corresponding Lie subalgebras

$$\mathfrak{n}_w = \mathfrak{n}_+ \cap Ad_w \mathfrak{n}_+, \qquad \mathfrak{n}'_w = \mathfrak{n}_+ \cap Ad_w \mathfrak{n}_-,$$
$$\bar{\mathfrak{n}}_w = \mathfrak{n}_- \cap \mathrm{Ad}_w\, \mathfrak{n}_-, \qquad \bar{\mathfrak{n}}'_w = \mathfrak{n}_- \cap \mathrm{Ad}_w\, \mathfrak{n}_+.$$

Clearly

$$\mathfrak{n}_+ = \mathfrak{n}_w \oplus \mathfrak{n}'_w\,, \qquad \mathfrak{n}_- = \bar{\mathfrak{n}}_w \oplus \bar{\mathfrak{n}}'_w.$$

Therefore, $\pi_{\mathfrak{n}_-} = \pi_{\bar{\mathfrak{n}}_w} + \pi_{\bar{\mathfrak{n}}'_w}$, where $\pi_{\bar{\mathfrak{n}}_w}, \pi_{\bar{\mathfrak{n}}'_w}$ are natural projections. Set $w = vv'$. Then, by Lemma 4.20(ii), $\nabla' f \perp \bar{\mathfrak{n}}'_w$, and hence

$$(\nabla' f, (\nabla' g)_-) = \left(\nabla' f, \pi_{\mathfrak{n}_-}(\nabla' g)\right) = (\nabla' f, \pi_{\bar{\mathfrak{n}}_w}(\nabla' g))\,.$$

Since $(\bar{\mathfrak{n}}_w)^\perp = \mathfrak{n}_-^\perp \oplus Ad_w \mathfrak{n}_-^\perp = \mathfrak{n}_- \oplus Ad_w \mathfrak{n}_-$, one has

$$(\nabla' f, \pi_{\bar{\mathfrak{n}}_w}(\nabla' g)) = (\pi_{\mathfrak{n}_w}(\nabla' f), \nabla' g)\,.$$

It is well-known that $l(vv') = l(v) + l(v')$ implies

$$\mathfrak{n}_{vv'} \subset \mathfrak{n}_v. \tag{4.36}$$

By Lemma 4.20(ii) and (4.36), $\nabla' g$ is orthogonal to $\mathfrak{n}_w \subset \mathfrak{n}_v$, and (4.35) follows.

Using (4.35), one can write

$$\begin{aligned}(\nabla' f, R(\nabla' g)) &= (\nabla' f, \nabla' g) - 2\,(\nabla' f, (\nabla' g)_-) - (\nabla' f, (\nabla' g)_0)\\ &= (\nabla' f, \nabla' g) - (\nabla' f, (\nabla' g)_0)\,.\end{aligned}$$

Since $(\nabla' g)_0 \in \mathfrak{h}$, this implies

$$(\nabla' f, R(\nabla' g)) = (\nabla' f, \nabla' g) - ((\nabla' f)_0, (\nabla' g)_0)\,.$$

Similarly,

$$(R(\nabla f), \nabla g) = -\,(\nabla f, \nabla g) + ((\nabla f)_0, (\nabla g)_0)\,.$$

Recall that $(\nabla f, \nabla g) = (\mathrm{Ad}_{x^{-1}} \nabla f, \mathrm{Ad}_{x^{-1}} \nabla g) = (\nabla' f, \nabla' g)$, and hence

$$2\{f, g\}_{\mathcal{G}} = ((\nabla f)_0, (\nabla g)_0) - ((\nabla' f)_0, (\nabla' g)_0)\,.$$

Now (4.33) follows from (4.34). □

As an immediate corollary of Proposition 4.19 we get the log-canonicity of the functions x_k with respect to $\{\cdot,\cdot\}_{\mathcal{G}}$.

COROLLARY 4.21. *The standard Poisson-Lie structure induces the following Poisson brackets for functions* $x_k = \Delta(k; \mathbf{i})$: *for* $j < k$,

$$\{x_j, x_k\}_{\mathcal{G}} = \Omega_{jk} x_j x_k = \frac{\left(u_{\le j}\omega_{|i_j|}, u_{\le k}\omega_{|i_k|}\right) - \left(v_{>j}\omega_{|i_j|}, v_{>k}\omega_{|i_k|}\right)}{2} x_j x_k.$$

PROOF. Recall that by (3.15), for $j < k$ we have $l(u_{\le k}) = l(u_{\le j}) + l(u_{\le j}^{-1} u_{\le k})$ and $l(v_{>j}) = l(v_{>k}) + l(v_{>k}^{-1} v_{>j})$. Therefore, we can apply (4.33) with $\xi = \omega_{|i_k|}$, $\eta = \omega_{|i_j|}$, $u = u_{\le j}$, $uu' = u_{\le k}$, $v = v_{>k}$, $vv' = v_{>j}$. □

To complete the proof of Theorem 4.18, we need to check that the matrix $\Omega = \Omega_{\mathbf{i}} = (\Omega_{jk})$ satisfies condition (4.4) in the proof of Theorem 4.5. Let us introduce some additional notation:

$$\mu_k = u_{\le k}\omega_{|i_k|}, \quad \nu_k = v_{>k}\omega_{|i_k|}, \quad \beta_{jk} = (\mu_j, \mu_k), \quad \gamma_{jk} = (\nu_j, \nu_k), \quad z_{jk} = \beta_{jk} - \gamma_{jk}.$$

Then $2\Omega = Z_+ - Z_-$, where $Z = (z_{jk})$ and, as usual, subscripts $\pm$ refer to strictly upper/strictly lower triangular parts of a matrix.

Let $m = l(u) + l(v) + r$, and let b'_{kl} denote the elements of the matrix $\widetilde{B}'_{\mathbf{i}}$.

LEMMA 4.22. *For any* $\mathbf{i}$*-exchangeable index* k*,*

$$\sum_{l=1}^{m} b'_{kl}\mu_l = \sum_{l=1}^{m} b'_{kl}\nu_l = 0. \tag{4.37}$$

PROOF. Fix an $\mathbf{i}$-exchangeable index k. We will assume that $\theta(k) = 1$. The other case can be treated in exactly the same way.

For any l, define l^- as in Section 2.2.3 and l^+ by $(l^+)^- = l$. Denote $p = |i_k|$, and for any $q \in [1, r]$, $q \neq p$, define

$$I_q = \{l \ : |i_l| = q, \ b'_{kl} \neq 0\}.$$

Let l_q and r_q be the leftmost and the rightmost elements in I_q, respectively. Note that by construction of $\Sigma'_{\mathbf{i}}$, the arrow is directed from k to l_q.

We will treat the cases $\theta(k^+) = \pm 1$ separately. First, suppose $\theta(k^+) = 1$. Then $\theta(r_q) = 1$, and the arrow in the graph $\Sigma'_{\mathbf{i}}$ is directed from r_q to k. Moreover, $r_q < k^+$, and for all $l \in [r_q + 1, k^+]$ such that $|i_l| = q$, one has $\theta(l) = -1$. Careful inspection reveals that I_q has the following properties:

(i) the number of elements in I_q is even;

(ii) the arrows in $\Sigma_{\mathbf{i}}$ are directed from k to even elements of I_q and from odd elements of I_q to k;

(iii) let j and j' be two consequent elements in I_q, then for all $l \in [j, j' - 1]$ such that $|i_l| = q$, $\theta(l) = \theta(j) = -\theta(j')$.

For any j put

$$\sigma_\mu^{qj} = \sum_{l \in I_q,\ l \le j} d_{kl}\mu_l, \qquad \sigma_\nu^{qj} = \sum_{l \in I_q,\ l > j} d_{kl}\nu_l,$$

where $D = (d_{kl})$ is the incidence matrix of the graph $\Sigma'_{\mathbf{i}}$. Let $\varkappa = \varkappa(j)$ be the number of elements in I_q that are less or equal than j.

LEMMA 4.23. *Let* $\theta(k) = \theta(k^+) = 1$*, then*

$$\sigma_\mu^{qj} = \begin{cases} 0, & \text{if } \varkappa \text{ is even}, \\ -u_{\le j}\omega_q, & \text{if } \varkappa \text{ is odd}, \end{cases} \qquad \sigma_\nu^{qj} = \begin{cases} (v_{>k^+} - v_{>j})\omega_q, & \text{if } \varkappa \text{ is even}, \\ v_{>k^+}\omega_q, & \text{if } \varkappa \text{ is odd}. \end{cases} \tag{4.38}$$

PROOF. Let $I_q = \{j_1 < \cdots < j_t\}$. Properties (ii) and (iii) of I_q imply

$$\sigma_\mu^{qj} = \sum_{s=1}^{\varkappa} (-1)^s \mu_{j_s}, \qquad \sigma_\nu^{qj} = \sum_{s=\varkappa+1}^{t} (-1)^s \nu_{j_s}.$$

Recall that by (1.5), elementary reflections s_i act on fundamental weights ω_k as

$$s_i\omega_k = \omega_k - \langle \omega_k | \alpha_i \rangle \alpha_i = \omega_k - \delta_{ik}\alpha_i. \tag{4.39}$$

We therefore conclude from property (iii) of I_q that

$$\mu_{2s} = u_{\le j_{2s}}\omega_q = u_{\le j_{2s-1}} \prod_{l=j_{2s-1}+1}^{j_{2s}} s_{|i_l|}^{\frac{1-\theta(l)}{2}} \omega_q = u_{\le j_{2s-1}}\omega_q = \mu_{2s-1}$$

and

$$\nu_{2s} = v_{>j_{2s}}\omega_q = v_{>j_{2s+1}} \prod_{l=j_{2s+1}}^{j_{2s}+1} s_{|i_l|}^{\frac{1+\theta(l)}{2}} \omega_q = v_{>j_{2s+1}}\omega_q = \nu_{2s+1}.$$

Now the first relation in (4.38) follows from property (i) immediately, whereas σ_ν^{qj} equals $\nu_{r_q} - \nu_{j_{\varkappa+1}}$ for $\varkappa$ even and ν_{r_q} for $\varkappa$ odd. Moreover, using properties of r_q, property (iii) and (4.39), one obtains $\nu_{r_q} = v_{>r_q}\omega_q = v_{>k^+}\omega_q$ and $\nu_{j_{\varkappa+1}} = v_{>j_{\varkappa+1}}\omega_q = v_{>j}\omega_q$, which completes the proof. □

Denote the sums in (4.37) by σ_μ and σ_ν, respectively. Then

$$\sigma_\mu = \mu_{k^-} - \mu_{k^+} - \sum_{q\ne p} a_{qp}\sigma_\mu^{qr_q}, \qquad \sigma_\nu = \nu_{k^-} - \nu_{k^+} - \sum_{q\ne p} a_{qp}\sigma_\nu^{q,l_q-1},$$

where $A = (a_{qp})$ is the Cartan matrix. Recall that $\varkappa(r_q) = |I_q|$ is even by property (i), and hence Lemma 4.23 yields

$$\sigma_\mu = \mu_{k^-} - \mu_{k^+} = u_{\le k^-}\left(1 - \prod_{l=k^-+1}^{k^+} s_{|i_l|}^{\frac{1-\theta(l)}{2}}\right)\omega_p = 0,$$

since the only reflection in the product above that acts nontrivially on ω_p is $s_{i_{k^+}}$, and it enters the product with degree 0.

On the other hand, $\varkappa(l_q - 1) = 0$ is even as well, and hence by the proof of Lemma 4.23, $\sigma_\nu^{q,l_q-1} = \nu_{k^+} - \nu_{l_q}$. Moreover, using properties of l_q and (4.39), one obtains $\nu_{l_q} = v_{>l_q}\omega_q = v_{>k}\omega_q$, and hence

$$\sigma_\nu^{q,l_q-1} = (v_{>k^+} - v_{>k})\,\omega_q.$$

For a similar reason, $\nu_{k^-} = v_{>k^-}\omega_p = v_{>k}s_p\omega_p$ and $\nu_k = v_{>k}\omega_p = v_{>k^+}s_p\omega_p$, thus

$$\sigma_\nu = (v_{>k^-} - v_{>k})\,\omega_p - (v_{>k^+} - v_{>k})\left(\omega_p + \sum_{q\ne p} a_{qp}\omega_q\right).$$

Taking into account (1.9), we get

$$\begin{aligned}
\sigma_\nu &= (v_{>k^-} - v_{>k})\,\omega_p - (v_{>k^+} - v_{>k})\,(-\omega_p + \alpha_p)\\
&= (v_{>k^-} - v_{>k})\,\omega_p + (v_{>k^+} - v_{>k})\,s_p\omega_p\\
&= (v_{>k^-} - v_{>k}s_p)\,\omega_p + (v_{>k^+}s_p - v_{>k})\,\omega_p = 0.
\end{aligned}$$

This completes the proof for the case $\theta(k^+) = 1$.

If $\theta(k^+) = -1$, then $\theta(r_q) = -1$, and the arrow in the graph $\Sigma_{\mathbf{i}}'$ is directed from k to r_q. Moreover, $r_q < k^+$, and for all $l \in [r_q + 1, k^+]$ such that $|i_l| = q$, one has $\theta(l) = 1$. A similar argument as above shows that I_q has the following properties:

(i) the number of elements in I_q is odd;

(ii) the arrows in $\Sigma_{\mathbf{i}}$ are directed from k to odd elements of I_q and from even elements of I_q to k;

(iii) let j and j' be two consequent elements in I_q, then for all $l \in [j+1, j']$ such that $|i_l| = q$, $\theta(l) = \theta(j') = -\theta(j)$.

In this case Lemma 4.23 is replaced by the following

LEMMA 4.24. *Let $\theta(k) = -\theta(k^+) = 1$, then*

(4.40)
$$\sigma_\mu^{qj} = \begin{cases} 0, & \text{if } \varkappa(j) \text{ is even,} \\ -u_{\le j}\omega_q, & \text{if } \varkappa(j) \text{ is odd,} \end{cases} \qquad \sigma_\nu^{qj} = \begin{cases} -v_{>j}\omega_q, & \text{if } \varkappa(j) \text{ is even,} \\ 0, & \text{if } \varkappa(j) \text{ is odd.} \end{cases}$$

PROOF. The proof is similar to that of Lemma 4.23 and is left to the reader. □

Consequently, $\sigma_\nu^{q,l_q-1} = -v_{>l_q}\omega_q = -v_{>k}\omega_q$ and $\sigma_\mu^{qr_q} = -u_{\le r_q}\omega_q = -u_{\le k^+}\omega_q$. Using (1.9) and (4.39) as above, we get

$$\sigma_\mu = \mu_{k^-} + \mu_{k^+} + \sum_{q\neq p} a_{qp} u_{\le k^+}\omega_q = u_{\le k^-}\omega_p + u_{\le k^+}\Big(\omega_p + \sum_{q\neq p} a_{qp}\omega_q\Big)$$
$$= u_{\le k^-}\omega_p - u_{\le k^+} s_p\omega_p = u_{\le k^-}\Big(\omega_p - \Big(\prod_{l=k^-+1}^{k^+-1} s_{|i_l|}^{\frac{1-\theta(l)}{2}}\Big) s_p s_p \omega_p\Big) = 0$$

and

$$\sigma_\nu = \nu_{k^-} + \nu_{k^+} + v_{>k}\sum_{q\neq p} a_{qp} u_{\le k^+}\omega_q = v_{>k^-}\omega_p + v_{>k}\Big(\omega_p + \sum_{q\neq p} a_{qp}\omega_q\Big)$$
$$= v_{>k^-}\omega_p - v_{>k}s_p\omega_p = v_{>k}s_p\omega_p - v_{>k}s_p\omega_p = 0.$$

□

Now we are ready to complete the proof of Theorem 4.18. We need to show that $(\widetilde{B}'_{\mathbf{i}}\Omega)_{kj} = d_k\delta_{kj}$ for any **i**-exchangeable index k. Lemma 4.22 implies

$$\sum_{l=1}^m b'_{kl}\gamma_{lj} = \Big(\sum_{l=1}^m b'_{kl}\mu_l, \mu_j\Big) = 0, \quad \sum_{l=1}^m b'_{kl}\beta_{lj} = \Big(\sum_{l=1}^m b_{kl}\nu_l, \nu_j\Big) = 0.$$

Besides,

$$\beta_{ll} = (\nu_l, \nu_l) = (\omega_{|i_l|}, \omega_{|i_l|}) = (\mu_l, \mu_l) = \gamma_{ll}.$$

Therefore,

$$(\widetilde{B}'_{\mathbf{i}}\Omega)_{kj} = \sum_{l=1}^m b'_{kl}\frac{\operatorname{sign}(j-l)}{2}(\beta_{lj} - \gamma_{lj}) = \sum_{l\le j} b'_{kl}\beta_{lj} + \sum_{l>j} b'_{kl}\gamma_{lj},$$

or

$$(\widetilde{B}'_{\mathbf{i}}\Omega)_{kj} = \Big(\sum_{l\le j} b'_{kl}\mu_l, \mu_j\Big) + \Big(\sum_{l>j} b'_{kl}\nu_l, \nu_j\Big). \tag{4.41}$$

Let us fix an **i**-exchangeable index k with $|i_k| = p$. Once again, it suffices to consider the case $\theta(k) = 1$. Since $b'_{kl} = 0$ for $l > k^+$, one gets $(\widetilde{B}'_{\mathbf{i}}\Omega)_{kj} = 0$ for $j \ge k^+$, and we only need to consider the case $j < k^+$.

Let $j \ge k$. First, assume that $\theta(k^+) = 1$. Denote $|i_j|$ by p^* and consider

$$\sigma_\mu^j = \sum_{l\le j} b'_{kl}\mu_l = \mu_{k^-} - \sum_{q\neq p} a_{qp}\sigma_\mu^{qj}, \qquad \sigma_\nu^j = \sum_{l>j} b'_{kl}\nu_l = -\nu_{k^+} - \sum_{q\neq p} a_{qp}\sigma_\nu^{qj}.$$

Since $\mu_j = u_{\le j}\omega_{p^*}$, $\nu_j = v_{>j}\omega_{p^*}$, Lemma 4.23 gives

$$(\sigma_\mu^{qj}, \mu_j) + (\sigma_\nu^{qj}, \nu_j) = (v_{>k^+}\omega_q, \nu_j) - (\omega_q, \omega_{p^*}).$$

Therefore,

$$\begin{aligned}(\widetilde{B}'_{\mathbf{i}}\Omega)_{kj} &= (\sigma^j_\mu, \mu_j) + (\sigma^j_\nu, \nu_j)\\ &= \Big(-\nu_{k^+} - v_{>k^+} \sum_{q\neq p} a_{qp}\omega_q, \nu_j \Big) + \Big(\sum_{q\neq p} a_{qp}\omega_q, \omega_{p^*} \Big) + (\mu_{k^-}, \mu_j).\end{aligned}$$

Mimicking the proof of Lemma 4.22, we can rewrite the last expression as

$$(\widetilde{B}'_{\mathbf{i}}\Omega)_{kj} = (v_{>k^+} s_p \omega_p, v_{>j}\omega_{p^*}) + (\alpha_p - 2\omega_p, \omega_{p^*}) + (u_{\le k^-}\omega_p, u_{\le j}\omega_{p^*}).$$

Since between k^- and j there are no indices l with $\theta(l) = -1$ and $|i_l| = p$, and between j and $k^+ - 1$ there are no indices l with $\theta(l) = 1$ and $|i_l| = p$, the first term and the third term are both equal to (ω_p, ω_{p^*}). Thus,

$$(\widetilde{B}'_{\mathbf{i}}\Omega)_{kj} = (\alpha_p, \omega_{p^*}) = \frac{1}{2}\delta_{kj}(\alpha_p, \alpha_p).$$

If $\theta(k^+) = -1$, a similar argument applies, with Lemma 4.24 replacing Lemma 4.23. The resulting formula for $(\widetilde{B}'_{\mathbf{i}}\Omega)_{kj}$ reads

$$\begin{aligned}(\widetilde{B}'_{\mathbf{i}}\Omega)_{kj} &= (\nu_{k^+}, \nu_j) + \Big(\sum_{q\neq p} a_{qp}\omega_q, \omega_{p^*} \Big) + (\mu_{k^-}, \mu_j)\\ &= (\nu_{k^+}, \nu_j) + (\alpha_p - 2\omega_p, \omega_{p^*}) + (\mu_{k^-}, \mu_j)\\ &= (\omega_p, \omega_{p^*}) + (\alpha_p - 2\omega_p, \omega_{p^*}) + (\omega_p, \omega_{p^*}) = (\alpha_p, \omega_{p^*}) = \frac{1}{2}\delta_{kj}(\alpha_p, \alpha_p).\end{aligned}$$

Now, let $j < k$. The proof in this case does not depend on $\theta(k^+)$. Similarly to (4.41), we can write

$$(\widetilde{B}'_{\mathbf{i}}\Omega)_{kj} = \Big(\sum_{l<j} b'_{kl}\nu_l, \nu_j \Big) - \Big(\sum_{l<j} b'_{kl}\mu_l, \mu_j \Big) = \sum_{l<j} b'_{kl}\left((\nu_l, \nu_j) - (\mu_l, \mu_j) \right).$$

Note that for each $q \in [1, r]$ there is at most one $l < k$ such that $|i_l| = q$ and $b'_{kl} \neq 0$. In notation of the proof of Lemma 4.22, $l = l_q$. Moreover, there is no other index l between l_q and k with $|i_l| = q$. This means that for $l_q < j < k$ one has

$$\begin{aligned}(\mu_{l_q}, \mu_j) &= (u_{\le l_q}\omega_q, u_{\le j}\omega_{p^*}) = (u_{\le j}\omega_q, u_{\le j}\omega_{p^*}) = (\omega_q, \omega_{p^*})\\ &= (v_{>j}\omega_q, v_{>j}\omega_{p^*}) = (v_{>l_q}\omega_q, v_{>j}\omega_{p^*}) = (\nu_{l_q}, \nu_j),\end{aligned}$$

and the corresponding contribution to $(\widetilde{B}'_{\mathbf{i}}\Omega)_{kj}$ equals zero.

This completes the proof of the theorem. □

4.4. Summary

- For any cluster algebra with an exchange matrix of full rank, we define and describe the set of Poisson structures compatible with a given cluster algebra structure. Moreover, we establish that in τ-coordinates obtained from cluster variables via a suitable monomial transformation, the Poisson bracket is given by a skew-symmetrization of the extended exchange matrix, see Theorem 4.5. In this setting, matrix mutations can be interpreted as transformations of Poisson brackets under coordinate changes.

- For Grassmannians, the Poisson structure induced by the Sklyanin Poisson bracket on GL_n is compatible with the natural cluster algebra structure on the open cell. In this case, the non-vanishing locus of stable cluster coordinates coincides with the union of generic symplectic leaves, see Proposition 4.11. We provide a complete description of generic symplectic leaves in Theorem 4.10.
- Furthermore, it is shown that the cluster algebra we constructed coincides with the ring of regular functions on the open cell in the Grassmannian, see Theorem 4.14. We extend the cluster algebra structure from this ring to the homogeneous coordinate ring of the Grassmannian and show that the resulting cluster algebra, in fact, coincides with the latter ring, see Theorem 4.17.
- Finally, we prove that the restriction of the standard Poisson-Lie structure on a simple Lie group to any double Bruhat cell is compatible with the cluster algebra structure described in Section 3.4.

Bibliographical notes

4.1. Exposition in this section closely follows [**GSV2**].

4.2. For cited results in theory of Poisson homogeneous spaces, see e.g. [**ReST**]. Section 4.2.1 is borrowed from [**GSV2**], as well as Theorem 4.14 and its proof. Symplectic leaves of the standard Poisson-Lie structure on GL_n and the corresponding Poisson homogeneous structure on the Grassmannian were comprehensively studied in [**BGY**] within a unifying framework of *matrix affine Poisson spaces.* Theorem 4.17 was first proved in a different way in [**Sc**] based on the Postnikov's combinatorial model for the Grassmannian derived in [**Po**]. Our proof is new. Projective normality of the Grassmannian in the Plücker embedding was proved by Severi in 1915, see [**HdP**], Chapter XIV, §7, Theorem 1.

4.3. A system of log-canonical coordinates for a double Bruhat cell was first constructed in [**KoZ**]. However, it consisted of *twisted* generalized minors, which do not extend to regular functions on the entire group. To the best of our knowledge, the results of this section have not appeared in the literature. A proof of inclusion (4.36) follows easily from Lemma 7 in Chapter IV.12 of [**Hu1**].

CHAPTER 5

The cluster manifold

Basic examples in Chapter 2 show that cluster algebras originate from (homogeneous) coordinate rings of some classical geometrical varieties, e.g., intersections of Schubert cells or double Bruhat cells.

In this Chapter we construct an algebraic variety $\mathcal{X}$ (which we call the *cluster manifold*) related to an abstract cluster algebra $\mathcal{A}(\widetilde{B})$.

Further, we extend calculations of the number of connected components in real reduced double Bruhat cells (see Section 2.2 and references therein) to a more general setting of geometric cluster algebras and compatible Poisson structures. We introduce an $\mathbb{F}^*$-action compatible with the cluster algebra $\mathcal{A}$ (here $\mathbb{F}$ is $\mathbb{C}$ or $\mathbb{R}$). Compatibility of the $\mathbb{F}^*$-action means that all exchange relations of $\mathcal{A}$ are preserved under this action. The union $\mathcal{X}^0$ of generic orbits with respect to this $\mathbb{F}^*$-action "almost coincides" with the union of generic symplectic leaves of a compatible Poisson structure in Spec($\mathcal{A}$). We compute the number of connected components of $\mathcal{X}^0$ for a cluster algebra over $\mathbb{R}$.

Finally we apply the previous result to the case of refined open Bruhat cells in the Grassmannian $G_k(n)$. Namely, the famous Sklyanin Poisson-Lie bracket on $SL_n(\mathbb{R})$ induces a Poisson bracket on the open Bruhat cell in $G_k(n)$. This Poisson bracket is compatible with the structure of a special cluster algebra, one of whose clusters consists only of Plücker coordinates. The corresponding $\mathbb{R}^*$-action determines the union of generic orbits, which is simply described as a subset of the Grassmannian defined by inequalities $x_{i,i+1,\dots,x+k-1} \neq 0$, where $x_{i,i+1,\dots,x+k-1}$ is the (cyclically dense) minor containing the ith, $(i+1)$-st, ..., $(i+k-1)$-st columns (here the indices are taken modulo n). We call this subset a *refined* open Bruhat cell in the Grassmannian $G_k(n)$; indeed, this subset is the intersection of n open Bruhat cells in general position. In the last part we compute the number of connected components of a refined open Bruhat cell in $G_k(n)$ over $\mathbb{R}$. This number is equal to $3 \cdot 2^{n-1}$ if $k \geq 3$ and $n \geq 7$.

5.1. Definition of the cluster manifold

A natural way to associate a variety with an algebra is suggested by algebraic geometry. One can naturally define the cluster manifold as Spec($\mathcal{A}(\widetilde{B})$), (or Proj($\mathcal{A}(\widetilde{B})$) in the (quasi)-homogeneous case), since it is the maximal manifold M satisfying the following two universal conditions:

(i) all the cluster functions are regular functions on M;

(ii) for any pair p_1, p_2 of two distinct points on M there exists an element $f \in \mathcal{A}(\widetilde{B})$ such that $f(p_1) \neq f(p_2)$.

However, as we have already seen, the Markov cluster algebra introduced in Example 3.12 is not finitely generated. This observation shows that Spec($\mathcal{A}(\widetilde{B})$)

might be a rather complicated object. Therefore we define the *cluster manifold* $\mathcal{X} = \mathcal{X}(\mathcal{A})$ as a "handy" nonsingular part of $\operatorname{Spec}(\mathcal{A}(\widetilde{B}))$.

We will describe $\mathcal{X}$ by means of charts and transition functions. Let us denote by $\mathbb{T}_n$ the universal covering of the exchange graph of $\mathcal{A}(\widetilde{B})$. Clearly, $\mathbb{T}_n$ is the n-regular tree introduced in Section 3.1. Vertices of $\mathbb{T}_n$ are equipped with ordered $(n+m)$-tuples of extended cluster variables and with $n\times(n+m)$ exchange matrices. Abusing notation, we will denote the $(n+m)$-tuple and the matrix at vertex v by $\widetilde{\mathbf{x}}^v$ and $\widetilde{B}^v$. Observe that the tuples at two vertices of $\mathbb{T}_n$ covering the same extended cluster, consist of the same elements and differ only in the order of the elements, while the matrices are obtained one from the other by an appropriate rearrangement of rows and columns. For each vertex v of $\mathbb{T}_n$ we define a chart, that is, an open subset $U^v \subset \mathcal{X}$ by

$$U^v = \operatorname{Spec}(\mathbb{F}[(x_1^v)^{\pm 1}, \dots, (x_n^v)^{\pm 1}, x_{n+1}^v, \dots, x_{n+m}^v])$$

(as before, $\mathbb{F}$ is a field of characteristic 0, and stable variables $x_{n+1}^v = x_{n+1}, \dots, x_{n+m}^v = x_{n+m}$ do not depend on v). An exchange relation in direction i between extended clusters $\widetilde{\mathbf{x}}$ and $\widetilde{\mathbf{x}}'$ induces an edge $(v, v') \in \mathbb{T}_n$, where v covers $\widetilde{\mathbf{x}}$ and v' covers $\widetilde{\mathbf{x}}'$. This edge defines an elementary transition map $U^v \to U^{v'}$ via

$$x_j^{v'} = x_j^v \quad \text{if } j \neq i, \qquad x_i^{v'} x_i^v = M_i^v + N_i^v, \tag{5.1}$$

where M_i^v and N_i^v are monomials in $x_1^v, \dots, \widehat{x_i^v}, \dots, x_{n+m}^v$ forming the right hand side of the exchange relation in direction i.

Note that any pair of vertices of $\mathbb{T}_n$ is connected by a unique path. Therefore, the transition map between the charts corresponding to two arbitrary vertices can be computed as the composition of the elementary transitions along this path. Finally, put

$$\mathcal{X} = \cup_{v \in \mathbb{T}_n} U^v.$$

It follows immediately from the definition that $\mathcal{X} \subset \operatorname{Spec}(\mathcal{A}(\widetilde{B}))$. More exactly, $\mathcal{X}$ contains the points $p \in \operatorname{Spec}(\mathcal{A}(\widetilde{B}))$ for which there exists a vertex v of $\mathbb{T}_n$ such that $x_1^v, \dots, x_n^v, x_{n+1}, \dots, x_{n+m}$ form a coordinate system in a neighborhood $U^v \ni p$.

EXAMPLE 5.1. Consider a cluster algebra $\mathcal{A}$ over $\mathbb{C}$ of rank 1 given by two extended clusters $\{x_1, x_2, x_3\}$ and $\{x_1', x_2, x_3\}$ subject to the relation: $x_1 x_1' = x_2^2 + x_3^2$. Clearly, $\operatorname{Spec}(\mathcal{A}) = \operatorname{Spec}(\mathbb{C}[x, y, z, t]/\{xy - z^2 - t^2 = 0\})$ is a singular affine hypersurface $H \subset \mathbb{C}^4$ given by the equation $xy = z^2 + t^2$ and containing a singular point $x = y = z = t = 0$. On the other hand, $\mathcal{X} = H \setminus \{x = y = z^2 + t^2 = 0\}$ is nonsingular.

In the general case, the following result stems immediately from the above definitions.

LEMMA 5.2. *The cluster manifold $\mathcal{X}$ is a nonsingular rational manifold.*

5.2. Toric action on the cluster algebra

Assume that an integer weight $\mathbf{w}^v = (w_1^v, \dots, w_{n+m}^v)$ is given at any vertex v of the tree $\mathbb{T}_n$. We define a *local toric action* at vertex v as the map $\mathbb{F}(\widetilde{\mathbf{x}}^v) \times \mathbb{F}^* \to \mathbb{F}(\widetilde{\mathbf{x}}^v)$ given on the generators of $\mathbb{F}(\widetilde{\mathbf{x}}^v)$ by the formula $\widetilde{\mathbf{x}}^v \mapsto \widetilde{\mathbf{x}}^v \cdot t^{\mathbf{w}^v}$, i.e.,

$$(x_1^v, \dots, x_{n+m}^v) \mapsto (x_1^v \cdot t^{w_1^v}, \dots, x_{n+m}^v \cdot t^{w_{n+m}^v}),$$

and extended naturally to $\mathbb{F}(\widetilde{\mathbf{x}}^v)$. We say that two local toric actions at v and u are *compatible* if the following diagram is commutative:

$$\begin{array}{ccc} \mathbb{F}(\widetilde{\mathbf{x}}^v) & \longrightarrow & \mathbb{F}(\widetilde{\mathbf{x}}^u) \\ {\scriptstyle t^{\mathbf{w}^v}}\big\downarrow & & {\scriptstyle t^{\mathbf{w}^u}}\big\downarrow \\ \mathbb{F}(\widetilde{\mathbf{x}}^v) & \longrightarrow & \mathbb{F}(\widetilde{\mathbf{x}}^u) \end{array}$$

where the horizontal arrows are induced by transitions along the path from u to v in $\mathbb{T}_n$. If all local toric actions are compatible, they determine a *global toric action* on $\mathcal{A}(\widetilde{B})$. This toric action is said to be an *extension* of the local toric action at v.

A toric action on the cluster algebra gives rise to a well-defined $\mathbb{F}^*$-action on $\mathcal{X}$. The corresponding flow is called a *toric flow*.

LEMMA 5.3. *Let $\mathbf{w}^v$ be an arbitrary integer weight. The local toric action at v defined by $\widetilde{\mathbf{x}}^v \mapsto \widetilde{\mathbf{x}}^v \cdot t^{\mathbf{w}^v}$ can be extended to a global toric action on $\mathcal{A}(\widetilde{B})$ if and only if $\widetilde{B}^v(\mathbf{w}^v)^T = 0$. Moreover, if such an extension exists, then it is unique.*

PROOF. Given a monomial $g = \prod_j x_j^{p_j}$, we define its *weighted degree* by $\deg g = \sum_j p_j w_j$. It is easy to see that local toric actions are compatible if and only if all the monomials in the right hand sides of all exchange relations for all clusters have the same weighted degree.

Consider an edge $(v, v') \in \mathbb{T}_n$ in direction i. The equality $\deg M_i^v = \deg N_i^v$ in the exchange relation (5.1) implies, by virtue of (3.1),

$$\sum_{b_{ik}^v > 0} b_{ik}^v w_k^v = \sum_{b_{ik}^v < 0} -b_{ik}^v w_k^v.$$

The above equalities written for all $i \in [1, n]$ are equivalent to the matrix equality $\widetilde{B}^v(\mathbf{w}^v)^T = 0$. Therefore, the latter condition is necessary for the existence of a global toric action.

Assume now that $\widetilde{B}^v(\mathbf{w}^v)^T = 0$. We claim that there is a unique weight $\mathbf{w}^{v'}$ that makes the local toric action at v' compatible with the one at v. Note first that identities $x_j^{v'} = x_j^v$ immediately incur $w_j^{v'} = w_j^v$ for $j \neq i$. Next, the exchange relation in direction i implies

$$w_i^{v'} = \sum_{b_{ik}^v > 0} b_{ik}^v w_k^v - w_i^v.$$

Hence, the compatible weight $\mathbf{w}^{v'}$ is unique (if it exists). To complete the proof it remains to check that $\widetilde{B}^v(\mathbf{w}^v)^T = 0$ implies $\widetilde{B}^{v'}(\mathbf{w}^{v'})^T = 0$.

For $k \neq i$, the kth entry of $\widetilde{B}^{v'}(\mathbf{w}^{v'})^T$ is

$$\left(\widetilde{B}^{v'}(\mathbf{w}^{v'})^T\right)_k = \sum_{j=1}^{n+m} b_{kj}^{v'} w_j^{v'}$$

$$= \sum_{j \neq i} b_{kj}^v w_j^v + \frac{1}{2} \sum_{j \neq i} \left(|b_{ki}^v| b_{ij}^v + b_{ki}^v |b_{ij}^v|\right) w_j^v - b_{ki}^v \left(\sum_{b_{il}^v > 0} b_{il}^v w_l^v - w_i^v \right)$$

$$= \sum_{j=1}^{n+m} b_{kj}^v w_j^v = 0.$$

Finally,

$$\left(\widetilde{B}^{v'}(\mathbf{w}^{v'})^T\right)_i = \sum_{j=1}^n b_{ij}^{v'} w_j^{v'} = -\sum_{j\neq i} b_{ij}^v w_j^v = 0,$$

since $b_{ii}^{v'} = 0$. Hence, $\widetilde{B}^{v'}(\mathbf{w}^{v'})^T = 0$. □

Note that the main source of examples of cluster algebras are coordinate rings of homogeneous manifolds. Toric actions on such cluster algebras are induced by the natural toric actions on these manifolds.

For example, the toric action on the Grassmannian $G_k(n)$ described in Section 4.2.1 is compatible with the cluster algebra $\mathcal{A}(G_k(n))$ defined in Section 4.2.2. By Lemma 5.3, it suffices to check that each short Plücker relation involved in the transformations of the initial cluster (4.30) is homogeneous with respect to this action. Write this relation in form (1.3); it is easy to see that an arbitrary index r either belongs to I', or belongs to $\{i, j, k, l\}$, or to neither of them. in the first case relation (1.3) is of degree 2, in the second case it is of degree 1, and in the third case, of degree 0. Similarly, action (4.22) in $G_k^0(n)$ is compatible with the cluster algebra $\mathcal{A}(G_k^0(n))$. This follows form the fact that the transformations of the initial cluster (4.24) are obtained from short Plücker relations by division of each monomial by the same power of $x_{[1,k]}$, which is, in its turn, homogeneous with respect to the toric action on $G_k(n)$. It follows that both $\mathcal{X}(\mathcal{A}(G_k(n)))$ and $\mathcal{X}(\mathcal{A}(G_k^0(n)))$ are endowed with a toric action compatible with the corresponding cluster algebra.

5.3. Connected components of the regular locus of the toric action

We are going to extend the results of Section 2.2.3 on the number of connected components in the real part of a reduced double Bruhat cell to general cluster manifolds. Let us denote by $\mathcal{X}_0 = \mathcal{X}_0(\mathcal{A})$ the open part of $\mathcal{X}$ given by the conditions of nonvanishing of stable variables: $x_i \neq 0$ for $i \in [n+1, n+m]$. Recall that stable variables are the same for all extended clusters, and hence $\mathcal{X}_0$ is well defined. We call $\mathcal{X}_0$ the *regular locus* of the toric action described in the previous section; $\mathcal{X}_0$ is a union of isomorphic orbits of the toric action. Moreover, we will see in the next section that in the presence of a compatible Poisson structure on $\mathcal{X}$, $\mathcal{X}_0$ is the union of generic symplectic leaves.

Denote $U_0^v = U^v \cap \mathcal{X}_0$, then we have a decomposition $\mathcal{X}_0 = \cup_{v\in\mathbb{T}_n} U_0^v$.

In what follows we assume that $\mathbb{F} = \mathbb{R}$. A natural generalization of Arnold's problem discussed in Section 2.2.1 is to find the number $\#(\mathcal{X}_0)$ of connected components of $\mathcal{X}_0$. In the current section we answer this question following the approach developed in Chapter 2.

Given a vertex $v \in \mathbb{T}_n$, we define an open subset $S(v) \subset \mathcal{X}_0$ by

$$S(v) = U_0^v \cup \bigcup_{(v,v')\in\mathbb{T}_n} U_0^{v'};$$

note that in the context of double Bruhat cells, $S(v)$ corresponds to $U_{\mathbf{i}} \bigcup (\cup_n U_{n,\mathbf{i}})$ studied in the proof of Theorem 2.16.

Recall that $U_0^v \simeq (\mathbb{R}^*)^{n+m}$. We can decompose U_0^v as follows. For every $\xi = (\xi_1, \dots, \xi_{n+m}) \in \mathbb{F}_2^{n+m}$, define $U_0^v(\xi)$ as the orthant $(-1)^{\xi_j} x_j^v > 0$ for all $j \in [1, n+m]$. Two orthants $U_0^v(\xi)$ and $U_0^v(\bar{\xi})$ are called *adjacent* if the following two conditions are fulfilled:

(i) there exists $i \in [1, n]$ such that $\xi_i \neq \bar{\xi}_i$, and $\xi_j = \bar{\xi}_j$ for all $j \neq i$;

(ii) there exists $p \in S(v)$ that belongs to the intersection of the closures of $U_0^v(\xi)$ and $U_0^v(\bar\xi)$ in $S(v)$.

The second condition can be restated as follows:

(ii′) there exists $p \in S(v)$ and a neighbor $\hat v$ of v in direction i such that x_i^v vanishes at p, $x_i^{\hat v}$ does not vanish at p, and $x_j^v = x_j^{\hat v}$ does not vanish at p for $j \neq i$.

LEMMA 5.4. *Let $(v, v') \in \mathbb{T}_n$, $\xi, \bar\xi \in \mathbb{F}_2^{n+m}$, and $U_0^v(\xi)$ be adjacent to $U_0^v(\bar\xi)$. Then $U_0^{v'}(\xi)$ and $U_0^{v'}(\bar\xi)$ are adjacent as well.*

PROOF. Assume that (v, v') is labeled by i. To simplify notation, we will denote coordinates x_j^v by x_j, and $x_j^{v'}$ by x'_j. Let us consider the case $\xi_j \neq \bar\xi_j$. Then $\hat v$ in condition (ii′) coincides with v', and hence

$$x_i x'_i = M_i + N_i, \tag{5.2}$$

where M_i, N_i are two monomials entering the right hand side of exchange relations. By condition (ii′), there exists a point $z^* \in U_0^{v'}$ such that $x_i(z^*) = 0$, $x'_i(z^*) \neq 0$. Hence, $M_i(z^*) + N_i(z^*) = 0$. Recall that $M_j + N_j$ does not depend on x_i. Choose a point $z^{**} \in U^v$ such that $x_j(z^{**}) = x_j(z^*)$ for $j \neq i$, $x'_i(z^{**}) = 0$, but $x_i(z^{**}) \neq 0$. Such a point z^{**} exists by virtue of (5.2). Therefore, condition (ii′) holds for $U_0^v(\xi)$ and $U_0^{v'}(\bar\xi)$, and they are adjacent.

Assume now that $\xi_j \neq \bar\xi_j$ for some $j \neq i$. As before, we have $M_j(z) + N_j(z) = 0$ for any $z \in U^v \cup U^{v'}$ such that $x_l(z) = x_l(z^*)$ for $l \neq j$. Consider the edge $(v', v'') \in \mathbb{T}_n$ in direction j; by the above assumption, $v'' \neq v$. Without loss of generality assume that x'_i does not enter N'_j. Then one has

$$M_j + N_j = N_j\left(\frac{M_j}{N_j} + 1\right) = N_j\left(\frac{M'_j}{N'_j} + 1\right)\Bigg|_{x'_i \leftarrow \frac{R}{x_i}}, \tag{5.3}$$

where $R = M_i + N_i|_{x_j=0}$. Therefore, for any z^{**} such that

$$\begin{aligned} x'_l(z^{**}) &= x_l(z^*) \qquad \text{if } l \neq i, j,\\ x'_i(z^{**}) &= R(z^*)/x_i(z^*),\\ x'_j(z^{**}) &= 0, \end{aligned}$$

one has

$$M'_j(z^{**}) + N'_j(z^{**}) = M_j(z^*) + N_j(z^*) = 0,$$

and hence one can choose $x''_j(z^{**}) \neq 0$. □

COROLLARY 5.5. *If $U_0^v(\xi)$ and $U_0^v(\bar\xi)$ are adjacent, then $U_0^{v'}(\xi)$ and $U_0^{v'}(\bar\xi)$ are adjacent for any vertex $v' \in \mathbb{T}_n$.*

PROOF. Since the tree $\mathbb{T}_n$ is connected, one can pick up the path connecting v and v'. Then the corollary follows immediately from Lemma 5.4. □

Let $\#_v$ denote the number of connected components in $S(v)$.

THEOREM 5.6. *The number $\#_v$ does not depend on v and is equal to $\#(\mathcal{X}_0)$.*

PROOF. Indeed, since $S(v)$ is dense in $\mathcal{X}_0$, one has $\#(\mathcal{X}_0) \leq \#_v$. Conversely, assume that there are points $z_1, z_2 \in \mathcal{X}_0$ that are connected by a path in $\mathcal{X}_0$. Therefore their small neighborhoods are also connected in $\mathcal{X}_0$, since $\mathcal{X}_0$ is a topological manifold. Since U_0^v is dense in $\mathcal{X}_0$, one can pick $\xi, \bar\xi \in \mathbb{F}_2^{n+m}$ such that the intersection of the first of the above neighborhoods with $U_0^v(\xi)$ and the intersection of the second one with $U_0^v(\bar{\xi})$ are both nonempty. Thus, $U_0^v(\xi)$ and $U_0^v(\bar{\xi})$ are connected in $\mathcal{X}_0$. Hence, there exist a loop γ in $\mathbb{T}_n$ with the initial point v, a subset $v_1, \dots, v_p$ of vertices of this loop, and a sequence $\xi^1 = \xi, \xi^2, \dots, \xi^{p+1} = \bar{\xi} \in \mathbb{F}_2^{n+m}$ such that $U_0^{v_l}(\xi^l)$ is adjacent to $U_0^{v_l}(\xi^{l+1})$ for all $l \in [1, p]$. Then by Corollary 5.5, $U_0^v(\xi^l)$ and $U_0^v(\xi^{l+1})$ are adjacent. Hence all $U_0^v(\xi^l)$ are connected with each other in $S(v)$. In particular, $U_0^v(\xi)$ and $U_0^v(\bar{\xi})$ are connected in $S(v)$. This proves the assertion. □

Fix a basis $\{e_i\}$ in $\mathbb{F}_2^{n+m}$. Let $\widehat{B} = (b_{ij})_{i,j=1}^{n+m}$ and $\widehat{D} = \operatorname{diag}\{d_1, \dots, d_{n+m}\}$ be the same as in Section 4.1.2, and let $\eta = \eta^v$ be a (skew)-symmetric bilinear form on $\mathbb{F}_2^{n+m}$ given by $\eta(e_i, e_j) = d_i b_{ij}$. Define a linear operator $\mathfrak{t}_i : \mathbb{F}_2^{n+m} \to \mathbb{F}_2^{n+m}$ by the formula $\mathfrak{t}_i(\xi) = \xi - \eta(\xi, e_i)e_i$, and let $\Gamma = \Gamma_v$ be the group generated by $\mathfrak{t}_i$, $i \in [1, n]$.

The following lemma is a minor modification of Theorem 2.10.

LEMMA 5.7. *The number of connected components $\#_v$ is equal to the number of Γ-orbits in $\mathbb{F}_2^{n+m}$.*

The proof follows the pattern already discussed in Section 2.2.3.

LEMMA 5.8. *Let ξ and $\bar{\xi}$ be two distinct vectors in $\mathbb{F}_2^{n+m}$. The closures of $U_0^v(\xi)$ and of $U_0^v(\bar{\xi})$ intersect in $S(v)$ if and only if $\bar{\xi} = \mathfrak{t}_i(\xi)$ for some $i \in [1, n]$.*

PROOF. The proof repeats the corresponding part of the proof of Theorem 2.16. □

Now we are ready to complete the proof of Lemma 5.7. Let Ξ be a Γ-orbit in $\mathbb{F}_2^{n+m}$, and let $U_0^v(\Xi) \subset S(v)$ be the union $\cup_{\xi\in\Xi} U_0^v(\xi)$. Each $U_0^v(\xi)$ is a copy of $\mathbb{R}_{>0}^{n+m}$, and is thus connected. Using the "if" part of Lemma 5.8, we conclude that $U_0^v(\Xi)$ is connected (since the closure of a connected set and the union of two non-disjoint connected sets are connected as well). On the other hand, by the "only if" part of the same lemma, all the sets $U_0^v(\Xi)$ are pairwise disjoint. Thus, they are the connected components of $S(v)$, and we are done.

Theorem 5.6 and Lemma 5.7 imply the following result.

THEOREM 5.9. *The number of connected components in $\mathcal{X}_0$ equals the number of Γ-orbits in $\mathbb{F}_2^{n+m}$.*

The number of Γ-orbits in $\mathbb{F}_2^{n+m}$ in many cases can be calculated explicitly.

A finite undirected graph is said to be E_6-*compatible* if it is connected and contains an induced subgraph with 6 vertices isomorphic to the Dynkin diagram E_6. A directed graph is said to be E_6-compatible if the corresponding undirected graph obtained by forgetting orientations of edges is E_6-compatible.

THEOREM 5.10. *Suppose that the diagram of B is E_6-compatible. Then the number of Γ-orbits in $\mathbb{F}_2^{m+n}$ is equal to*

$$2^m \cdot (2 + 2^{\dim(\mathbb{F}_2^B \cap \ker \eta)}),$$

where $\mathbb{F}_2^B$ is the n-dimensional subspace of F_2^{n+m} induced by cluster variables.

The proof of this result falls beyond the scope of this book.

5.4. Cluster manifolds and Poisson brackets

In this section we assume that the rank of the exchange matrix $\widetilde{B}$ is equal to n. Our goal is to interpret $\mathcal{X}_0$ in terms of symplectic leaves of a Poisson bracket.

LEMMA 5.11. *Let* rank $\widetilde{B} = n$, *then the cluster manifold* $\mathcal{X}$ *possesses a Poisson bracket compatible with the cluster algebra* $\mathcal{A}(\widetilde{B})$.

PROOF. Definitions and Theorem 4.5 imply immediately the existence of a compatible Poisson bracket defined on rational functions on $\mathcal{X}$. This bracket is extended naturally to a compatible Poisson bracket on $C^\infty(\mathcal{X})$. □

Let $\{\cdot,\cdot\}$ be one of these Poisson brackets. The smooth manifold $\mathcal{X}$ is foliated into a disjoint union of symplectic leaves of $\{\cdot,\cdot\}$. In this section we are interested in generic symplectic leaves.

We start with the description of genericity conditions. Recall that a *Casimir element* of $\{\cdot,\cdot\}$ is a function that is in involution with all the other functions on $\mathcal{X}$. All rational Casimir functions form a subfield $\mathbb{F}_C$ in the field of rational functions $\mathbb{F}(\mathcal{X})$. Fix some generators $q_1,\dots,q_s$ of $\mathbb{F}_C$, where s is the corank of $\{\cdot,\cdot\}$. They define a map $Q\ :\mathcal{X}\to\mathbb{F}^s$, $Q(x)=(q_1(x),\dots,q_s(x))$. Let $\mathcal{L}$ be a symplectic leaf. We say that $\mathcal{L}$ is *generic* if there exist s vector fields u_i in a neighborhood of $\mathcal{L}$ such that

(i) at every point $x^*\in\mathcal{L}$, the vector $u_i(x^*)$ is transversal to the surface $q_i(x)=q_i(x^*)$, i.e., $\nabla_{u_i}q_i(x^*)\neq 0$;

(ii) the translation along u_i for a sufficiently small time t determines a diffeomorphism between $\mathcal{L}$ and a close symplectic leaf $\mathcal{L}_t$.

THEOREM 5.12. $\mathcal{X}_0$ *is a regular Poisson manifold foliated into a disjoint union of generic symplectic leaves of the Poisson bracket* $\{\cdot,\cdot\}$.

PROOF. First note that for any $x\in\mathcal{X}_0$ there exists a chart U^v such that x_j^v does not vanish at x for all $j\in[1,n+m]$. Therefore we can consider functions $\log x_j^v$ as a local coordinate system on U^v. The Poisson structure $\{\cdot,\cdot\}$ written in these coordinates becomes a constant Poisson structure. Since the rank of a constant Poisson structure is the same at every point, we conclude that $\{\cdot,\cdot\}$ has a constant rank on $\mathcal{X}_0$. In other words, $\mathcal{X}_0$ is a *regular* Poisson submanifold of $\mathcal{X}$.

If a Poisson structure on $\mathcal{X}$ is *generically symplectic*, i.e. when its rank equals the dimension of $\mathcal{X}$ at a generic point of $\mathcal{X}$, the observation above shows that every point x of $\mathcal{X}_0$ is generic. Moreover, note that the complement $\mathcal{X}\setminus\mathcal{X}_0$ consists of degenerate symplectic leaves of smaller dimensions. Hence, if the Poisson structure is generically symplectic then $\mathcal{X}_0$ is the union of its generic symplectic leaves.

Let us assume that the rank of $\{\cdot,\cdot\}$ equals $r<n$. In this case, there exist $s=n-r$ Casimir functions that generate the field $\mathbb{F}_C$. The following proposition describes $\mathbb{F}_C$.

LEMMA 5.13. $\mathbb{F}_C=\mathbb{F}(\mu_1,\dots,\mu_s)$, *where* μ_j *has a monomial form*

$$\mu_j=\prod_{i=m+1}^{n} x_i^{\alpha_{ji}}$$

for some integer α_{ji}.

PROOF. Fix a vertex $v \in \mathbb{T}_n$. For the sake of simplicity, we will omit the dependence on v when it does not cause confusion. Define the τ-coordinates as in Section 4.1.2 and note that each τ_i is distinct from 0 and from ∞ in U_0. Therefore, $\log \tau_i$ form a coordinate system in U_0, and $\{\log \tau_p, \log \tau_q\} = \omega^\tau_{pq}$, where $\Omega^\tau = (\omega^\tau_{pq})$ is the coefficient matrix of $\{\cdot,\cdot\}$ in the basis τ. The Casimir functions of $\{\cdot,\cdot\}$ that are linear in $\{\log \tau_i\}$ are given by the left nullspace $N_\mathfrak{l}(\Omega^\tau)$ in the following way. Since Ω^τ is an integer matrix, its left nullspace contains an integral lattice L. For any vector $u = (u_1, \dots, u_{n+m}) \in L$, the sum $\sum_i u_i \log \tau_i$ is in involution with all the coordinates $\log \tau_j$. Hence the product $\prod_{i=1}^{n+m} \tau_i^{u_i}$ belongs to $\mathbb{F}_C$; moreover $\mathbb{F}_C$ is generated by the monomials $\mu^u = \prod_{i=1}^{n+m} \tau_i^{u_i}$ for s distinct vectors $u \in L$.

Let us calculate $\log \mu^u = u \log \tau^T$, where $\log \tau = (\log \tau_1, \dots, \log \tau_{n+m})$. Recall that by (4.1), $\log \tau^T = (\widetilde{B} + K) \log \widetilde{\mathbf{x}}^T$, where K is a diagonal matrix whose first n diagonal entries are equal to zero, and $\log \widetilde{\mathbf{x}} = (\log x_1, \dots, \log x_{n+m})$. So $u \log \tau^T = \alpha \log \widetilde{\mathbf{x}}^T$, where $\alpha = u(\widetilde{B} + K)$. Further, consider the decompositions

$$\widetilde{B} = \begin{pmatrix} B_1 & B_2 \\ -B_2^T D & B_4 \end{pmatrix}, \qquad \Omega^\tau = \begin{pmatrix} \Omega_1 & \Omega_2 \\ -\Omega_2^T & \Omega_4 \end{pmatrix},$$

where B_1 and Ω_1 are $n \times n$ submatrices (and hence $B_1 = B$). By Theorem 4.5, $B_1 = \Lambda^{-1}\Omega_1 D$ and $B_2 = \Lambda^{-1}\Omega_2$; moreover it is easy to check that Λ and Ω_1 commute. Let $u = (u^1 \;\; u^2)$ and $\alpha = (\alpha^1 \;\; \alpha^2)$ be the corresponding decompositions of u and α. Since $u\Omega^\tau = 0$, one has $u^1\Omega_1 - u^2\Omega_2^T = 0$. Therefore $\alpha^1 = u^1 B_1 - u^2 B_2^T D = (u^1\Omega_1 - u^2\Omega_2^T)\Lambda^{-1} D = 0$, and hence $\mu^u = \prod_{i=m+1}^{n} x_i^{\alpha_i}$. Finally, $U_0 = U_0^v$ is open and dense in U^v, so the above defined meromorphic Casimir functions can be extended to the whole U^v, and hence to $\mathcal{X}$. □

Following Lemma 5.13, choose an integer vector u in the left nullspace $N_\mathfrak{l}(\Omega^\tau)$ and take the corresponding monomials μ^u. Observe that if $u = (u^1 \;\; u^2) \in N_\mathfrak{l}(\Omega^\tau)$ and $u' = (D^{-1}u^1 \;\; u^2)$, then $(u')^T \in N_\mathfrak{r}(B)$. Therefore, by Lemma 5.3, u' as above defines a toric flow on $\mathcal{X}$. To accomplish the proof it is enough to show that the toric flow corresponding to the vector u' is transversal to the level surface $\{y \in \mathcal{X}_0 : \mu^u(y) = \mu^u(x)\}$, and that a small translation along the trajectory of the toric flow transports one symplectic leaf to another one.

We will first show that if $x(t)$ is a trajectory of the toric flow corresponding to u' with the initial value $x(1) = x$ and the initial velocity vector

$$\nu = dx(t)/dt|_{t=1} = (u'_1 x_1, \dots, u'_{n+m} x_{n+m})$$

then $d\mu^u(\nu)/dt \neq 0$. Indeed, by Lemma 5.13,

$$d\mu^u(\nu)/dt = \sum_{i=n+1}^{n+m} \alpha_i \frac{\prod_{j=n+1}^{n+m} x_j^{\alpha_j}}{x_i} \cdot u_i x_i = \mu^u \alpha^2 (u^2)^T.$$

Since $x \in \mathcal{X}_0$, one has $\mu^u(x) \neq 0$. To find $\alpha^2(u^2)^T$ recall that by the proof of Lemma 5.13, $\alpha^2 = u^1 B_2 + u^2 B_4 + u^2 K'$, where K' is the submatrix of K whose entries lie in the last n rows and columns. Clearly, $u^2 B_4 (u^2)^T = 0$, since B_4 is skew-symmetric. Next,

$$u^1 B_2 (u^2)^T = u^1 \Lambda^{-1} \Omega_2 (u^2)^T = u^1 \Lambda^{-1} \Omega_1^T (u^1)^T = 0,$$

since $u^2\Omega_2^T = u^1\Omega_1$ and $\Lambda^{-1}\Omega_1^T$ is skew-symmetric. Thus, $\alpha^2(u^2)^T = u^2 K' (u^2)^T \neq 0$, since K' can be chosen to be a diagonal matrix with positive elements on the diagonal, see the proof of Lemma 4.3.

Consider now another basis vector $\bar{u} \in N_{\mathfrak{l}}(\Omega^\tau)$ and the corresponding Casimir function $\mu^{\bar{u}}$. It is easy to see that $d\mu^{\bar{u}}(\nu)/dt = \mu^{\bar{u}}\bar{\alpha}^2(u^2)^T$. Note that the latter expression does not depend on the point x, but only on the value $\mu^{\bar{u}}(x)$ and on the vectors $\bar{\alpha}$ and u. Therefore the value of the derivative $d\mu^{\bar{u}}/dt$ is the same for all points x lying on the same symplectic leaf, and the toric flow transforms one symplectic leaf into another one. □

EXAMPLE 5.14. It is not true, in general, that $\mathcal{X}_0$ coincides with the union of all "generic" symplectic leaves. A simple counterexample is provided by the cluster algebra of rank 1 containing two clusters $\{x_1, x_2, x_3\}$ and $\{x_1', x_2, x_3\}$ subject to the relation: $x_1'x_1 = x_2^2x_3^2 + 1$. By Theorem 4.5 we can choose a compatible Poisson bracket on $\mathcal{X}$ as follows: $\{x_1, x_2\} = x_1x_2$, $\{x_1, x_3\} = x_1x_3$, $\{x_2, x_3\} = 0$. Equivalently, $\{x_1', x_2\} = -x_1'x_2$, $\{x_1', x_3\} = -x_1'x_3$, $\{x_2, x_3\} = 0$. Generic symplectic leaves of this Poisson structure are solutions of the equation $Ax_2+Bx_3 = 0$ where $(A : B)$ is a homogeneous coordinate on P^1. Thus, the set of all generic symplectic leaves is P^1. In particular, two leaves $(1 : 0)$ and $(0 : 1)$ (equivalently, subsets $x_2 = 0, x_3 \neq 0, x_1x_1' = 1$ and $x_3 = 0, x_2 \neq 0, x_1x_1' = 1$) are generic symplectic leaves in $\mathcal{X}$. According to the definition of $\mathcal{X}_0$ these leaves are not contained in $\mathcal{X}_0$.

5.5. The number of connected components of refined Schubert cells in real Grassmannians

Let $S_k^{\mathbb{R}}(n)$ be the real part of the locus $S_k(n)$ defined in Proposition 4.11. Since $S_k^{\mathbb{R}}(n)$ is an intersection of n Schubert cells in general position, we call it a *real refined Schubert cell.* According to Proposition 4.11, it is the union of generic symplectic leaves of the real toric action (4.22). Below we explain how to find the number of connected components of a real refined Schubert cell.

THEOREM 5.15. *The number of connected components of a real refined Schubert cell $S_k^{\mathbb{R}}(n)$ is equal to $3 \cdot 2^{n-1}$ if $k > 3$, $n - k > 3$.*

PROOF. Note that regular orbits of action (4.22) in $G_k^0(n)$ and $\mathcal{X}_0(\mathcal{A}(G_k^0(n)))$ are given by the same nonvanishing conditions for the dense minors as $S_k(n)$. Let U be given by (4.31), then $U \subset \mathcal{X}(G_k^0(n) \subset G_k^0(n)$. By the proof of Theorem 4.14 we know that the complement to U in $G_k^0(n)$ has codimension 2. codimension Therefore, $\mathcal{X}_0(\mathcal{A}(G_k^0(n))) \hookrightarrow S_k(n)$, and its complement has codimension 2 in $S_k(n)$. It follows that the number of connected component of $S_k^{\mathbb{R}}(n)$ equals the number of connected components of the real part of $\mathcal{X}_0(\mathcal{A}(G_k^0(n)))$.

By Theorem 5.9 we know that the number of connected components of the real part of $\mathcal{X}_0(\mathcal{A}(G_k^0(n)))$ equals the number of orbits of Γ-orbits in $\mathbb{F}_2^{k(n-k)}$, where Γ is defined by the graph shown on Fig. 4.2 and the cluster variables correspond to all vertices except for the first column and the last row. Since in the case $k \geq 4$, $n \geq 8$ the subgraph spanned by these vertices is evidently E_6-compatible, Theorem 5.10 implies that to prove the statement it is enough to show that $\mathbb{F}_2^B \cap \ker \eta = 0$; in other words, that there is no nontrivial vector in $\ker \eta$ with vanishing stable components.

Indeed, let us denote such a vector by $h \in \mathbb{F}_2^B$, and let δ_{ij} be the (i, j)-th basis vector in $\mathbb{F}_2^{k(n-k)}$. Note that the condition $\eta(h, \delta_{k,n-k}) = 0$ implies that $h_{k-1,n-k} = 0$. Further, assuming $h_{k-1,n-k} = 0$ we see that the condition $\eta(h, \delta_{k,n-k-1}) = 0$ implies $h_{k-1,n-k-1} = 0$ and so on. Since $\eta(h, \delta_{kj}) = 0$ for any $j \in [1, n - k]$, we conclude that $h_{k-1,j} = 0$ for any $j \in [1, n-k]$. Proceeding by induction we prove that $h_{ij} = 0$ for any $i \in [1, k]$ and any $j \in [1, n-k]$. Hence $h = 0$. Note that the number of stable variables equals $n - 1$, which accomplishes the proof of the statement. □

By the standard duality between $G_k(n)$ and $G_{n-k}(n)$ we can make a conclusion that the number of connected components for $S_k^{\mathbb{R}}(n)$ equals the number of connected components for $S_{n-k}^{\mathbb{R}}(n)$. Therefore, taking in account Theorem 5.15, in order to find the number of connected components for real refined Schubert cells for all Grassmannians we need to consider only two remaining cases: $G_2(n)$ and $G_3(n)$. To accomplish our investigation let us mention the following result.

PROPOSITION 5.16. *The number of connected components of a real refined Schubert cell equals to* $(n-1) \cdot 2^{n-2}$ *for* $G_2(n)$*,* $n \geq 3$*, and equals to* $3 \cdot 2^{n-1}$ *for* $G_3(n)$ *for* $n \geq 6$*.*

5.6. Summary

- We define a cluster manifold for a general cluster algebra of geometric type. It is a smooth (generally speaking, non-compact) rational manifold endowed with toric actions compatible with the cluster algebra. These actions correspond to integer vectors in the right nullspace of the exchange matrix, see Lemma 5.3.
- The number of connected components of the regular locus of the toric action over $\mathbb{R}$ can be computed as the number of orbits of a linear group action in a finite dimensional vector space over the finite field $\mathbb{F}_2$, see Theorem 5.9.
- If the exchange matrix is of the full rank, the cluster manifold can be endowed with a compatible Poisson structure, while the nonsingular locus of the toric action is foliated into a union of generic symplectic leaves of this structure, see Theorem 5.12.
- As an application of these ideas, we calculate the number of connected components of a real refined Schubert cell, see Theorem 5.15.

Bibliographical notes

Exposition in this chapter closely follows [**GSV2**]. Relation (5.3) in the proof of Lemma 5.4 is borrowed from [**FZ2**]. Theorem 5.10 is proved in [**SSVZ**]; for a further generalization of this result see [**Se**]. Proposition 5.16 is proved in [**GSV2**] based on techniques from [**GSV1**].

CHAPTER 6

Pre-symplectic structures compatible with the cluster algebra structure

In Chapter 4 we have discussed Poisson properties of cluster algebras of geometric type with exchange matrices of full rank. Now we are going to relax the maximal rank condition for exchange matrix. Example 4.2 shows that in degenerate case a compatible Poisson structure may not exist. The main idea is that the relevant geometric object in this case is a pre-symplectic structure compatible with the cluster algebra structure. We also make efforts to restore cluster algebra exchange relations given such a pre-symplectic structure as "canonical" transformations satisfying some "natural" conditions.

6.1. Cluster algebras of geometric type and pre-symplectic structures

6.1.1. Characterization of compatible pre-symplectic structures. Let ω be a *pre-symplectic structure* (that is, a closed differential 2-form) on an $(n+m)$-dimensional rational manifold. We say that functions $g_1, \dots, g_{n+m}$ are *log-canonical* with respect to ω if

$$\omega = \sum_{i,j=1}^{n+m} \omega_{ij} \frac{dg_i}{g_i} \wedge \frac{dg_j}{g_j},$$

where ω_{ij} are constants. The matrix $\Omega^g = (\omega_{ij})$ is called the *coefficient matrix* of ω (with respect to g); evidently, Ω^g is skew-symmetric. We say that a pre-symplectic structure ω on a rational manifold is *compatible* with the cluster algebra $\mathcal{A}(Z)$ if all clusters in $\mathcal{A}(Z)$ are log-canonical with respect to ω.

EXAMPLE 6.1. Consider the same seed as in Example 4.2 above. Recall that it consists of a cluster $\mathbf{x} = (x_1, x_2, x_3)$ and a 3×3 skew-symmetric matrix $\widetilde{B} = B = \begin{pmatrix} 0 & 1 & -1 \\ -1 & 0 & 1 \\ 1 & -1 & 0 \end{pmatrix}$, and that the exchange relations imply

$$x_1' = \frac{x_2 + x_3}{x_1}, \qquad x_2' = x_2, \qquad x_3 = x_3.$$

Let ω be a closed 2-form compatible with the coefficient free cluster algebra $\mathcal{A}(B)$. Then, by the definition of compatibility,

$$\omega = \lambda \frac{dx_1}{x_1} \wedge \frac{dx_2}{x_2} + \mu \frac{dx_2}{x_2} \wedge \frac{dx_3}{x_3} + \nu \frac{dx_1}{x_1} \wedge \frac{dx_3}{x_3}.$$

Besides,

$$\frac{dx_1}{x_1} = -\frac{dx_1'}{x_1'} + \frac{dx_2}{x_2 + x_3} + \frac{dx_3}{x_2 + x_3},$$

and hence

$$\omega = -\lambda \frac{dx'_1}{x'_1} \wedge \frac{dx_2}{x_2} + \left(\mu + \frac{\lambda x_3 - \nu x_2}{x_2 + x_3}\right) \frac{dx_2}{x_2} \wedge \frac{dx_3}{x_3} - \nu \frac{dx'_1}{x'_1} \wedge \frac{dx_3}{x_3}.$$

By the compatibility, $\mu + \frac{\lambda x_3 - \nu x_2}{x_2 + x_3}$ is a constant, which is only possible when $\lambda = -\nu$. In a similar way one gets $\lambda = \mu$, and hence the coefficient matrix of a compatible 2-form with respect to $\mathbf{x}$ is given by $\Omega^{\mathbf{x}} = \lambda B$.

Recall that a square matrix A is *reducible* if there exists a permutation matrix P such that PAP^T is a block-diagonal matrix, and *irreducible* otherwise; $\rho(A)$ is defined as the maximal number of diagonal blocks in PAP^T. The partition into blocks defines an obvious equivalence relation $\sim$ on the rows (or columns) of A.

The following result, which can be considered as a natural analog of Theorem 4.5, shows that the answer in Example 6.1 is not just a coincidence.

THEOREM 6.2. *Assume that $\widetilde{B}$ is D-skew-symmetrizable and does not have zero rows. Then all pre-symplectic structures compatible with $\mathcal{A}(\widetilde{B})$ form a vector space of dimension $\rho(B) + \binom{m}{2}$, where $B = \widetilde{B}[n;n]$. Moreover, the coefficient matrices of these pre-symplectic structures with respect to $\widetilde{\mathbf{x}}$ are characterized by the equation $\Omega^{\widetilde{\mathbf{x}}}[n; n+m] = \Lambda D\widetilde{B}$, where $\Lambda = \operatorname{diag}(\lambda_1, \dots, \lambda_n)$ with $\lambda_i = \lambda_j$ whenever $i \sim j$. In particular, if B is irreducible, then $\Omega^{\widetilde{\mathbf{x}}}[n; n+m] = \lambda D\widetilde{B}$.*

PROOF. Indeed, let ω be a 2-form compatible with $\mathcal{A}(B)$. Then

$$\omega = \sum_{j,k=1}^{n+m} \omega_{jk} \frac{dx_j}{x_j} \wedge \frac{dx_k}{x_k} = \sum_{j,k=1}^{n+m} \omega'_{jk} \frac{dx'_j}{x'_j} \wedge \frac{dx'_k}{x'_k},$$

where x'_j is given by (3.1) and ω'_{jk} are the coefficients of ω with respect to $\widetilde{\mathbf{x}}'$. Recall that the only variable in the extended cluster $\widetilde{\mathbf{x}}'$ different from the corresponding variable in $\widetilde{\mathbf{x}}$ is x_i, and

$$\frac{dx'_i}{x'_i} = -\frac{dx_i}{x_i} + \sum_{b_{ik}>0} \frac{b_{ik}}{1 + \prod_{k=1}^{n+m} x_k^{-b_{ik}}} \frac{dx_k}{x_k} - \sum_{b_{ik}<0} \frac{b_{ik}}{1 + \prod_{k=1}^{n+m} x_k^{b_{ik}}} \frac{dx_k}{x_k}.$$

Thus, for any $j \in [1, n+m]$ we immediately get

$$\omega'_{ij} = -\omega_{ij}. \tag{6.1}$$

Next, consider any pair $j, k \neq i$ such that both b_{ij} and b_{ik} are nonnegative, and at least one of the two is positive. Then

$$\omega_{jk} = \omega'_{jk} + \frac{\omega'_{ik} b_{ij} + \omega'_{ji} b_{ik}}{1 + \prod_{k=1}^{n+m} x_k^{-b_{ik}}}.$$

This equality can only hold if $\omega'_{ik} b_{ij} + \omega'_{ji} b_{ik} = 0$, which by (6.1) is equivalent to $\omega_{ij} b_{ik} = \omega_{ik} b_{ij}$. If both b_{ij} and b_{ik} are positive, this gives

$$\frac{\omega_{ij}}{b_{ij}} = \frac{\omega_{ik}}{b_{ik}} = \mu_i. \tag{6.2}$$

Otherwise, if, say, $b_{ij} = 0$, one gets $\omega_{ij} = 0$. Besides, in any case

$$\omega'_{jk} = \omega_{jk}. \tag{6.3}$$

Similarly, if both b_{ij} and b_{ik} are nonpositive, and at least one of the two is negative, then

$$\omega_{jk} = \omega'_{jk} - \frac{\omega'_{ik} + \omega'_{ji} b_{ik}}{1 + \prod_{k=1}^{n+m} x_k^{b_{ik}}},$$

and hence the same relations as above hold.

Finally, let $b_{ij} \cdot b_{ik} < 0$, say, $b_{ij} > 0$ and $b_{ik} < 0$; then

$$\omega_{jk} = \omega'_{jk} + \frac{\omega'_{ik} b_{ij} \prod_{k=1}^{n+m} x_k^{b_{ik}} - \omega'_{ji} b_{ik}}{1 + \prod_{k=1}^{n+m} x_k^{b_{ik}}},$$

which again leads to (6.2); the only difference is that in this case

$$\omega'_{jk} = \omega_{jk} + \omega_{ik} b_{ij}. \tag{6.4}$$

We have thus obtained that $\Omega^{\widetilde{\mathbf{x}}}[n; n+m] = \operatorname{diag}(\mu_1, \dots, \mu_n)\widetilde{B}$. Recall that $\Omega^{\widetilde{\mathbf{x}}}$ is skew-symmetric; however, any skew-symmetrizer for $\widetilde{B}$ can be written as ΛD, where $\Lambda = \operatorname{diag}(\lambda_1, \dots, \lambda_n)$ and $\lambda_i = \lambda_j$ whenever $i \sim j$, which completes the proof. □

It is worth mentioning that relations (6.1), (6.3),(6.4) are equivalent to the matrix mutation rules from Definition 3.5.

6.1.2. Secondary cluster manifold and the Weil-Petersson form. In this section we restrict ourself to the skew-symmetric case with no stable variables, i.e., $m = 0$ and the exchange matrix is a square $n \times n$ skew-symmetric matrix. Consider the cluster manifold $\mathcal{X}$ introduced and studied in Chapter 5; recall that by Lemma 5.2, $\mathcal{X}$ is rational. Let ω be one of the forms described in Theorem 6.2. Note that ω is degenerate whenever $r = \operatorname{rank} B < n$. Kernels of ω form an integrable distribution (since it becomes linear in the coordinates $\log x$), so one can define the quotient mapping π^ω. Observe that this mapping does not depend on the choice of ω, provided ω is generic (that is, the corresponding diagonal matrix Λ is nonsingular), since the kernels of all generic ω coincide. The image of the quotient mapping is called the *secondary cluster manifold* and denoted $\mathcal{Y}$.

PROPOSITION 6.3. *The secondary cluster manifold is a rational manifold of dimension* r.

PROOF. Indeed, note that by construction $\mathcal{X}$ is covered by a collection of open dense charts $\{U_i\}$. Each chart U_i is isomorphic to a torus $U_i \simeq (\mathbb{C} \setminus 0)^n$, which is fibered by kernels of ω in such a way that the quotient $U_i^\omega \simeq (\mathbb{C} \setminus 0)^r$ is well defined. Since every chart U_i is dense in $\mathcal{X}$, the intersection of any two charts $U_i \cap U_j$ is an open dense subset of $\mathcal{X}$ as well. We claim that a well defined global quotient space $\mathcal{Y}$ is obtained by gluing open charts U_i^ω. To see that let us show that for any i no two distinct points of U_i^ω are identified by the gluing procedure.

Note first that it is enough to prove that no two distinct fibers of U_i that intersect a fixed dense subset V of U_i are identified. Then the claim follows by continuity. Define V as the intersection $\cap_j U_j$. Clearly, V is dense in U_i as the intersection of countably many open dense subsets of U_i. Assume that we can find a sequence of charts $U_{j_0} = U_i, U_{j_1}, \dots, U_{j_s} = U_i$ and two distinct fibers f_1 and f_2 of U_i such that $f_1 \cap V \neq \varnothing$, $f_2 \cap V \neq \varnothing$, and f_1 is identified with f_2 after consecutive identifications of fibers in U_{j_k} and $U_{j_{k+1}}$, $k = 0, \dots, s-1$. Choose a point $p \in f_1 \cap V$. By the definition of V, $p \in U_j$ for all j, and hence the fiber of U_j identified with f_1 via the consecutive gluing procedure is the one that contains p.

By construction, if a fiber f of the whole $\mathcal{X}$ has a nonempty intersection with a chart U_j, then $f \cap U_j$ coincides with f up to real codimension 2. It follows that if two points $p_1, p_2 \in U_j \cap U_l$ belong to the same fiber of U_l (i.e., represent the same element of U_l^ω), then they both belong to the same fiber of U_j (i.e., represent the same element of U_j^ω). Therefore, the consecutive identification of the fiber of U_{j_k} containing $p \in V$ with the corresponding fiber of $U_{j_{k+1}}$ can not identify points of two distinct fibers in U_i, meaning that $\mathcal{Y}$ is well defined.

Finally, we note that any chart U_i^ω is an open dense subset of $\mathcal{Y}$, which implies rationality of $\mathcal{Y}$. □

Clearly, the symplectic reduction $\pi_*^\omega \omega$ of ω is well defined; moreover, $\pi_*^\omega \omega$ is a symplectic form on $\mathcal{Y}$.

Let us introduce coordinates on the secondary cluster manifold $\mathcal{Y}$ that are naturally related to the initial coordinates $\mathbf{x}$ on $\mathcal{X}$. To do this, we use the τ-coordinates introduced in Chapter 4. Recall that by (4.1),

$$\tau_j = \prod_{k=1}^{n} x_k^{b_{jk}}, \qquad j \in [1, n].$$

It is easy to see that elements of the τ-cluster are no longer functionally independent. To get a functionally independent subset one has to choose at most r entries in such a way that the corresponding rows of B are linearly independent.

Clearly, any choice of r functionally independent coordinates τ provides log-canonical coordinates for the 2-form $\pi_*^\omega \omega$. Besides, $\pi_*^\omega \omega$ is compatible with the cluster algebra transformations. Moreover, if γ is any other 2-form on $\mathcal{Y}$ compatible with the cluster algebra transformations, then its pullback is a compatible 2-form on the preimage $\mathcal{X}$. Thus, we get the following result.

COROLLARY 6.4. *Let B be irreducible, then there exists a unique, up to a constant factor, closed 2-form on the secondary cluster manifold $\mathcal{Y}$ compatible with $\mathcal{A}(B)$. Moreover, this form is symplectic.*

The unique symplectic form described in the above corollary is called the *Weil-Petersson* form associated with the cluster algebra $\mathcal{A}(B)$. The reasons for this name will become clear in Section 6.2.3 below.

Let us discuss the relation between the Poisson brackets and closed 2-forms compatible with the cluster algebra structure. Fix a multiindex I of length r such that the rank of the corresponding $r \times n$ submatrix equals r and consider the restriction $\mathcal{A}_I(B)$ of $\mathcal{A}(B)$ obtained by declaring all the variables x_i, $i \notin I$, stable. By Theorem 4.5, there exists a linear space of Poisson structures compatible with $\mathcal{A}_I(B)$. To distinguish a unique, up to a constant factor, Poisson structure, we consider the restriction of the form ω to an affine subspace $x_i = \text{const}$, $i \notin I$. This restriction is a symplectic form, since the kernels of ω are transversal to the above subspace, and it is compatible with $\mathcal{A}_I(B) \subset \mathcal{A}(B)$. Its dual is therefore a Poisson structure on this affine subspace, compatible with $\mathcal{A}_I(B)$. By the same transversality argument, we can identify each affine subspace $x_i = \text{const}$, $i \notin I$, with the secondary cluster manifold, and hence each of the obtained Poisson structures can be identified with the dual to the Weil-Petersson symplectic form.

In Chapter 4 we have considered an alternative description of the Poisson structure dual to $\pi^\omega_* \omega$. It follows from the proof of Theorem 4.5 that it is the unique (up to a nonzero scalar factor) Poisson structure $\{\cdot,\cdot\}$ on $\mathcal{Y}$ satisfying the following condition: for any Poisson structure on $\mathcal{X}$ compatible with any reduced cluster algebra $\mathcal{A}_I(B)$, the projection $p\colon \mathcal{X} \to \mathcal{Y}$ is Poisson. The basis τ_i is log-canonical with respect to $\{\cdot,\cdot\}$, and $\{\tau_i, \tau_j\} = \lambda z_{ij}\tau_i\tau_j$ for some constant λ.

6.2. Main example: Teichmüller space

Let Σ be a smooth closed orientable 2-dimensional surface of genus g, S be a finite set of marked points, $|S| = s > 2 - 2g$. For technical reasons we also exclude the case $g = 0$, $s = 3$.

6.2.1. Teichmüller space, horocycles, decorated Teichmüller space. Let $\mathcal{T}_g^s$ denote the *Teichmüller space* of complex structures on Σ degenerating at marked points modulo the action of the connected component of identity in the group of diffeomorphisms of Σ. We require a tubular neighborhood of each marked point to be isomorphic as a complex manifold to a punctured disc; in what follows we call these marked points *punctures*. By the uniformization theorem, Σ is identified with the quotient of the complex upper half-plane $\mathbb{H}$ modulo the action of a Fuchsian group Γ. A *horocycle* centered at a point p on the real axis $\partial\mathbb{H}$ is a circle orthogonal to any geodesic passing through p.

If p is a puncture of Σ, then the stabilizer of p is generated by a parabolic element $\varphi \in \Gamma$. Let γ be a geodesic in $\mathbb{H}$ through p, then $\varphi(\gamma)$ is a geodesic through p as well. Denote the interval of a horocycle h centered at p between geodesics γ and $\varphi(\gamma)$ by $\hat{h}$. Note that h is orthogonal to both γ and $\varphi(\gamma)$. After taking the quotient modulo Γ, the interval $\hat{h}$ becomes a closed curve on Σ around p—the image of the horocycle on Σ that we, abusing notation, will also denote $\hat{h}$. The *length* of $\hat{h}$ with respect to the hyperbolic metric is denoted by $L(h)$. Finally, two horocycles h_1 and h_2 centered at the same puncture p are called *conjugate* if $L(h_1)\cdot L(h_2) = 1$.

The *decorated* Teichmüller space $\widetilde{\mathcal{T}}_g^s$ classifies hyperbolic metrics on Σ with a chosen horocycle centered at each of the marked points; $\widetilde{\mathcal{T}}_g^s$ is a fiber bundle over $\mathcal{T}_g^s$ whose fiber is $\mathbb{R}_+^s$.

Let us recall here a construction of coordinates on the decorated Teichmüller space. An *ideal arc* is a simple (possibly closed) curve in Σ considered up to isotopy relative S with endpoints at marked points such that its relative interior is disjoint from S. A trivial ideal arc is the isotopy class of a marked point. Two ideal arcs are *compatible* if there exist two curves in the corresponding isotopy classes that do not intersect outside S. An *ideal triangulation* Δ is a maximal collection of distinct pairwise compatible nontrivial ideal arcs. It is well-known that if two ideal arcs are compatible, then the corresponding geodesics do not intersect outside S. These geodesics are called the *edges* of Δ. The edges of Δ cut Σ into *ideal triangles*, which are images of ideal triangles in the complex upper-half plane under the uniformization map; observe that two sides of an ideal triangle on Σ may coincide (see the punctured monogon on Fig. 6.1).

Any edge $e \in \mathrm{Edge}(\Delta)$ has an infinite hyperbolic length, however the hyperbolic length of its segment between a pair of horocycles centered at the endpoints of e is finite. Let us denote by $l(e)$ the signed length of this segment, i.e., we take the

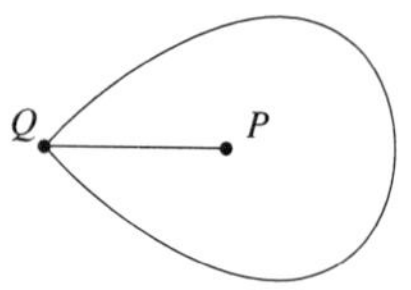

FIGURE 6.1. Punctured monogon: a triangle with two coinciding sides

length of e with the positive sign if the horocycles do not intersect, and with the negative sign if they do, and define $f(e) = \exp(l(e)/2)$.

The following result is well known.

THEOREM 6.5. *For any ideal geodesic triangulation Δ and any choice of horocycles around the punctures, the functions $f(e)$, $e \in \mathrm{Edge}(\Delta)$, define a homeomorphism between the decorated Teichmüller space $\widetilde{\mathcal{T}}_g^s$ and $\mathbb{R}_+^{6g-6+3s}$.*

The coordinates described in Theorem 6.5 are called the *Penner coordinates* on $\widetilde{\mathcal{T}}_g^s$.

6.2.2. Whitehead moves and the cluster algebra structure. Given a triangulation of Σ, one can obtain a new triangulation via a sequence of simple transformations called *flips*, or *Whitehead moves*, see Fig. 6.2 (a very simple particular case of Whitehead moves was already discussed in Section 2.1.3). Note that some of the sides a, b, c, d can coincide, while p is distinct from any of them.

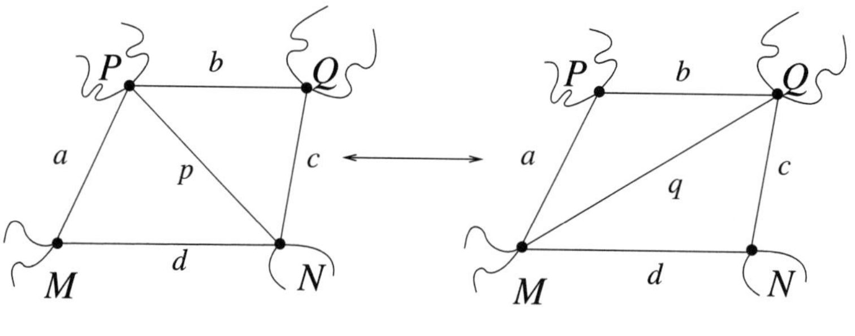

FIGURE 6.2. Whitehead move

Under such a transformation the Penner coordinates undergo the following change described by the famous *Ptolemy relation*:

$$f(p)f(q) = f(a)f(c) + f(b)f(d). \tag{6.5}$$

Note that this transformation looks very much like a cluster algebra transformation; we would like to make this statement more precise. Let us call an ideal triangulation of Σ *nice* if all the edges of any triangle are pairwise distinct. We call a nice triangulation *perfect* if, moreover, all vertices have at least three incident edges (it is assumed that loops are counted with multiplicity 2). We will consider Penner coordinate systems only for nice triangulations. Note that every edge in a perfect triangulation borders two distinct triangles with different edges. So, a and c can coincide with each other, but not with any of b and d; similarly, b and d can coincide as well.

Therefore in this situation the Ptolemy relation holds for any edge. In general, flips do not preserve the perfectness of a triangulation; however, the result of a flip of a perfect triangulation is a nice one.

Given a perfect triangulation Δ, we construct the following coefficient-free cluster algebra $\mathcal{A}(\Delta)$: the Penner coordinates are cluster coordinates, and the transformation rules are defined by Ptolemy relations. The transformation matrix $B(\Delta)$ is determined by Δ in the following way.

Let $\nu_P(a,b)$ be the number of occurrences of the edge b immediately after a in the counterclockwise order around vertex P. For any pair of edges $a, b \in \mathrm{Edge}(\Delta)$, put

$$B(\Delta)_{ab} = \sum_{P \in \mathrm{Vert}(\Delta)} (\nu_P(a,b) - \nu_P(b,a)), \tag{6.6}$$

and define $\mathcal{A}(\Delta) = \mathcal{A}(B(\Delta))$.

For example, the left part of Fig. 6.3 shows a triangulation of the sphere with four marked points. Here $\nu_P(a,b) = \nu_P(b,a) = 1$ and $\nu(a,b) = \nu(b,a) = 0$ at all other marked points, hence $B(\Delta)_{ab} = 0$. The right part of Fig. 6.3 shows a triangulation of the torus with one marked point. Here $\nu_P(a,b) = 0$, $\nu_P(b,a) = 2$, and hence $B(\Delta)_{ab} = -2$.

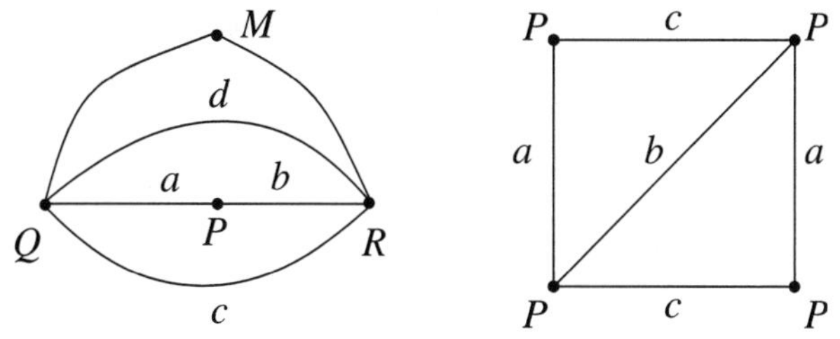

FIGURE 6.3. To the definition of $B(\Delta)$

Let $\Delta \mapsto \Delta'$ be a flip shown on Fig. 6.2. By the definition of $\mathcal{A}(\Delta)$ we see that cluster variables in $\mathcal{A}(\Delta)$ corresponding to the cluster obtained from the initial one by the transformation T: $f(p) \mapsto f(q)$ coincide with Penner coordinates for the flipped triangulation Δ'. To check that the flip is indeed a cluster algebra transformation, it is enough to check that the transformation matrix B' is determined by the adjacency of edges in Δ' in the same way as $B = B(\Delta)$ above.

To prove this, recall that the new transformation matrix B' is obtained from B by the matrix mutation rules from Definition 3.5. Assume first that the degrees of both endpoints of p are at least three; the Whitehead move in this case is shown on Fig. 6.2.

Note that the image of the diagonal p under the flip is the diagonal q. Using Definition 3.5 we obtain $B'_{ad} = B_{ad} - 1$, $B'_{ba} = B_{ba} + 1$, $B'_{cb} = B_{cb} - 1$, $B'_{dc} = B_{dc} + 1$, $B'_{xq} = -B_{xp}$, where x runs over $\{a, b, c, d\}$.

For instance, substituting the values $B_{da} = B_{ap} = B_{pd} = B_{cp} = B_{pb} = B_{bc} = 1$, $B_{ab} = B_{cd} = 0$ into the mutation rule we obtain $B'_{ab} = B'_{bq} = B'_{qa} = B'_{dq} = B'_{qc} = B'_{cd} = 1$ and $B'_{ad} = B'_{bc} = 0$ which corresponds to the adjacency rules of Fig. 6.2. Similarly, one can easily check that the adjacency rules hold for other triangulations containing the rectangle $ABCD$.

This shows that Penner coordinates for two perfect triangulations related by a Whitehead move form two sets of cluster coordinates for adjacent clusters in $\mathcal{A}(\Delta)$.

A problem arises in the case of non-perfect triangulations, as shown in Fig. 6.4, where edge PR is replaced by a loop QQ.

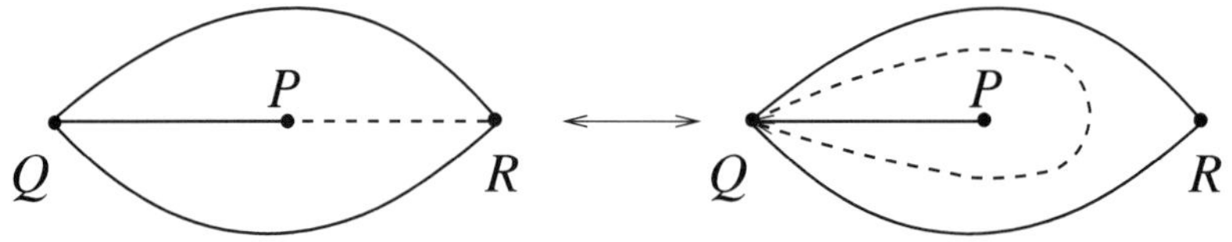

FIGURE 6.4. Prohibited Whitehead move

Indeed, Ptolemy relation (6.5) implies

$$f(PR)f(QQ) = f(PQ)(f(Q \frown R) + f(Q \smile R)). \tag{6.7}$$

However, by the definition of cluster algebras, the right part of the cluster algebra relation must be an irreducible polynomial, and in fact the cluster algebra relation would be

$$f(PR)f(QQ) = f(Q \frown R) + f(Q \smile R). \tag{6.8}$$

We thus see that the flip with respect to the edge BC breaks the cluster algebra rules, meaning that Penner coordinates after the flip differ from the coordinates prescribed by the cluster algebra rules.

To overcome this problem we are going to prohibit flips of certain diagonals in non-perfect triangulations. Namely, we call a Whitehead move *allowed* only if both endpoints of the flipped diagonal before the move have degrees at least three. As we have already mentioned, the transformation rule for Penner coordinates coincides with the transformation rule for the corresponding cluster algebra, and we are guaranteed that Penner coordinates of triangulations obtained by a sequence of such flips form clusters in the same cluster algebra $\mathcal{A}(\Delta)$.

Observe that all nice triangulations might be obtained one from another by a sequence of allowed flips. This can be proved by an immediate modification of the known proofs for a similar statement in the case of ideal triangulations, or all triangulations; another proof follows from the fact that nice triangulations label maximal simplices of triangulations of the Teichmüller space. Therefore, cluster algebras $\mathcal{A}(\Delta)$ defined by all perfect triangulations Δ of a fixed punctured surface are isomorphic to the same cluster algebra, which we denote by $\mathcal{A}(\Sigma)$.

The clusters of $\mathcal{A}(\Sigma)$ corresponding to nice triangulations are called *Teichmüller clusters*. Let $\mathcal{X}(\Sigma)$ be the cluster manifold for the cluster algebra $\mathcal{A}(\Sigma)$. It is easy to see that the decorated Teichmüller space $\tilde{\mathcal{T}}_g^s$ coincides with the positive part of any toric chart defined by a Teichmüller cluster.

6.2.3. The Weil-Petersson form. Recall that if Δ is a perfect triangulation, then flips of all edges are allowed. Let us define the *star* of Δ as the subset of clusters in $\mathcal{A}(\Sigma)$ formed by Δ itself and all the clusters obtained by flips with respect to all edges of Δ; we denote the star of Δ by $\star(\Delta)$. It follows immediately from the proof of Theorem 6.2f and the connectedness of the graph of Δ that there exists a unique (up to a constant) closed 2-form on $\mathcal{X}(\Sigma)$ compatible with all clusters in

$\star(\Delta)$; moreover, this form is compatible with the whole algebra $\mathcal{A}(\Sigma)$. Recall that this form is given in coordinates $f(e)$, $e \in \text{Edge}(\Delta)$, by the following expression:

$$\omega = \text{const} \cdot \sum_{b \curvearrowleft a} \frac{df(a) \wedge df(b)}{f(a)f(b)}, \tag{6.9}$$

where $b \curvearrowleft a$ means that edge b follows immediately after a in the counterclockwise order around some vertex of Δ.

It is well-known that the Teichmüller space $\mathcal{T}_g^s$ is a symplectic manifold with respect to the *Weil-Petersson* symplectic form W. Let us recall here the definition of W. The cotangent space $T^*_\Sigma \mathcal{T}_g^s$ is the space of holomorphic quadratic differentials $\varphi(z)dz^2$. Define the Weil-Petersson nondegenerate co-metric by the following formula:

$$\langle \varphi_1(z)dz^2, \varphi_2(z)dz^2 \rangle = \frac{i}{2} \int_\Sigma \frac{\varphi_1(z)\overline{\varphi_2(z)}}{\lambda(z)} dz \wedge \overline{dz},$$

where $\lambda(z)|dz^2|$ is the hyperbolic metric on Σ. The Weil-Petersson metric is defined by duality. It is known to be Kähler, hence its imaginary part is a symplectic 2-form W, which is called the Weil-Petersson symplectic form.

As we have already mentioned above, the decorated Teichmüller space is fibered over $\mathcal{T}_g^s$ with a trivial fiber $\mathbb{R}^s_{>0}$. The projection π: $\widetilde{\mathcal{T}}_g^s \to \mathcal{T}_g^s$ is given by forgetting the horocycles.

Given a geodesic triangulation, we introduce *Thurston shear coordinates* on the Teichmüller space. Namely, we associate to each edge of the triangulation a real number in the following way. Choose an edge e and two triangles incident to it and lift the resulting rectangle to the upper half plane. Vertices of this geodesic quadrilateral lie on the real axis. Therefore, we obtain four points on the real axis, or more precisely, on $\mathbb{R}P^1$. Among the constructed four points there are two distinguished ones, which are the endpoints of the edge e we have started with. Using the Möbius group action on the upper half plane, we can shift these points to zero and infinity, respectively, and one of the remaining points to -1. Finally, we assign to e the logarithm of the coordinate of the fourth point (the fourth coordinate itself is a suitable cross-ratio of those four points).

To show that we indeed obtained coordinates on the Teichmüller space, it is enough to reconstruct the surface. We will glue the surface out of ideal hyperbolic triangles. The lengths of the sides of ideal triangles are infinite, and therefore we can glue two triangles in many ways which differ by shifting one triangle w.r.t. another one along the side. The ways of gluing triangles can be parametrized by the cross-ratio of four vertices of the obtained quadrilateral (considered as points of $\mathbb{R}P^1$).

The projection ρ is written in these natural coordinates as follows:

$$g(e) = \log f(e_1) + \log f(e_3) - \log f(e_2) - \log f(e_4), \tag{6.10}$$

where $\{g(e)\colon e \in \text{Edge}(\Delta)\}$ and $\{f(e)\colon e \in \text{Edge}(\Delta)\}$ are Thurston shear coordinates on the Teichmüller space and Penner coordinates on the decorated Teichmüller space, respectively, with respect to the same triangulation Δ; notation e_i is explained in Fig. 6.5.

We see that $\exp(\pi)$ coincides with the transition from cluster variables to τ-coordinates in the cluster algebra $\mathcal{A}(\Sigma)$. Therefore, the pullback of the Weil-Petersson symplectic form determines a degenerate 2-form of corank s on the decorated Teichmüller space. For any nice triangulation Δ its expression in terms of

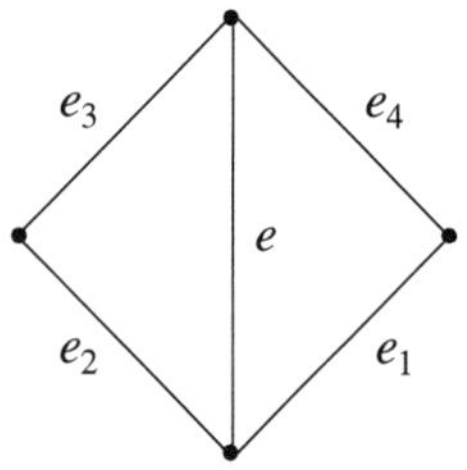

FIGURE 6.5. Notation for neighboring edges

Penner coordinates $f(e)$, $e \in \mathrm{Edge}(\Delta)$, is as follows:

$$(6.11)\qquad \omega = \frac{1}{2} \sum_{e \in \mathrm{Edge}(\Delta)} \left(dx_{e_1} \wedge dx_e - dx_{e_2} \wedge dx_e + dx_{e_3} \wedge dx_e - dx_{e_4} \wedge dx_e \right),$$

where $x_e = \log f(e)$.

Comparing this expression with (6.9) we can summarize the previous discussion as follows.

THEOREM 6.6. (i) *The decorated Teichmüller space $\widetilde{\mathcal{T}}_g^s$ is the subset of the cluster manifold $\mathcal{X}(\Sigma)$ given by the positive part of the toric chart defined by any Teichmüller cluster.*

(ii) *The projection to the secondary cluster manifold is the forgetful map from the decorated Teichmüller space to the Teichmüller space.*

(iii) *The corank of the corresponding transformation matrix equals s.*

(iv) *The unique compatible symplectic structure induced on the Teichmüller space is the classical Weil-Petersson symplectic structure.*

6.2.4. Denominators of transformation functions and modified geometric intersection numbers. In this Section we provide an interpretation of denominators of transition functions in terms of certain intersection numbers on the surface Σ. Let us express all cluster variables in $\mathcal{A}(\Delta)$ as rational functions in the cluster variables of the initial cluster. Recall that all the clusters in $\mathcal{A}(\Delta)$ are related to nice triangulations of Σ, while cluster variables correspond to edges of a triangulation. Abusing notation, we denote an edge of a triangulation and the corresponding cluster variable by the same letter. For a cluster variable p and an initial cluster variable x we denote by $\delta_x(p)$ the exponent of x in the denominator of p written as an irreducible rational function in the initial cluster variables. Besides, let a and b be two distinct edges possibly belonging to different nice triangulations. We define the *modified geometric intersection number* $[a, b]$ as the sum of contributions of all inner intersection points of a and b (that is, the ones distinct from their ends). The contribution of an intersection point equals 1 in all the cases except for the following one: if b is a loop based at a point P, $A1$ and A_2 are intersection points of a and b consecutive along a, and the triangle PA_1A_2 is contractible in Σ, then the contributions of both A_1 and A_2 to $[a, b]$ equal $1/2$; if this is the case we say that A_1 and A_2 are *conjugate* and that b is in the *exceptional* position w.r.t. a, otherwise it is in the *regular* position. Observe that in general, $[a, b] \neq [b, a]$.

For example, consider the edges a and b presented on Fig. 6.6; they form two triangles with one vertex at P and two others being intersection points of

a and b consecutive along a. In the left part of the figure, both these triangles are not contractible, and hence in this case b is in a regular position w.r.t. a and $[a, b] = [b, a] = 4$. In the middle part, the smaller triangle is contractible, while the larger one is not, hence b is in the exceptional position w.r.t. a, there exists one pair of conjugate points, and $[a, b] = 3$, $[b, a] = 4$. Finally, in the right part of the figure both triangles are contractible. Here again b is in the exceptional position w.r.t. a, but the number of conjugate pairs is 2, and hence $[a, b] = 2$, $[b, a] = 4$.

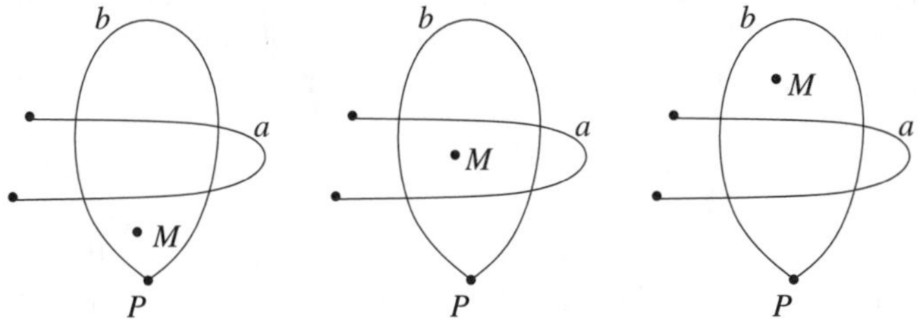

FIGURE 6.6. To the definition of the modified geometric intersection number

We extend the definition of the modified geometric intersection number by setting $[a, a] = -1$ for any edge a.

THEOREM 6.7. $\delta_x(p) = [p, x]$.

PROOF. The claim is trivial for $p = x$, so in what follows we assume that $p \neq x$. Consider two triangles bordering on p (see Fig. 6.7) and denote by q the edge obtained by flipping p. We want to prove the following two relations:

$$\delta_x(p) + \delta_x(q) = \max\{\delta_x(a) + \delta_x(c), \delta_x(b) + \delta_x(d)\}. \tag{6.12}$$

$$[p, x] + [q, x] = \max\{[a, x] + [c, x], [b, x] + [d, x]\}. \tag{6.13}$$

To prove (6.12), consider an arbitrary cluster variable, say p, as a function of the initial cluster variable x, provided all the other initial cluster variables are fixed to 1. Then $p = P(x)/x^k$, where $P(x)$ is a polynomial in x with a nonzero constant term, and $k \geq 0$. It is easy to notice that the above constant term is positive. Indeed, by Theorem 6.5, $p > 0$ for any choice of positive values of the initial cluster coordinates. On the other hand, the sign of p coincides with the sign of the constant term under consideration, provided x is sufficiently small. Therefore, $\delta_x(ac + bd) = \max\{\delta_x a + \delta_x c, \delta_x b + \delta_x d\}$, and (6.12) follows immediately from this fact and the exchange relation $pq = ac + bd$.

To prove (6.13), we consider all the inner intersection points of x with the sides a, b, c, d. These points break x into consecutive segments, each either lying entirely inside the quadrangle $abcd$ (*internal* segments), or entirely outside this quadrangle (*external* segments). We are going to partition the set of all internal segments into pairs and singletons such that the following two claims hold true.

Claim 1. For each part (pair or singleton) x_i of x,

$$[p, x_i] + [q, x_i] = \max\{[a, x_i] + [c, x_i], [b, x_i] + [d, x_i]\}$$

(here and in what follows we extend the notation for the modified geometric intersection number to parts of x).

Claim 2. If there exists a part x_i such that $[a, x_i] + [c, x_i] > [b, x_i] + [d, x_i]$, then $[a, x_j] + [c, x_j] \geq [b, x_j] + [d, x_j]$ for any other part x_j.

Evidently, (6.13) will follow from the above two claims by summing up the contributions of all parts.

Assume first that x is not a loop. In this case the partition consists only of singletons. We distinguish three types of inner segments. The segments of the first type intersect two adjacent sides of the quadrangle and exactly one of the diagonals p and q (see Fig. 6.7). Thus, for any such segment x_i one has $[p, x_i] + [q, x_i] = [a, x_i] + [c, x_i] = [b, x_i] + [d, x_i] = 1$. Therefore, Claim 1 for the segments of the first type holds true. The segments of the second type intersect two opposite sides of the quadrangle and both diagonals. Thus, for any such segment x_i one has $[p, x_i] + [q, x_i] = \max\{[a, x_i] + [c, x_i], [b, x_i] + [d, x_i]\} = 2$, and hence Claim 1 holds true. Finally, the segments of the third type start at a vertex and intersect the opposite diagonal and one of the opposite sides of the quadrangle. They are treated in the same way as the segments of the second type, the only difference being that their contribution to both sides of (6.13) equals 1. To prove Claim 2 it suffices to notice that all segments of the second and the third types contribute to the same pair of the opposite sides, otherwise x would have a self-intersection.

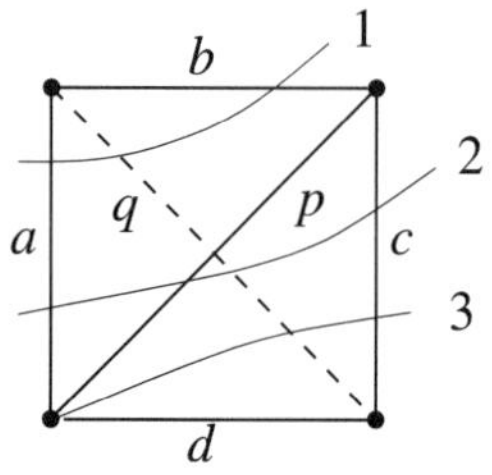

FIGURE 6.7. Types of segments

Assume now that x is a loop based at a point lying outside the quadrangle. If x is in the regular position w.r.t. the sides of the quadrangle, then it is also in the regular position w.r.t. its diagonals, and we proceed exactly as in the previous case. Otherwise, x is in the exceptional position w.r.t. at least one of the sides of the quadrangle. Consider a pair of conjugate points on the boundary of the quadrangle and the corresponding pair of internal segments of x, one starting at one of the conjugate points and the other ending at the other conjugate point. We say that these segments are *conjugate*. The partition consists of pairs of conjugate internal segments and of remaining single internal segments.

Consider a part x_i consisting of an arbitrary pair of conjugate internal segments of x. We can distinguish four possible cases.

Case 1. Both segments in x_i are of the second type. Therefore, x is in the exceptional position w.r.t. both diagonals, and hence $[p, x_i] + [q, x_i] = \max\{[a, x_i] + [c, x_i], [b, x_i] + [d, x_i]\} = 2$, so Claim 1 holds true.

Case 2. Both segments in x_i are of the first type, and their other endpoints belong to the same side of the quadrangle. Therefore, x is in the exceptional position w.r.t. one of the diagonals, and hence $[p, x_i] + [q, x_i] = [a, x_i] + [c, x_i] = [b, x_i] + [d, x_i] = 1$, so Claim 1 holds true.

Case 3. Both segments in x_i are of the first type, and their other endpoints belong to distinct sides of the quadrangle. Therefore, x is in the regular position w.r.t. both diagonals, and hence $[p, x_i] + [q, x_i] = \max\{[a, x_i] + [c, x_i], [b, x_i] + [d, x_i]\} = 2$, so Claim 1 holds true.

Case 4. One of the segments in x_i is of the first type, and the other one is of the second type. Therefore, x is in the exceptional position w.r.t. one of the diagonals, and hence $[p, x_i] + [q, x_i] = \max\{[a, x_i] + [c, x_i], [b, x_i] + [d, x_i]\} = 2$, so Claim 1 holds true.

Internal segments of x not included in conjugate pairs are treated as before; observe that they all are of the first or of the second type.

Claim 2 follows once again from the absence of self-intersections.

Finally, let x be a loop based at a vertex of the quadrangle. All pairs of internal conjugate segments of the first and the second types are treated as in the previous case. If both segments in a pair x_i are of the third type (and thus are incident to the same vertex of the quadrangle and intersect the same side of the quadrangle), x is in the exceptional position w.r.t. one of the diagonals and hence $[p, x_i] + [q, x_i] = \max\{[a, x_i] + [c, x_i], [b, x_i] + [d, x_i]\} = 1$, so Claim 1 holds true.

Internal segments of x of the first and the second types not included in conjugate pairs are treated as before. The remaining case to be considered occurs when two segments of the third type are incident to the same vertex of the quadrangle, and their endpoints lie on its distinct sides. In this case these two segments together form a part x_i, though they are not conjugate. It is easy to see that x is in the exceptional position w.r.t. one of the diagonals, and hence $[p, x_i] + [q, x_i] = [a, x_i] + [c, x_i] = [b, x_i] + [d, x_i] = 1$, so Claim 1 holds true. Claim 2 is treated as before.

Thus, we have proved that both $\delta_x(p)$ and $[p, x]$ satisfy the same relation with respect to flips. To get the statement of the theorem it suffices to notice that it obviously holds when p and x belong to two adjacent clusters. □

We illustrate the theorem with the following example, see Fig. 6.8. Consider the triangulation of the sphere with 6 marked points presented on the upper left part of the figure (two additional marked points and four edges lie inside the dashed part of the figure). After the first move we have $\bar{x} = (ac + bd)/x$, and hence $\delta_x(\bar{x}) = 1$, which corresponds to the unique intersection point of the edges x and $\bar{x}$. After the second move we have $\bar{a} = (abc + b^2d + cex)/ax$, and hence $\delta_x(\bar{a}) = 1$, which corresponds to the unique intersection point of the edges x and $\bar{a}$. After the third move we have $\bar{d} = (acg + bdg + bfx)/dx$, and hence $\delta_x(\bar{d}) = 1$, which corresponds to the unique intersection point of the edges x and $\bar{d}$. Finally, after the fourth move we have $\bar{b} = (2abcdg + b^2dg + a^2c^2g + abcfx + cdegx)/abdx$, and hence $\delta_x(\bar{b}) = 1$, which corresponds to the two conjugate intersection points of x and $\bar{b}$.

6.2.5. Geometric description of the algebra $\mathcal{A}(\Sigma)$. Recall that in Section 6.2.2 we have prohibited flips that do not preserve the perfectness of the triangulation. The reason for this was that the corresponding Ptolemy relations (6.7) deviate from the cluster algebra framework. Here we explain how to overcome this difficulty by introducing an additional geometric structure.

Consider a punctured monogon pictured on Fig. 6.1. Assume that Penner coordinates $f(PQ)$ and $f(QQ)$ are calculated with respect to some pair of horocycles, and let $f^*(PQ)$ be the Penner coordinate obtained when the horocycle around P is

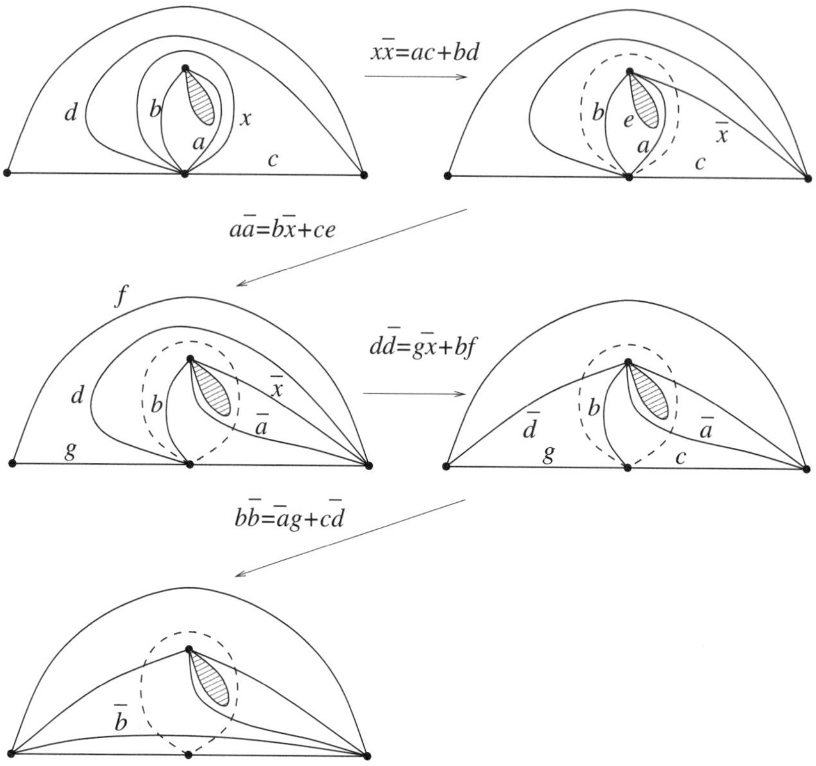

FIGURE 6.8. Denominators and the modified geometric intersection number

replaced by the conjugate one (defined in Section 6.2.1 above). One can show that in this case

$$f(QQ) = f(PQ)f^*(PQ). \tag{6.14}$$

Therefore, (6.7) can be rewritten as

$$f(PR)f^*(PQ) = f(Q \frown R) + f(Q \smile R),$$

which means that the cluster algebra relation (6.8) will be preserved with $f(QQ)$ replaced by $f^*(PQ)$. In other words, the prohibited Whitehead move presented on Fig. 6.4 should be replaced by a new one that, instead of introducing a loop QQ with the corresponding variable $f(QQ)$, makes use of the conjugate horocycle at the point P and introduces an additional variable $f^*(PQ)$ corresponding to the edge PQ.

To be more precise, we fix at each puncture an ordered pair of distinct conjugate horocycles. Therefore, any geodesic arc e joining two distinct marked points P and Q defines four different variables $f_{ij}(e)$, $i, j \in \{-1, 1\}$, depending on the choice of a horocycle at each of its endpoints. Similarly, any geodesic loop e based at a marked point Q defines two different variables $f_{ii}(e)$, $i \in \{-1, 1\}$.

Given an ideal triangulation Δ, define $S'(\Delta) \subseteq S$ as the subset of marked points having degree at least 2 in Δ. Clearly, for a nice triangulation $S'(\Delta) = S$.

We say that a geodesic arc $e \in \text{Edge}(\Delta)$ is a *special edge* of Δ if one of its endpoints does not belong to $S'(\Delta)$. Note that the same geodesic arc on Σ may be a special edge in one triangulation and a regular edge in another one. Further, we say that a geodesic loop is a *noose* if it cuts off a single marked point (see Fig. 6.1). Clearly, any triangulation containing a noose contains also the special edge joining the base of the loop with the point inside it. The situation when two special edges correspond to the same noose never occurs, since it corresponds to the case of a sphere with three marked points, which was prohibited from the very beginning.

For an ideal triangulation Δ, a *spin* σ is defined as a function $\sigma\colon S'(\Delta) \to \{-1, 1\}$. The pair (Δ, σ) is called a *spun triangulation*. A *spun geodesic arc* is a pair (e, ξ), where e is a geodesic arc and ξ maps the endpoints of e to $\{-1, 1\}$. We say that (e, ξ) is a *spun edge* of (Δ, σ) if e is an edge of Δ distinct from a noose and the values of σ and ξ coincide at the endpoints of e lying in $S'(\Delta)$. Therefore, to each regular edge $e \in \text{Edge}(\Delta)$ distinct from a noose corresponds a spun edge (e, ξ) of (Δ, σ) with ξ being the restriction of σ to the endpoints of e. To each special edge $e = PQ \in \text{Edge}(\Delta)$, $P \notin S'(\Delta)$, correspond two spun edges (e, ξ) and $(e, \bar{\xi})$ with $\xi(Q) = \bar{\xi}(Q) = \sigma(Q)$ and $\xi(P) = -\bar{\xi}(P)$. There are no spun edges corresponding to nooses.

Given a spun triangulation (Δ, σ), we define *spun Whitehead moves* (or *spun flips*) for all its spun edges. Let $(e = PQ, \xi)$ be a spun edge of (Δ, σ). If the degrees of both P and Q in Δ are at least 3, then the spun flip coincides with the usual one, and the spin σ remains unchanged. If the degree at one of the endpoints, say at P, equals 2, then the spun flip is the prohibited move presented on Fig. 6.4; observe that after the spun flip P does not belong to $S'(\Delta)$, and hence σ is not defined at P. Finally, if the degree at one of the endpoints, say at P, equals 1, then the spun flip is the inverse of the prohibited move. Observe that after the spun flip P belongs to $S'(\Delta)$, and hence we have to define σ at P; this is done by setting $\sigma(P) = -\xi(P)$.

Define the graph $G(\Sigma)$ as follows: the vertices of $G(\Sigma)$ are all spun triangulations (Δ, σ), and two vertices are connected by an edge if they differ by a spun flip. A simple modification of the standard argument proves the following result.

PROPOSITION 6.8. *The graph $G(\Sigma)$ is connected for $s > 1$ and consists of two isomorphic components for $s = 1$.*

Finally, define the adjacency matrix $B(\Delta, \sigma)$. Let (a, ξ), (b, ζ) be a pair of spun edges in (Δ, σ); this means that ξ and ζ coincide with σ at any point in $S'(\Delta)$. Put

$$B(\Delta, \sigma)_{(a,\xi),(b,\zeta)} = \begin{cases} B(\Delta)_{ab} & \text{if } a \neq b, \\ 0 & \text{otherwise,} \end{cases}$$

where $B(\Delta)_{ab}$ is defined by (6.6). Clearly, for a nice triangulation Δ the matrices $B(\Delta)$ and $B(\Delta, \sigma)$ coincide.

THEOREM 6.9. (i) *Clusters of $\mathcal{A}(\Sigma)$ are labeled by the spun triangulations.*

(ii) *Cluster variables of $\mathcal{A}(\Sigma)$ are labeled by spun geodesic arcs distinct from nooses.*

(iii) *The exchange graph of $\mathcal{A}(\Sigma)$ coincides with a connected component of $G(\Sigma)$.*

(iv) *The exchange matrix for the cluster labeled by a spun triangulation (Δ, σ) coincides with the adjacency matrix $B(\Delta, \sigma)$.*

PROOF. A spun triangulation defines a cluster in the following way: to each spun edge (e, ξ) corresponds a cluster variable $f(e, \xi) = f_{\xi(P),\xi(Q)}(e)$. It follows from the above discussion that adjacent clusters correspond bijectively to the spun triangulations obtained from (Δ, σ) by spun Whitehead moves, and that the exchange relations are governed by the adjacency matrix $B(\Delta, \sigma)$. Indeed, the only situations that are new as compared to the proof of Theorem 6.6(i) arise when we apply the spun flip to an arc adjacent to a noose or to a loop adjacent to two nooses. Both these situations are pictured in Fig. 6.9.

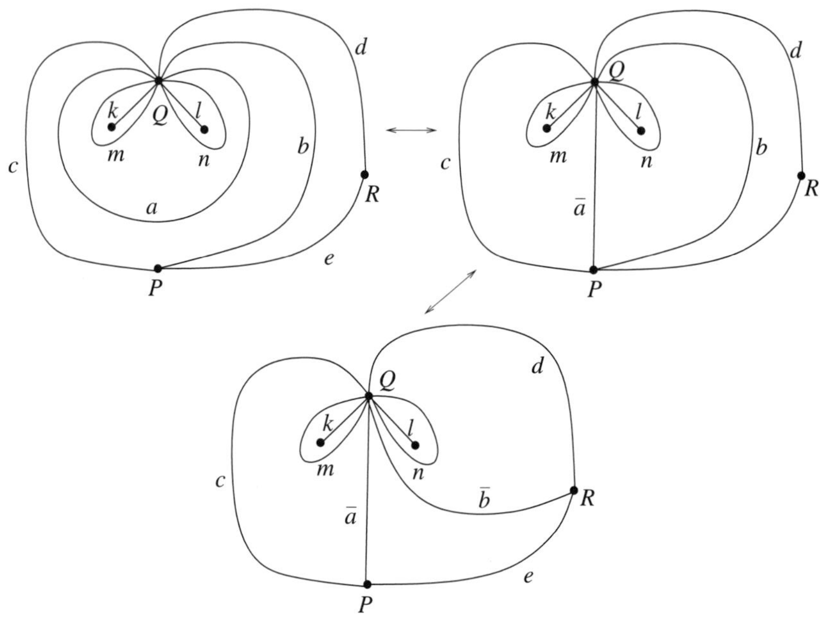

FIGURE 6.9. Spun flips adjacent to a noose

In the first of them, the usual Ptolemy relation reads $f(a)f(\bar{a}) = f(b)f(m) + f(c)f(n)$. Recall that $f(m) = f(k)f^*(k)$ and $f(n) = f(l)f^*(l)$, hence for any fixed spin σ,

$$f(a, \sigma)f(\bar{a}, \sigma) = f(b, \sigma)f(k_+)f(k_-) + f(c, \sigma)f(l_+)f(l_-),$$

where $k_+ = (k, 1, \sigma(Q))$, $k_- = (k, -1, \sigma(Q))$, $l_+ = (l, 1, \sigma(Q))$, $l_- = (l, -1, \sigma(Q))$. This corresponds to the fact that, by the definition of $B(\Delta, \sigma)$,

$$B(\Delta, \sigma)_{(a,\sigma),(b,\sigma)} = B(\Delta, \sigma)_{(a,\sigma),k_+} = B(\Delta, \sigma)_{(a,\sigma),k_-} = 1,$$
$$B(\Delta, \sigma)_{(a,\sigma),(c,\sigma)} = B(\Delta, \sigma)_{(a,\sigma),l_+} = B(\Delta, \sigma)_{(a,\sigma),l_-} = -1.$$

Similarly, in the second situation, the usual Ptolemy relation reads $f(b)f(\bar{b}) = f(\bar{a})f(d) + f(e)f(n)$, which leads to

$$f(b, \sigma)f(\bar{b}, \sigma) = f(\bar{a}, \sigma)f(d, \sigma) + f(e, \sigma)f(l_+)f(l_-),$$

in agreement with

$$B(\Delta, \sigma)_{(b,\sigma),(\bar{a},\sigma)} = B(\Delta, \sigma)_{(b,\sigma),(d,\sigma)} = 1,$$
$$B(\Delta, \sigma)_{(b,\sigma),(e,\sigma)} = B(\Delta, \sigma)_{(b,\sigma),l_+} = B(\Delta, \sigma)_{(b,\sigma),l_-} = -1.$$

Therefore, claims (i), (iii) and (iv) hold true, and the only statement which remains to be proven is (ii). More exactly, since we have assigned a cluster variable to each spun geodesic arc, it remains to show that distinct spun geodesic arcs define distinct cluster variables.

For two spun geodesic arcs (e,ξ) and (e',ξ') define the *spun intersection number* $[(e,\xi),(e',\xi')]$ by

$$[(e,\xi),(e',\xi')] = [e,e'] + \sum_P |\xi(P) - \xi'(P)| l(e), \tag{6.15}$$

where $[e,e']$ is the modified geometric intersection number defined in the previous section, the sum is taken over all common endpoints of e and e', and $l(e)$ equals 1 if e is a loop and $1/2$ otherwise. To complete the proof, we need the following statement is a straightforward generalization of Theorem 6.7 and can be proved similarly.

PROPOSITION 6.10. *The exponent of $f(e,\xi)$ in the denominator of $f(e',\xi')$ written via any cluster containing (e,ξ) equals $[(e',\xi'),(e,\xi)]$.*

Assume now that (e,ξ) and (e',ξ') are two distinct spun geodesic arcs such that $f(e,\xi) = f(e',\xi')$. Then the exponent of a $f(e,\xi)$ in the denominator of $f(e',\xi')$ written via any cluster containing (e,ξ) equals -1, and hence by Proposition 6.10, $[(e',\xi'),(e,\xi)] = -1$. It follows immediately from (6.15) and the definition of the modified geometric intersection number that the spun intersection number is non-negative unless (e,ξ) and (e',ξ') coincide. The obtained contradiction proves the statement. □

6.3. Restoring exchange relations

In this section we suggest an axiomatic approach to exchange relations (3.1). Namely, we start from a list of properties which these relations should have and derive from this list the corresponding expressions. In what follows we confine ourselves to the skew-symmetric case.

Consider an infinite bi-indexed sequence of maps

$$T_{n,k} : (B; x_1, \dots, x_n) \mapsto (B'; x'_1, \dots, x'_n), \qquad n \in [1,\infty), \quad i \in [1,n],$$

where $B, B' \in so_n(\mathbb{Z})$, $x_i, x'_i \in \mathcal{F}$. We assume that this sequence satisfies the following conditions:

(i) (*locality*) $x'_j = x_j$ for all $j \neq k$, x'_k depends on $\mathbf{x} = (x_1, \dots, x_n)$ and $\mathbf{b}_k = (b_{k1}, \dots, b_{kn})$;

(ii) (*involutivity*) $T^2_{n,k} = \mathrm{id}$;

(iii) (*polynomiality*) x'_k is a nontrivial polynomial in x_j, $j \neq k$ not divisible by any of x_j;

(iv) (*compatibility*) the 2-forms $\omega = \sum b_{ij} \frac{dx_i}{x_i} \wedge \frac{dx_j}{x_j}$ and $\omega' = \sum b'_{ij} \frac{dx'_i}{x'_i} \wedge \frac{dx'_j}{x'_j}$ coincide;

(v) (S_n-*invariance*) $T_{n,\sigma(k)}(\sigma B; x_{\sigma(1)}, \dots, x_{\sigma(n)}) = (\sigma B'; x'_{\sigma(1)}, \dots, x'_{\sigma(n)})$ for any permutation $\sigma \in S_n$, where σB stands for the matrix obtained from B by applying σ to its rows and columns;

(vi) (*universality*) for any $j \in [1, n+1]$,

$$T_{n,k}(B^j; x_1, \dots, \widehat{x}_j, \dots, x_{n+1}) = T_{n+1,k^j}(B; x_1, \dots, x_j, \dots, x_{n+1})|_{x_j=1},$$

where B^j is obtained from B by deleting the jth row and the jth column, $k^j = k$ if $k < j$ and $k^j = k+1$ if $k \geq j$.

It is easy to see that the identical map, as well as the map that changes the sign of x_k and preserves B, trivially satisfy conditions (1)–(6).

THEOREM 6.11. *Let $T_{n,k}$ be a sequence of nontrivial maps satisfying conditions* (i)–(vi). *Then x'_k is given by*

$$x'_k = \frac{v(\mathbf{x};\mathbf{b}_k)^d P(z(\mathbf{x};\mathbf{b}_k))}{x_k}, \tag{6.16}$$

where

$$z(\mathbf{x};\mathbf{b}_k) = \prod_{i=1}^{n} x_i^{b_{ki}}, \qquad v(\mathbf{x};\mathbf{b}_k) = \prod_{b_{ki}<0} x_i^{b_{ki}},$$

and P is a reciprocal polynomial of degree d.

PROOF. Let us write down ω' using the locality condition:

$$\omega' = \sum b'_{ij}\frac{dx'_i}{x'_i} \wedge \frac{dx'_j}{x'_j} = \sum_{i,j\neq k} b'_{ij}\frac{dx_i}{x_i} \wedge \frac{dx_j}{x_j} + \sum_{i\neq k} b'_{ik}\frac{dx_i}{x_i} \wedge \frac{dx'_k}{x'_k}. \tag{6.17}$$

Since by locality, $x'_k = F_k(\mathbf{x};\mathbf{b}_k)$, we get

$$\frac{dx'_k}{x'_k} = \sum_{j=1}^{n} \frac{\partial F_k(\mathbf{x};\mathbf{b}_k)}{\partial x_j}\frac{x_j}{F_k(\mathbf{x};\mathbf{b}_k)} \cdot \frac{dx_j}{x_j}. \tag{6.18}$$

Therefore, compatibility implies

$$b'_{ik}\frac{\partial F_k(\mathbf{x};\mathbf{b}_k)}{\partial x_k}\frac{x_k}{F_k(\mathbf{x};\mathbf{b}_k)} = b_{ik}, \quad i \in [1,n].$$

If there exists $i \in [1,n]$ such that $b_{ik} \neq 0$ then $b'_{ik} \neq 0$ as well, and hence the above first-order equation is solved to

$$F_k(\mathbf{x};\mathbf{b}_k) = \Phi_k(\mathbf{x};\mathbf{b}_k)x_k^{c_k}$$

with $c_k = b_{ik}/b'_{ik}$ and Φ_k not depending on x_k. Since $x'_k \in \mathcal{F}$, c_k should be an integer; by involutivity, this integer is either 1 or -1. Observe that by polynomiality, Φ_k is a polynomial in $x_1, \dots, \widehat{x}_k, \dots, x_n$.

Assume first that $c_k = 1$. Then $b'_{ik} = b_{ik}$ for $i \in [1,n]$, and hence $T^2_{n,k}$ takes x_k to $\Phi_k^2(\mathbf{x};\mathbf{b}_k)x_k$. By involutivity, this means that $\Phi_k \equiv \pm 1$, which contradicts the nontriviality assumption.

Therefore, $c_k = -1$, and hence $b'_{ik} = -b_{ik}$ for $i \in [1,n]$. Note that by S_n-invariance, neither c_k nor Φ_k depend on k, and hence

$$x'_k = \frac{\Phi(\mathbf{x};\mathbf{b}_k)}{x_k}. \tag{6.19}$$

In case $b_{ik} = 0$ for all $i \in [1,n]$ we extend B by adding one additional row and column and setting $b_{n+1,k} = 1$. For the obtained matrix all the above reasoning is valid. To get back to the initial matrix it suffices to use the universality condition.

Substituting (6.19) into (6.18) and (6.17) and using compatibility, we get

$$b'_{ij} + \frac{b'_{ik}x_j\dfrac{\partial \Phi(\mathbf{x};\mathbf{b}_k)}{\partial x_j} - b'_{jk}x_i\dfrac{\partial \Phi(\mathbf{x};\mathbf{b}_k)}{\partial x_i}}{\Phi(\mathbf{x};\mathbf{b}_k)} = b_{ij}, \qquad i,j \in [1,n], \quad i,j \neq k. \tag{6.20}$$

The polynomial Φ can be rewritten as

$$\Phi(\mathbf{x};\mathbf{b}_k)=\sum_{\mathbf{r}\geq 0} c(\mathbf{r};\mathbf{b}_k)x_1^{r_1}\cdots x_n^{r_n},$$

where $\mathbf{r}=(r_1,\dots,r_n)$. Taking into account that $b'_{lk}=-b_{lk}$ for any $l\in[1,n]$, we can rewrite (6.20) as

$$\frac{\sum_{\mathbf{r}\geq 0}(b_{jk}r_i-b_{ik}r_j)c(\mathbf{r};\mathbf{b}_k)x_1^{r_1}\cdots x_n^{r_n}}{\sum_{\mathbf{r}\geq 0} c(\mathbf{r};\mathbf{b}_k)x_1^{r_1}\cdots x_n^{r_n}}=b_{ij}-b'_{ij},\qquad i,j\in[1,n],\quad i,j\neq k.$$

Since the right hand side of the above equality does not depend on $\mathbf{x}$, the same holds true for the left hand side. This is only possible if all points $\mathbf{r}$ for which $c(\mathbf{r};\mathbf{b}_k)\neq 0$ lie on the same affine line parallel to $\mathbf{b}_k$. We thus get

$$\Phi(\mathbf{x};\mathbf{b}_k)=M(\mathbf{x};\mathbf{b}_k)P(z(\mathbf{x};\mathbf{b}_k);\mathbf{b}_k), \tag{6.21}$$

where $M(\mathbf{x};\mathbf{b}_k)$ is a monomial in $(x_1,\dots,\widehat{x}_k,\dots,x_n)$ and P is a polynomial in one variable with a nonzero constant term. Polynomiality condition implies immediately that the only possible choice for the monomial M is $v(\mathbf{x};\mathbf{b}_k)^{d(\mathbf{b}_k)}$, where $d(\mathbf{b}_k)$ is the degree of P.

Lets us prove that the polynomial P in (6.21) does not depend on $\mathbf{b}_k$. Indeed, extend the matrix B by two rows and columns and set $b_{k,n+1}=-1$, $b_{k,n+2}=1$ (the values of other additional entries of the extended matrix are irrelevant). Denote by $\bar{\mathbf{b}}_k$ the extended row of B and by $\bar{\mathbf{x}}$ the corresponding $(n+2)$-tuple of variables. Then, by the above reasoning,

$$\Phi(\bar{\mathbf{x}};\bar{\mathbf{b}}_k)=v(\bar{\mathbf{x}};\bar{\mathbf{b}}_k)^{d(\bar{\mathbf{b}}_k)}P(z(\bar{\mathbf{x}};\bar{\mathbf{b}}_k);\bar{\mathbf{b}}_k),$$
$$\Phi(x_k,x_{n+1},x_{n+2};\mathbf{e})=v(x_k,x_{n+1},x_{n+2};\mathbf{e})^{d(\mathbf{e})}P(z(x_k,x_{n+1},x_{n+2};\mathbf{e});\mathbf{e})$$

with $\mathbf{e}=(0,-1,1)$. Besides,

$$v(x_k,x_{n+1},x_{n+2};\mathbf{e})=v(\bar{\mathbf{x}};\bar{\mathbf{b}}_k)|_{x_1=\dots=x_n=1}=x_{n+1},$$
$$z(x_k,x_{n+1},x_{n+2};\mathbf{e})=z(\bar{\mathbf{x}};\bar{\mathbf{b}}_k)|_{x_1=\dots=x_n=1}=x_{n+2}/x_{n+1},$$

and hence by universality,

$$x_{n+1}^{d(\mathbf{e})}P(x_{n+2}/x_{n+1};\mathbf{e})=x_{n+1}^{d(\bar{\mathbf{b}}_k)}P(x_{n+2}/x_{n+1};\bar{\mathbf{b}}_k),$$

which amounts to $P(z;\mathbf{e})=P(z;\bar{\mathbf{b}}_k)$.

On the other hand,

$$\Phi(\mathbf{x};\mathbf{b}_k)=v(\mathbf{x};\mathbf{b}_k)^{d(\mathbf{b}_k)}P(z(\mathbf{x};\mathbf{b}_k);\mathbf{b}_k)$$

and

$$v(\mathbf{x};\mathbf{b}_k)=v(\bar{\mathbf{x}};\bar{\mathbf{b}}_k)|_{x_{n+1}=x_{n+2}=1},\qquad z(\mathbf{x};\mathbf{b}_k)=z(\bar{\mathbf{x}};\bar{\mathbf{b}}_k)|_{x_{n+1}=x_{n+2}=1}.$$

Therefore, universality implies

$$v(\mathbf{x};\mathbf{b}_k)^{d(\bar{\mathbf{b}}_k)}P(z(\mathbf{x};\mathbf{b}_k);\bar{\mathbf{b}}_k)=v(\mathbf{x};\mathbf{b}_k)^{d(\mathbf{b}_k)}P(z(\mathbf{x};\mathbf{b}_k);\mathbf{b}_k),$$

which amounts to $P(z;\mathbf{b}_k)=P(z;\bar{\mathbf{b}}_k)$.

We thus get that $P(z;\mathbf{b}_k)=P(z;\mathbf{e})$ for any $\mathbf{b}_k$. Therefore, $P(z;\mathbf{b}_k)$ does not depend on $\mathbf{b}_k$, and in what follows we write just $P(z)$. It remains to prove that P is reciprocal, that is, $P(z)=z^dP(1/z)$, where d is the degree of P.

Consider a 3×3 matrix B with the first row $\mathbf{b}_1=(0,-1,1)$. By the above reasoning, $\Phi(x_1,x_2,x_3;0,-1,1)=x_2^dP(x_3/x_2)$. Besides, the S_3-invariance condition for $\sigma = (132)$ yields $\Phi(x_1, x_3, x_2; 0, 1, -1) = \Phi(x_1, x_2, x_3; 0, -1, 1)$. We therefore get $x_3^d P(x_2/x_3) = x_2^d P(x_3/x_2)$, and the reciprocity follows. □

It is easy to see that the simplest case of transformations (6.16) obtained for $P(z) = 1 + z$ is equivalent to exchange relations (3.1).

6.4. Summary

- We define a pre-symplectic structure compatible with a cluster algebra structure. Unlike the compatible Poisson structure, a compatible pre-symplectic structure exists even in the case of a degenerate exchange matrix. We provide a complete description of all compatible pre-symplectic structures in Theorem 6.2. In the non-degenerate case, the compatible pre-symplectic structure is dual to the compatible Poisson structure. The compatible pre-symplectic structure reduces to the symplectic (Weil-Petersson) form on the secondary cluster manifold, see Corollary 6.4.
- As an example of the general construction we give a cluster algebra interpretation to the classical Weil-Petersson form on the Teichmüller space. In this case, the Penner coordinates associated with nice geodesic triangulations of a surface Σ serve as cluster variables of a cluster algebra $\mathcal{A}(\Sigma)$, and cluster transformations are induced by Whitehead moves. Furthermore, we express exponents in denominators of cluster variables as modified geometric intersection numbers for geodesic arcs, see Theorem 6.7.
- Nice geodesic triangulations correspond to a subset of all clusters in $\mathcal{A}(\Sigma)$. We provide a geometric description of $\mathcal{A}(\Sigma)$ based on the notion of the spin of a triangulation, see Theorem 6.9.
- Finally, we use the notion of compatible pre-symplectic structure to describe, in the skew-symmetric case, a set of axioms that allows one to restore cluster algebra exchange relations, see Theorem 6.11.

Bibliographical notes

6.1. Theorem 6.2 is proved in [**GSV4**]. It expands a similar result in [**GSV3**] proved for skew-symmetric matrices. The Weil–Petersson form associated with an arbitrary cluster algebra and the secondary cluster manifold were introduced in [**GSV3**].

6.2. For more details on uniformization, Fuchsian groups and hyperbolic geometry, see Chapter 9 in [**Sp**] and Chapter 4 in [**Ka**]. Penner coordinates on the decorated Teichmüller space are introduced in [**Pe1**]. The proof of Theorem 6.5 and Ptolemy relation for Penner coordinates can be found in the same paper. Connectivity for the set of ideal triangulations is proved in [**Ha**], and for the set of all triangulations in [**Bu**]. For the correspondence between nice triangulations and maximal simplices of triangulations of the Teichmüller space see [**Iv**] and references therein. For the definition of the classical Weil–Petersson form, see e.g. [**Pe2**]. Thurston shear coordinates are described, e.g., in [**Th, Pe1, Fo**]. Relation (6.10) is borrowed from [**Pe1, Fo**]. Expression (6.11) can be found, e.g., in [**FoR**]. Theorems 6.6 and 6.7 are proved in [**GSV3**]. Relation (6.14) can be derived from Corollary 3.4 in [**Pe1**]. Theorem 6.9 is a slight modification of one of the results in [**FST**], where tagged triangulations were used instead of spun triangulations.

6.3. The results in this section are new. A similar but weaker result involving Poisson brackets instead of closed 2-forms can be found in [**GSV2**].

CHAPTER 7

On the properties of the exchange graph

In this Chapter we prove, under certain nondegeneracy conditions, two conjectures about the structure of the exchange graph of a cluster algebra formulated in Section 3.5.

7.1. Covering properties

In this Section we consider in more detail general cluster algebras over an arbitrary semifield $\mathcal{P}$ as defined in Remark 3.13. It will be more convenient to consider, instead of the $2n$-tuple $\mathbf{p}$, an n-tuple $\mathbf{y} = (y_1, \dots, y_n)$, $y_i \in \mathcal{P}$, defined by $y_j = p_j^+/p_j^-$ for all j. Consequently, a seed is defined as a triple $\Sigma = (\mathbf{x}, \mathbf{y}, B)$. It is easy to see that the two settings are equivalent since the coefficients $p_j^\pm$ are recovered via

$$p_j^+ = \frac{y_j}{y_j \oplus 1}, \qquad p_j^- = \frac{1}{y_j \oplus 1}.$$

Using this new notation one can rewrite exchange relations (3.2) as

$$x_k x_k' = \frac{y_k}{y_k \oplus 1} \prod_{b_{ki}>0} x_i^{b_{ki}} + \frac{1}{y_k \oplus 1} \prod_{b_{ki}<0} x_i^{-b_{ki}}. \tag{7.1}$$

Correspondingly, relations (3.4) now become

$$y_j' = \begin{cases} y_k^{-1} & \text{if } j = k, \\ y_j y_k^{b_{jk}} (y_k \oplus 1)^{-b_{jk}} & \text{if } j \neq k \text{ and } b_{jk} > 0, \\ y_j (y_k \oplus 1)^{-b_{jk}} & \text{if } j \neq k \text{ and } b_{jk} \leq 0. \end{cases} \tag{7.2}$$

In what follows we assume that the initial cluster $\mathbf{x}$ and the initial exchange matrix B are fixed, and the n-tuple $\mathbf{y}$ varies; denote by $\mathfrak{A}(B)$ the family of all cluster algebras thus obtained. Let B_{pr} be the $n \times 2n$ matrix whose principal part equals B and the remaining part is the $n \times n$ identity matrix. The corresponding algebra of geometric type $\mathcal{A}(B_{\mathrm{pr}}) \in \mathfrak{A}(B)$ is called the algebra with *principal coefficients* and is denoted by $\mathcal{A}_{\mathrm{pr}}(B)$. Note that for this algebra initial coefficients $y_1, \dots, y_n$ coincide with the stable variables $x_{n+1}, \dots, x_{2n}$.

Recall that the exchange graph of a cluster algebra was defined in Section 3.1 as the quotient of the n-regular tree $\mathbb{T}$ modulo the equivalence relation that identifies two seeds obtained one from the other by an arbitrary permutation of cluster variables and the corresponding permutation of the rows and columns of B. Let $\mathcal{A}$ and $\mathcal{A}'$ be two cluster algebras in $\mathfrak{A}(B)$. The exchange graph of $\mathcal{A}'$ *covers* the exchange graph of $\mathcal{A}$ if the equivalence of two seeds in $\mathcal{A}'$ implies the equivalence of the corresponding two seeds in $\mathcal{A}$. It follows immediately from the definition that the exchange graph of any cluster algebra in $\mathfrak{A}(B)$ covers the exchange graph of the coefficient-free cluster algebra $\mathcal{A}_{\mathrm{cf}}(B)$ over the one-element semifield $\{1\}$ defined by the seed $(\mathbf{x}, 1, B)$. The corresponding map takes all elements of $\mathcal{P}$ to 1.

THEOREM 7.1. *The exchange graph of any cluster algebra $\mathcal{A} \in \mathfrak{A}(B)$ is covered by the exchange graph of the cluster algebra with principal coefficients $\mathcal{A}_{\mathrm{pr}}(B)$.*

PROOF. Define $X_{i;t}(x_1, \dots, x_n; y_1, \dots, y_n)$ as a rational function expressing the cluster variable $x_{i;t}$ in $\mathcal{A}_{\mathrm{pr}}(B)$; further, define rational functions $F_{i;t}$ by

$$F_{i;t}(y_1, \dots, y_n) = X_{i;t}(1, \dots, 1; y_1, \dots, y_n).$$

Observe that by (3.1) $X_{i;t}$, and hence $F_{i;t}$, can be written as subtraction-free rational expressions in $x_1, \dots, x_n, y_1, \dots, y_n$. For any subtraction-free rational expression $G(z_1, \dots, z_k)$ and arbitrary elements $u_1, \dots, u_k \in \mathcal{P}$ denote by $G|_{\mathcal{P}}(u_1, \dots, u_k)$ the evaluation of G at $u_1, \dots, u_k$; this operation is well-defined since any subtraction-free identity in the field of rational functions remains valid in any semifield.

LEMMA 7.2. *Let $\mathcal{A}$ be a cluster algebra over an arbitrary semifield $\mathcal{P}$ defined by a seed $(\mathbf{x}, \mathbf{y}, B)$ at the initial vertex t_0. Then the cluster variables at a vertex t can be expressed as follows:*

$$x_{i;t} = \frac{X_{i;t}|_{\mathcal{F}}(x_1, \dots, x_n; y_1, \dots, y_n)}{F_{i;t}|_{\mathcal{P}}(y_1, \dots, y_n)}, \qquad i \in [1, n]. \tag{7.3}$$

PROOF. Let $(\mathbf{x}_t, \mathbf{y}_t, B^t)$ be the seed of $\mathcal{A}$ at a vertex $t \in \mathbb{T}$. Introduce a new n-tuple $\widehat{\mathbf{y}}_t = (\widehat{y}_{1;t}, \dots, \widehat{y}_{n;t}) \in \mathcal{F}^n$ by

$$\widehat{y}_{j;t} = y_{j;t} \prod_{k=1}^{n} x_{k;t}^{b^t_{jk}}. \tag{7.4}$$

Observe that for the algebra with principal coefficients the initial n-tuple $\widehat{\mathbf{y}}$ coincides with the n-tuple of τ-coordinates defined in Section 4.1.2 (see (4.1)). The following proposition is a straightforward generalization of Lemma 4.4.

PROPOSITION 7.3. *Assume that $\widehat{\mathbf{y}}$ and $\widehat{\mathbf{y}}'$ are calculated at a pair of vertices adjacent in direction k, then they are related by (7.2) with $\oplus$ replaced by the usual addition in $\mathcal{F}$.*

PROOF. The proof is similar to the proof of Lemma 4.4 and is based on the coefficient relations (7.2) and exchange relations (7.1) which can be rewritten as

$$x_k x'_k = \frac{\widehat{y}_k + 1}{y_k \oplus 1} \prod_{b_{ki}<0} x_i^{-b_{ki}}. \tag{7.5}$$

□

Define $Y_{i;t}(y_1, \dots, y_n)$ as a rational subtraction-free expression for the coefficient $y_{i;t}$. Note that the functions $Y_{i;t}$ are defined for general coefficients, while $X_{i;t}$ and $F_{i;t}$ are defined for principal coefficients. The evaluation of $Y_{i;t}$ over a semifield $\mathcal{P}$ gives the value of $y_{i;t}$ in the cluster algebra over $\mathcal{P}$. In particular, for the cluster algebra with principal coefficients this evaluation equals $\prod_{j=1}^{n} y_j^{b^t_{i,n+j}}$. Using this notation, Proposition 7.3 can be restated as

$$\widehat{y}_{j;t} = Y_{j;t}|_{\mathcal{F}}(\widehat{y}_1, \dots, \widehat{y}_n). \tag{7.6}$$

We can now prove (7.3) by induction on the distance between t and t_0. The base of the induction is trivial since for $t = t_0$ both sides of (7.3) are equal to x_i.

Assume now that $x_{i;t}$, $i \in [1,n]$, satisfy (7.3) and that t' is adjacent to t in direction k. We have to prove that $x_{k;t'}$ satisfies (7.3) as well.

Indeed, by Proposition 7.3, exchange relation (7.5) can be rewritten as

$$x_{k;t'} = \frac{(Y_{k;t}+1)|_{\mathcal{F}}(\widehat{y}_1,\dots,\widehat{y}_n)}{(Y_{k;t}+1)|_{\mathcal{P}}(y_1,\dots,y_n)} x_{k;t}^{-1} \prod_{b_{ki}<0} x_{i;t}^{-b_{ki}}. \tag{7.7}$$

Applying the above formula to the algebra $\mathcal{A}_{\mathrm{pr}}(B)$ we obtain

$$X_{k;t'} = (Y_{k;t}+1)(\widehat{y}_1,\dots,\widehat{y}_n) \prod_{b^t_{k,n+i}<0} y_i^{-b^t_{k,n+i}} X_{k;t}^{-1} \prod_{b_{ki}<0} X_{i;t}^{-b_{ki}}. \tag{7.8}$$

Further specializing all x_i in (7.8) to 1 and evaluating the resulting expression in $\mathcal{P}$ we get

$$F_{k;t'}|_{\mathcal{P}} = (Y_{k;t}+1)|_{\mathcal{P}}(y_1,\dots,y_n) \prod_{b^t_{k,n+i}<0} y_i^{-b^t_{k,n+i}} F_{k;t}|_{\mathcal{P}}^{-1} \prod_{b_{ki}<0} F_{i;t}|_{\mathcal{P}}^{-b_{ki}}.$$

Therefore,

$$\begin{aligned}\frac{X_{k;t'}|_{\mathcal{F}}}{F_{k;t'}|_{\mathcal{P}}} &= \frac{(Y_{k;t}+1)|_{\mathcal{F}}(\widehat{y}_1,\dots,\widehat{y}_n)}{(Y_{k;t}+1)|_{\mathcal{P}}(y_1,\dots,y_n)} \frac{F_{k;t}|_{\mathcal{P}}}{X_{k;t}|_{\mathcal{F}}} \prod_{b_{ki}<0} \left(\frac{X_{i;t}|_{\mathcal{F}}}{F_{i;t}|_{\mathcal{P}}}\right)^{-b_{ki}} \\ &= \frac{(Y_{k;t}+1)|_{\mathcal{F}}(\widehat{y}_1,\dots,\widehat{y}_n)}{(Y_{k;t}+1)|_{\mathcal{P}}(y_1,\dots,y_n)} x_{k;t}^{-1} \prod_{b_{ki}<0} x_{i;t}^{-b_{ki}} = x_{k;t'}\end{aligned}$$

by (7.7). □

We can now complete the proof of the theorem. Indeed, assume that the seeds at the vertices t_1 and t_2 of $\mathbb{T}$ are equivalent in $\mathcal{A}_{\mathrm{pr}}(B)$. This means that there exists a permutation $\sigma \in S_n$ such that $X_{i;t_2} = X_{\sigma(i);t_1}$ for all $i \in [1,n]$. Therefore, $F_{i;t_2} = F_{\sigma(i);t_1}$ by the definition of F, and hence by Lemma 7.2, $x_{i;t_2} = x_{\sigma(i);t_1}$ in $\mathcal{A}$. To prove that $y_{i;t_2} = y_{\sigma(i);t_1}$ in $\mathcal{A}$, observe that specializing $x_1 = \dots = x_n = 1$ in (7.4) and (7.6) gives

$$Y_{i;t}|_{\mathcal{F}}(y_1,\dots,y_n) = y_{i;t} \prod_{j=1}^{n} F_{j;t}^{b^t_{ij}}$$

For the algebra with principal coefficients this translates to

$$Y_{i;t} = \prod_{b^t_{i,n+k}<0} y_i^{-b^t_{i,n+k}} \prod_{j=1}^{n} F_{j;t}^{b^t_{ij}}.$$

The latter equality yields $Y_{i;t_2} = Y_{\sigma(i);t_1}$ for all $i \in [1,n]$, and the result follows. □

7.2. The vertices and the edges of the exchange graph

By the definition, the vertices of the exchange graph are seeds considered up to an arbitrary permutation and if two such seeds are adjacent then they have exactly $n-1$ common cluster variables. In this Section we prove that under certain conditions, Conjecture 3.46 holds true, namely, the vertices of the exchange graph are, in fact, clusters considered up to an arbitrary permutation, and two such clusters are adjacent if and only if they have exactly $n-1$ common variables.

THEOREM 7.4. *Let a cluster algebra $\mathcal{A} \in \mathfrak{A}(B)$ satisfy one of the following two conditions:*

(i) $\mathcal{A}$ is of geometric type;

(ii) $\mathcal{A}$ is arbitrary and B is nondegenerate.

Then every seed in $\mathcal{A}$ is uniquely defined by its cluster (in other words, Conjecture 3.46(i) holds true for $\mathcal{A}$).

PROOF. Observe that Conjecture 3.46(i) is equivalent to the following one: suppose that the seeds at the vertices t_1 and t_2 of $\mathbb{T}$ satisfy relations

$$x_{i;t_2} = x_{\sigma(i);t_1} \tag{7.9}$$

for some $\sigma \in S_n$ and any $i \in [1, n]$, then

$$y_{i;t_2} = y_{\sigma(i);t_1} \tag{7.10}$$

for $i \in [1, n]$ and

$$b_{ij}^{t_2} = b_{\sigma(i)\sigma(j)}^{t_1} \tag{7.11}$$

for any $i, j \in [1, n]$.

Consider case (i) and assume first that the initial extended matrix $\widetilde{B}$ does not have zero rows. Then by Theorem 6.2 we can pick a closed 2-form ω compatible with $\mathcal{A}$ so that at any vertex t of $\mathbb{T}$ holds

$$B^t = D^{-1}\Omega_t^{\widetilde{\mathbf{x}}}[n; n+m] \tag{7.12}$$

where D is a skew-symmetrizer of B. Since (7.9) implies $\omega_{ij}^{\widetilde{\mathbf{x}}_{t_2}} = \omega_{\sigma(i)\sigma(j)}^{\widetilde{\mathbf{x}}_{t_1}}$, we get

$$b_{\sigma(i)\sigma(j)}^{t_1} = d_{\sigma(i)}^{-1}\omega_{\sigma(i)\sigma(j)}^{\widetilde{\mathbf{x}}_{t_1}} = d_{\sigma(i)}^{-1}\omega_{ij}^{\widetilde{\mathbf{x}}_{t_2}} = d_i d_{\sigma(i)}^{-1} b_{ij}^{t_2}. \tag{7.13}$$

Let us show that $d_i = d_{\sigma(i)}$. Indeed, relation (7.13) implies that σD is a skew-symmetrizer for B along with D. If B is irreducible then any two skew-symmetrizers coincide up to the multiplication by a positive constant, hence $d_i = \lambda d_{\sigma(i)}$ for some $\lambda > 0$. Iterating this equality we get $\lambda^k = 1$ for some integer $k > 1$, and hence $\lambda = 1$. If B is reducible then the same holds for any matrix mutation equivalent to B with the same partition into blocks, and hence σ cannot permute cluster variables from different blocks. Inside each block the previous reasoning remains valid. We thus proved that (7.11) holds true for any $i, j \in [1, n]$. Similarly, taking into account (7.12) and $d_i = d_{\sigma(i)}$ we get $b_{i,n+j}^{t_2} = b_{\sigma(i),n+j}^{t_1}$ for any $i, j \in [1, n]$. It remains to recall that for cluster algebras of geometric type the n-tuple $\mathbf{y}_t$ is completely defined by the initial n-tuple $\mathbf{y}$ and the matrix B^t.

Assume now that $\widetilde{B}$ has k zero rows. In this case $\mathcal{A}$ is a direct product of k copies of the simplest cluster algebra of rank 1 (see Example 3.9 with $b_1 = \cdots = b_{m+1} = 0$) and a cluster algebra defined by a $(n-k) \times (n+m)$-submatrix of $\widetilde{B}$ with no zero rows, for which the above reasoning applies.

Consider now case (ii). Evidently, (7.9) for the algebra $\mathcal{A}$ implies the same relation for the coefficient-free cluster algebra $\mathcal{A}_{\mathrm{cf}}(B)$. Since the latter is of geometric type, the result of case (i) applies, and hence relation (7.11) holds true. Therefore τ-coordinates defined by (4.1) at the vertices t_1 and t_2 coincide up to σ. Observe that for the coefficient-free cluster algebra, the n-tuple of τ-coordinates at a vertex t of $\mathbb{T}$ coincides with $\widehat{\mathbf{y}}_t$. Therefore, by (7.6), $Y_{i;t_2}(\tau_1, \dots, \tau_n) = Y_{\sigma(i);t_1}(\tau_1, \dots, \tau_n)$. Since B is nondegenarate, Lemma 4.3 implies that the transformation $\mathbf{x} \mapsto \tau$ is a bijection, and hence $Y_{i;t_2}(z_1,\dots,z_n) = Y_{\sigma(i);t_1}(z_1,\dots,z_n)$ for any set of variables $z_1,\dots,z_n$. Therefore, (7.10) holds for any cluster algebra $\mathcal{A} \in \mathfrak{A}(B)$. $\square$

THEOREM 7.5. *Let every seed in a cluster algebra $\mathcal{A} \in \mathfrak{A}(B)$ be uniquely defined by its cluster. Then two clusters are adjacent in the exchange graph of $\mathcal{A}$ if and only if they have exactly $n-1$ common variables. In particular, Conjecture 3.46(ii) holds true for $\mathcal{A}$ satisfying one of the conditions* (i) *or* (ii) *of Theorem 7.4.*

PROOF. Denote by $x_1,\dots,x_{n-1}$ the common variables in the two clusters and by $\dot{x}$ and $\ddot{x}$ the remaining variables; the clusters themselves will be denoted $\dot{\mathbf{x}}$ and $\ddot{\mathbf{x}}$, respectively. By Theorem 3.14, $\ddot{x}$ can be written as a Laurent polynomial in $x_1,\dots,x_{n-1},\dot{x}$. Since each cluster transformation is birational, $\dot{x}$ enters this polynomial with exponent 1 or -1; we write this as $\ddot{x} = \mathcal{L}_0 + L_1\dot{x}^{\pm 1}$ where L_0 and L_1 are Laurent polynomials in $x_1,\dots,x_{n-1}$. Denote by $\ddot{x}'$ the cluster variable in the cluster adjacent to $\ddot{\mathbf{x}}$ that replaces $\ddot{x}$, then

$$\ddot{x}' = \frac{M+N}{L_0 + L_1\dot{x}^{\pm 1}}, \tag{7.14}$$

where M and N are monomials in $x_1,\dots,x_{n-1}$. Since $\ddot{x}'$ is a Laurent polynomial in $x_1,\dots,x_{n-1},x$, we immediately get $L_0 = 0$.

We have to consider two cases: $\ddot{x} = L_1\dot{x}$ and $\ddot{x} = L_1/\dot{x}$. In the first case we write down $\dot{x} = L_1^{-1}\ddot{x}$ and apply once again Theorem 3.14 to see that L_1 is a Laurent monomial M_+. In the second case we use (7.14) to get $\ddot{x}' = M_-\dot{x}$ for some Laurent monomial M_-. It remains to prove that both M_+ and M_- are identically equal to 1. In the first case this would mean $\ddot{\mathbf{x}}$ and $\dot{\mathbf{x}}$ have n common cluster variables, a contradiction. In the second case this would mean that $\dot{\mathbf{x}}$ is adjacent to $\ddot{\mathbf{x}}$ as required.

Assume that a variable x_i enters M_+ with a negative exponent $-k$, $k > 0$. Let x_i' denote the cluster variable that replaces x_i in the cluster adjacent to $\dot{\mathbf{x}}$ in the corresponding direction. Then $x_i x_i' = M_i + N_i$, where M_i and N_i are monomials in $x_1,\dots,x_{n-1},\dot{x}$. Therefore,

$$\ddot{x} = \frac{\dot{x}(x_i')^k M_+'}{(M_i+N_i)^k},$$

where M_+' is a Laurent monomial. We may assume that B does not have zero rows; for the case of nondegenerate B this condition holds trivially, and the geometric type is treated as before. Hence, the denominator in the above expression is a nontrivial polynomial, which contradicts the Laurent property. To handle the case of a positive exponent $k > 0$ we write $\dot{x} = M_+^{-1}\ddot{x}$ and proceed in a similar way using the variable x_i'' that replaces x_i in the cluster adjacent to $\ddot{\mathbf{x}}$ in the corresponding direction. We thus obtained that M_+ is a constant. It is an easy exercise to prove that the constant is equal to 1. The case of M_- is handled similarly. $\square$

REMARK 7.6. Note that we do not claim that the whole cluster complex is determined by set-theoretic combinatorics of subsets (clusters) of the ground set of all cluster variables, although we strongly believe that it is true. Theorem 7.5 states only that the exchange graph of a cluster algebra is determined by this combinatorics. In other words, we can restore which maximal simplices are adjacent by a codimension one face.

7.3. Exchange graphs and exchange matrices

In this Section we prove that all cluster algebras in $\mathfrak{A}(B)$ share the same exchange graph, provided B is nondegenerate, thus establishing Conjecture 3.45 for nondegenerate initial exchange matrices.

THEOREM 7.7. *Let B be nondegenerate, then the exchange graphs of all cluster algebras in $\mathfrak{A}(B)$ coincide.*

PROOF. In view of Theorem 7.1, it suffices to prove that the exchange graphs for $\mathcal{A}\mathrm{pr}(B)$ and $\mathcal{A}_{\mathrm{cf}}(B)$ coincide.

Define rational functions $G_{i;t}$ by $G_{i;t}(x_1,\dots,x_n) = X_{i;t}(x_1,\dots,x_n;1,\dots,1)$; it stems immediately from (7.3) that $G_{i;t}$ express the variables $x_{i;t}$ via the initial variables in the coefficient-free algebra $\mathcal{A}_{\mathrm{cf}}(B)$. By Theorem 7.4, we have to prove the following implication: if

$$G_{i;t_2} = G_{\sigma(i);t_1} \tag{7.15}$$

for some $\sigma \in S_n$, some $t_1, t_2 \in \mathbb{T}$ and any $i \in [1,n]$ then

$$X_{i;t_2} = X_{\sigma(i);t_1} \tag{7.16}$$

for any $i \in [1,n]$. The proof is based on the following lemma.

LEMMA 7.8. *For any $t \in T$ there exist Laurent monomials $M_{i;t}$ such that*

$$\begin{aligned} &X_{i;t}(x_1,\dots,x_n;y_1,\dots,y_n) \\ &= M_{i;t}(z_1,\dots,z_n)G_{i;t}(x_1M_1(z_1,\dots,z_n),\dots,x_nM_n(z_1,\dots,z_n)), \qquad i\in[1,n], \end{aligned} \tag{7.17}$$

where $M_i = M_{i;t_0}$ and $z_i = y_i^{1/\det B}$.

PROOF. Consider toric actions on $\mathcal{A}_{\mathrm{pr}}(B)$ similar to those introduced in Section 5.2. A slight modification of Lemma 5.3 guarantees that if $\mathbf{w}^1 = (w_1^1,\dots,w_{2n}^1)$, $\dots$, $\mathbf{w}^n = (w_1^n,\dots,w_{2n}^n)$ are integer weights at the initial vertex t_0 then the local toric action at t_0 given by $\widetilde{\mathbf{x}} \mapsto \widetilde{\mathbf{x}} \cdot t_1^{\mathbf{w}^1}\dots t_n^{\mathbf{w}^n}$ can be extended to a global toric action. Define the weights $\mathbf{w}^i$ as follows: the first n entries of $\mathbf{w}^i$ constitute the ith row of the matrix B^{-1} multiplied by $\det B$, while its last n entries constitute the ith row of the $n\times n$-matrix $\mathrm{diag}(-\det B,\dots,-\det B)$. The corresponding monomials M_i are defined by $M_i(t_1,\dots,t_n) = \prod_{j=1}^n t_j^{w_j^i}$. Then the compatibility condition of Section 5.2 gives

$$\begin{aligned} X_{i;t}(x_1t_1^{w_1^1}\dots t_n^{w_1^n},\dots,x_nt_1^{w_n^1}\dots t_n^{w_n^n};y_1t_1^{-\det B},\dots,y_nt_n^{-\det B}) \\ = N_{i;t}(t_1,\dots,t_n)X_{i;t}(x_1,\dots,x_n;y_1,\dots,y_n). \end{aligned}$$

Relation (7.17) follows from the above condition with $t_i = z_i$ and $M_{i;t} = N_{i;t}^{-1}$. □

From (7.15) and (7.17) we get $X_{i;t_2} = M_{i;t_2}M_{\sigma(i);t_1}^{-1}X_{\sigma(i);t_1}$. Let us prove that the monomial $M(y_1,\dots,y_n) = M_{i;t_2}M_{\sigma(i);t_1}^{-1}$ is, in fact, trivial.

Indeed, assume that there exists a variable, say, y_1, that enters M with a negative exponent. Consider the $(n+1)\times 2n$-matrix $\widehat{B}^{t_1}$ obtained from $(B_{\mathrm{pr}})^{t_1}$ by adding the $(n+1)$th row $(d_1b_{1,n+1}^{t_1},\dots,d_nb_{n,n+1}^{t_1},0,\dots,0)$, where $d_1,\dots,d_n$ are the diagonal entries of D. Clearly, $\mathcal{A}_{\mathrm{pr}}(B)$ is the restriction of $\mathcal{A}(\widehat{B}^{t_1})$ to the first n variables; therefore, $x_{i;t_2} = x_{n+1;t_1}^kM'(y_2,\dots,y_n)x_{\sigma(i);t_1}$ in $\mathcal{A}(\widehat{B}^{t_1})$, where $x_{n+1;t_1}$

is naturally identified with y_1. The rest of the proof proceeds exactly as the proof of the second part of Conjecture 3.46 in Theorem 7.5. □

7.4. Summary

- We discuss general cluster algebras (not necessary of geometric type) and study the dependence of the exchange graph on coefficients. We define principal coefficients and prove that the exchange graph of a cluster algebra with principal coefficients covers the exchange graph of any other cluster algebra with the same exchange matrix, see Theorem 7.1.
- Using properties of compatible 2-forms, we prove that every seed of a cluster algebra $\mathcal{A}$ is completely determined by the collection of cluster variables in this seed, i.e., by its cluster, in the following two cases: (i) $\mathcal{A}$ is a cluster algebra of geometric type, or (ii) the exchange matrix of $\mathcal{A}$ is nondegenerate, see Theorem 7.4.
- We deduce then that if every seed of a cluster algebra $\mathcal{A}$ of rank n is determined by its cluster, then the exchange graph of $\mathcal{A}$ is determined by clusters of $\mathcal{A}$. Namely, two clusters are adjacent in the exchange graph if and only if the clusters share exactly $n-1$ common variables, see Theorem 7.5. In particular, this holds for both cluster algebras of geometric type and cluster algebras with a nondegenerate exchange matrix.
- Finally, we prove that if the exchange matrix is nondegenerate, then any two cluster algebras sharing this exchange matrix have isomorphic exchange graphs, see Theorem 7.7.

Bibliographical notes

7.1. All results in this Section are borrowed from [**FZ7**]. The fact that any subtraction-free identity in the field of rational functions remains valid in any semifield is proved in [**BFZ1**], Lemma 2.1.6.

7.2. Theorems 7.4 and 7.5 are proved in [**GSV4**]. Theorem 7.4(i) for cluster algebras with no stable variables with a skew-symmetric exchange matrix satisfying an additional acyclicity condition was proved in [**BMRT**]. The proof of Theorem 7.4(ii) is similar to the proof of Proposition 2.7 in [**FoG**].

7.3. Theorem 7.7 is proved in [**GSV4**].

CHAPTER 8

Perfect planar networks in a disk and Grassmannians

We have already seen in Section 2.2.2 that to describe connected components in the real part of a reduced double Bruhat cell $L^{u,v}(\mathbb{R}) \subset SL_n(\mathbb{R})$, $u, v \in S_n$, one makes use of a factorization of upper triangular matrices into a product of elementary bidiagonal matrices of a special form. Each elementary matrix depends on one parameter t_i called a factorization parameter. This factorization is not unique; all possible factorizations are labeled by reduced words for the pair (u, v) (see Section 1.2.3). Each reduced word gives rise to its own toric chart, and the union of all such charts covers $L^{u,v}(\mathbb{R})$ up to a codimension two subset. In Section 2.2.3 we introduce regular functions M_i on $L^{u,v}(\mathbb{R})$, which are expressed monomially in terms of t_i, and use them to find the number of connected components of $L^{u,v}(\mathbb{R})$.

Most importantly, there are elementary transformations of reduced words that do not change the resulting product of simple transpositions. An elementary transformation of a reduced word leads to the elementary transformation of the matrix factorization. We change factorization parameters in such a way that the resulting product remains unchanged; this condition determines a birational transformation of factorization parameters. The change of factorization parameters leads to a cluster algebra type change of the corresponding system of regular functions M_i. Poisson structures compatible with the arising cluster algebra include the famous Sklyanin bracket, see Chapter 4.

Factorizations discussed above can be visualized via representing an elementary bidiagonal matrix by a simple weighted directed planar graph. The product of elementary bidiagonal matrices is given by the concatenation of graphs corresponding to factors. This tool has been used a lot in the study of totally positive matrices. It turns out that Poisson brackets mentioned above also enjoy a natural graphical interpretation.

In this Chapter we develop a similar construction for Grassmannians. In order to achieve our goal we utilize the notion of perfect planar networks. Such a network is a directed planar trivalent graph satisfying certain conditions. The edges of the network are equipped with weights. The boundary measurement map provides a parametrization of cells in a certain cell decomposition of the Grassmannian by edge weights, more exactly, by edge weights modulo gauge transformations. The edge weights can be viewed as analogs of factorization parameters discussed above. A more convenient parametrization invariant under gauge transformations is provided by face weights. Face weights are restored from the boundary measurements, and are, in a sense, similar to the functions τ_i (see Remark 2.14).

Next, we introduce a natural family of Poisson brackets on the space of edge weights. These brackets induce, via the boundary measurement map, a two-parameter family of Poisson structures on $\mathrm{Mat}_{k,n-k}$. The map from edge weights to matrices is then extended to a map into the Grassmannian $G_k(n)$ that is Poisson with respect to a certain two-parameter family of Poisson bracket on the Grassmannian. Every Poisson bracket in this family is compatible with the cluster algebra on the Grassmannian described in Chapter 4.

Furthermore, for a special class of networks the boundary measurement map extends to a Poisson map to GL_n equipped with a family of R-matrix Poisson–Lie structures. The corresponding R-matrices form a two-dimensional linear space, which contain the standard R-matrix. The resulting families of Poisson structures on $G_k(n)$ and on GL_n do not depend on a particular network. Every Poisson bracket in the former family is Poisson homogeneous with respect to the natural right action of GL_n equipped with the R-matrix Poisson–Lie bracket with matching parameters.

8.1. Perfect planar networks and boundary measurements

8.1.1. Networks, paths and weights. Let $G = (V, E)$ be a directed planar graph drawn inside a disk (and considered up to an isotopy) with the vertex set V and the edge set E. Exactly n of its vertices are located on the boundary circle of the disk. They are labelled counterclockwise $b_1, \dots, b_n$ and called *boundary vertices*; occasionally we will write b_0 for b_n and b_{n+1} for b_1. Each boundary vertex is marked as a source or a sink. A *source* is a vertex with exactly one outcoming edge and no incoming edges. *Sinks* are defined in the same way, with the direction of the single edge reversed. The number of sources is denoted by k, and the corresponding set of indices, by $I \subset [1, n]$; the set of the remaining $m = n - k$ indices is denoted by J. All the internal vertices of G have degree 3 and are of two types: either they have exactly one incoming edge, or exactly one outcoming edge. The vertices of the first type are called (and shown on figures) *white*, those of the second type, *black*.

Let $x_1, \dots, x_d$ be independent variables. A *perfect planar network* $N = (G, w)$ is obtained from G as above by assigning an *edge weight* $w_e \in \mathbb{Z}(x_1, \dots, x_d)$ to each edge $e \in E$. In what follows we occasionally write "network" instead of "perfect planar network". Each network defines a rational map $w : \mathbb{R}^d \to \mathbb{R}^{|E|}$; the *space of edge weights* $\mathcal{E}_N$ is defined as the intersection of the image of $(\mathbb{R} \setminus 0)^d$ under w with $(\mathbb{R} \setminus 0)^{|E|}$. In other words, a point in $\mathcal{E}_N$ is a graph G as above with edges weighted by nonzero reals obtained by specializing the variables $x_1, \dots, x_d$ in the expressions for w_e to nonzero values.

An example of a perfect planar network is shown in Fig. 8.1. It has two sources: b_1 and b_2, and two sinks b_3 and b_4. Each edge e_i is labelled by its weight. The weights depend on four independent variables x_1, x_2, x_3, x_4 and are given by

$$w_1 = x_1^2/(x_2+1), \qquad w_2 = x_2, \qquad w_3 = x_2 + 1, \qquad w_4 = x_1 + x_3,$$
$$w_5 = x_3, \qquad w_6 = x_3, \qquad w_7 = x_3, \qquad w_8 = x_4,$$
$$w_9 = 1, \qquad w_{10} = 1, \qquad w_{11} = 1.$$

The space of edge weights is the 4-dimensional subvariety in $(\mathbb{R} \setminus 0)^{11}$ given by equations $w_1 w_3 = (w_4 - w_5)^2$, $w_3 = w_2 + 1$, $w_5 = w_6 = w_7$, $w_9 = w_{10} = w_{11} = 1$ and condition $w_3 \neq 1$.

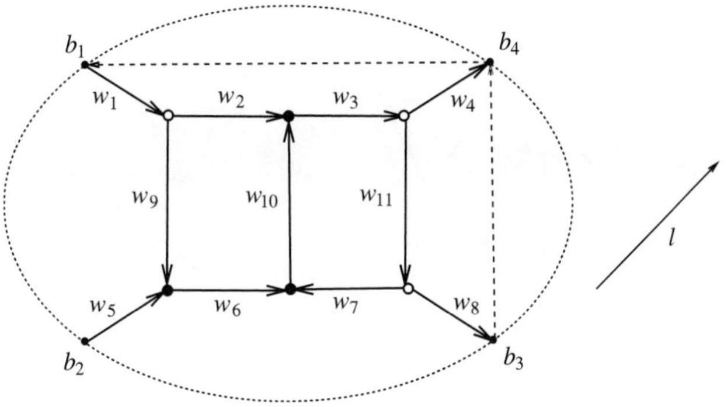

FIGURE 8.1. A perfect planar network in a disk

A *path* P in N is an alternating sequence $(v_1, e_1, v_2, \dots, e_r, v_{r+1})$ of vertices and edges such that $e_i = (v_i, v_{i+1})$ for any $i \in [1, r]$. Sometimes we omit the names of the vertices and write $P = (e_1, \dots, e_r)$. A path is called a *cycle* if $v_{r+1} = v_1$ and a *simple cycle* if additionally $v_i \neq v_j$ for any other pair $i \neq j$.

To define the weights of the paths we need the following construction. Consider a closed oriented polygonal plane curve C. Let e' and e'' be two consequent oriented segments of C, and let v be their common vertex. We assume for simplicity that for any such pair (e', e''), the cone spanned by e' and e'' is not a line; in other words, if e' and e'' are collinear, then they have the same direction. Observe that since C is not necessary simple, there might be other edges of C incident to v (see Figure 8.2 below). Let l be an arbitrary oriented line. Define $c_l(e', e'') \in \mathbb{Z}/2\mathbb{Z}$ in the following way: $c_l(e', e'') = 1$ if the directing vector of l belongs to the interior of the cone spanned by e' and e'' and $c_l(e', e'') = 0$ otherwise (see Figure 8.2 for examples). Define $c_l(C)$ as the sum of $c_l(e', e'')$ over all pairs of consequent segments in C. One can check that $c_l(C)$ does not depend on l, provided l is not collinear to any of the segments in C. The common value of $c_l(C)$ for different choices of l is denoted by $c(C)$ and called the *concordance number* of C. In fact, $c(C)$ equals $\bmod 2$ the rotation number of C; the definition of the latter is similar, but more complicated.

In what follows we assume without loss of generality that N is drawn in such a way that all its edges are straight line segments and all internal vertices belong to the interior of the convex hull of the boundary vertices. Given a path P between a source b_i and a sink b_j, we define a closed polygonal curve C_P by adding to P the path between b_j and b_i that goes counterclockwise along the boundary of the convex hull of all the boundary vertices of N. Finally the weight of P is defined as

$$w_P = (-1)^{c(C_P)-1} \prod_{e \in P} w_e.$$

The weight of an arbitrary cycle in N is defined in the same way via the concordance number of the cycle.

If edges e_i and e_j in P coincide and $i < j$, the path P can be decomposed into two parts: the path $P' = (e_1, \dots, e_{i-1}, e_i = e_j, e_{j+1}, \dots, e_r)$ and the cycle $C^0 = (e_i, e_{i+1}, \dots, e_{j-1})$. Clearly, $c(C_P) = c(C_{P'}) + c(C^0)$, and hence

$$w_P = -w_{P'} w_{C^0}. \tag{8.1}$$

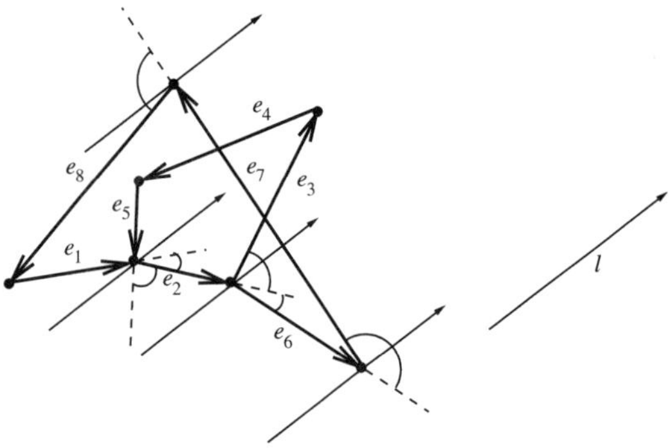

FIGURE 8.2. To the definition of the concordance number: $c_l(e_1, e_2) = c_l(e_5, e_2) = 0$; $c_l(e_2, e_3) = 1, c_l(e_2, e_6) = 0$; $c_l(e_6, e_7) = 1, c_l(e_7, e_8) = 0$

EXAMPLE 8.1. Consider the path $P = (e_1, e_2, e_3, e_{11}, e_7, e_{10}, e_3, e_{11}, e_8)$ in Figure 8.1. Choose l as shown in the Figure; l is neither collinear with the edges of P nor with the relevant edges of the convex hull of boundary vertices (shown by dotted lines). Clearly, $c_l(e', e'') = 0$ for all pairs of consecutive edges of C_P except for the pairs (e_{10}, e_3) and $(e_8, \bar{e})$, where $\bar{e}$ is the additional edge joining b_3 and b_4. So, $c(C_P) = 0$, and hence $w_P = -w_1 w_2 w_3^2 w_7 w_8 w_{10} w_{11}^2$. The same result can be obtained by decomposing P into the path $P' = (e_1, e_2, e_3, e_{11}, e_8)$ and the cycle $C^0 = (e_3, e_{11}, e_7, e_{10})$.

REMARK 8.2. Instead of closed polygonal curves C_P, one can use curves C_P^* obtained by adding to P the path between b_j and b_i that goes clockwise along the boundary of the convex hull of all the boundary vertices of N. It is a simple exercise to prove that the concordance numbers of C_P and C_P^* coincide. Therefore, the weight of a path can be defined also as

$$w_P = (-1)^{c(C_P^*)-1} \prod_{e \in P} w_e.$$

8.1.2. Boundary measurements. Given a perfect planar network as above, a source b_i, $i \in I$, and a sink b_j, $j \in J$, we define the *boundary measurement* $M(i, j)$ as the sum of the weights of all paths starting at b_i and ending at b_j. Clearly, the boundary measurement thus defined is a formal infinite series in variables w_e, $e \in E$. However, this series possesses certain nice properties.

Recall that a formal power series $g \in \mathbb{Z}[[w_e, e \in E]]$ is called a rational function if there exist polynomials $p, q \in \mathbb{Z}[w_e, e \in E]$ such that $p = qg$ in $\mathbb{Z}[[w_e, e \in E]]$. In this case we write $g = p/q$. For example, $1 - z + z^2 - z^3 + \cdots = (1 + z)^{-1}$ in $\mathbb{Z}[[z]]$. Besides, we say that g admits a *subtraction-free* rational expression if it can be written as a ratio of two polynomials with nonnegative coefficients. For example, $x^2 - xy + y^2$ admits a subtraction-free rational expression since it can be written as $(x^3 + y^3)/(x + y)$.

PROPOSITION 8.3. *Let N be a perfect planar network in a disk, then each boundary measurement in N is a rational function in the weights w_e admitting a subtraction-free rational expression.*

PROOF. We prove the claim by induction on the number of internal vertices. The base of induction is the case when there are no internal vertices at all, and hence each edge connects a source and a sink; in this case the statement of the proposition holds trivially.

Assume that N has r internal vertices. Consider a specific boundary measurement $M(i,j)$. The claim concerning $M(i,j)$ is trivial if b_j is the neighbor of b_i. In the remaining cases b_i is connected by an edge e_0 to its only neighbor in G, which is either white or black.

Assume first that the neighbor of b_i is a white vertex v. Create a new network $\widetilde{N}$ by deleting b_i and the edge (b_i,v) from G, splitting v into 2 sources $b_{i'_v}, b_{i''_v}$ (so that $i-1 \prec i'_v \prec i''_v \prec i+1$ in the counterclockwise order) and replacing the edges $e_1=(v,v')$ and $e_2=(v,v'')$ by $(b_{i'_v},v')$ and $(b_{i''_v},v'')$, respectively, both of weight 1 (see Figure 8.3). Clearly, to any path P from b_i to b_j corresponds either a path P' from $b_{i'_v}$ to b_j or a path P'' from $b_{i''_v}$ to b_j. Moreover, $c(C_P)=c(C_{P'})=c(C_{P''})$. Therefore

$$M(i,j)=w_{e_0}(w_{e_1}\widetilde{M}(i'_v,j)+w_{e_2}\widetilde{M}(i''_v,j)),$$

where $\widetilde{M}$ means that the measurement is taken in $\widetilde{N}$. Observe that the number of internal vertices in $\widetilde{N}$ is $r-1$, hence the claim follows from the above relation by induction.

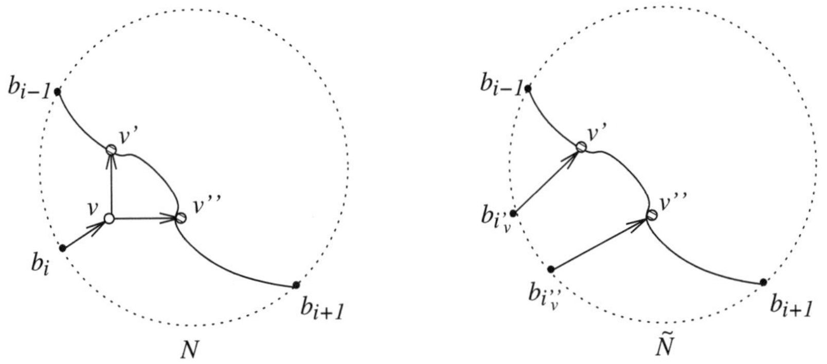

FIGURE 8.3. Splitting a white vertex

Assume now that the neighbor of b_i is a black vertex u. Denote by u_+ the unique vertex in G such that $(u,u_+)\in E$, and by u_- the neighbor of u distinct from u_+ and b_i. Create a new network $\widehat{N}$ by deleting b_i and the edge (b_i,u) from G, splitting u into one new source b_{i_u} and one new sink b_{j_u} (so that either $i-1\prec i_u\prec j_u\prec i+1$ or $i-1\prec j_u\prec i_u\prec i+1$ in the counterclockwise order) and replacing the edges $e_+=(u,u_+)$ and $e_-=(u_-,u)$ by new edges $\hat{e}_+$ and $\hat{e}_-$ in the same way as in the previous case, see Figure 8.4.

Let us classify the paths from b_i to b_j according to the number of times they traverse e_+. Similarly to the previous case, the total weight of the paths traversing e_+ only once is given by $w_{e_0}w_{e_+}\widehat{M}(i_v,j)$, where $\widehat{M}$ means that the measurement is taken in $\widehat{N}$. Any path P traversing e_+ exactly twice can be represented as $P=(e_0,e_+,P_1,e_-,e_+,P_2)$ for some path $\widehat{P}_1=(\hat{e}_+,P_1,\hat{e}_-)$ from i_u to j_u in $\widehat{N}$ and a path $\widehat{P}_2=(\hat{e}_+,P_2)$ from i_u to b_j in $\widehat{N}$. Clearly, $C_1=(e_+,P_1,e_-)$ is a cycle in N, and $c(C_1) = c(C_{\widehat{P}^1})$. Besides, $c(C_{P'}) = c(C_{\widehat{P}_2})$ for $P' = (e_0, e_+, P_2)$. Therefore, by (8.1), $w_P = -w_{C_1} w_{P'}$. Taking into account that $w_{C_1} = w_{e_-} w_{e_+} w_{\widehat{P}_1}$ and $w_{P'} = w_{e_0} w_{e_+} w_{\widehat{P}_2}$, we see that the total contribution of all paths traversing e_+ exactly twice to $M(i,j)$ equals

$$-w_{e_0} w_{e_+} \widehat{M}(i_u, j) w_{e_-} w_{e_+} \widehat{M}(i_u, j_u).$$

In general, the total contribution of all paths traversing e_+ exactly $s+1$ times to $M(i,j)$ equals

$$(-1)^s w_{e_0} w_{e_+} \widehat{M}(i_u, j)(w_{e_-} w_{e_+} \widehat{M}(i_u, j_u))^s.$$

Therefore, we find

$$M(i,j) = \frac{w_{e_0} w_{e_+} \widehat{M}(i_u, j)}{1 + w_{e_-} w_{e_+} \widehat{M}(i_u, j_u)} \tag{8.2}$$

and the claim follows by induction, since the number of internal vertices in $\widehat{N}$ is $r-1$. □

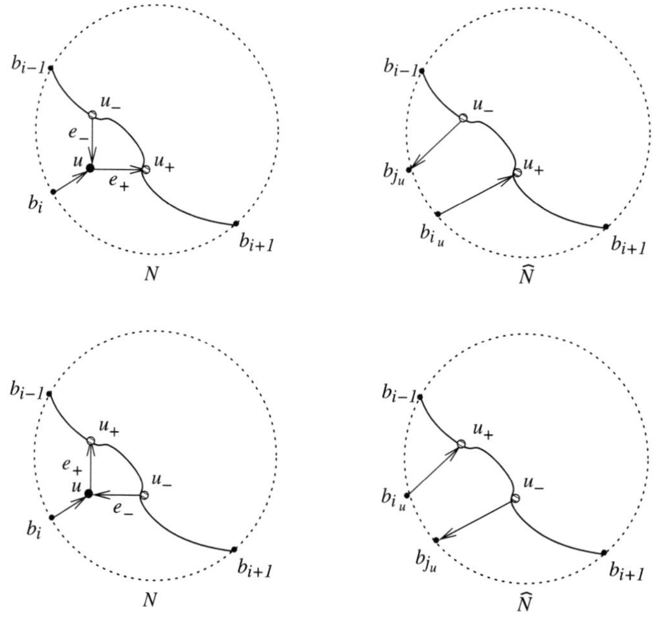

FIGURE 8.4. Splitting a black vertex: cases $i-1 \prec i_u \prec j_u \prec i+1$ (upper part) and $i-1 \prec j_u \prec i_u \prec i+1$ (lower part)

Boundary measurements can be organized into a $k \times m$ *boundary measurement matrix* M_N in the following way. Let $I = \{i_1 < i_2 < \cdots < i_k\}$ and $J = \{j_1 < j_2 < \cdots < j_m\}$. We define $M_N = (M_{pq})$, $p \in [1,k]$, $q \in [1,m]$, where $M_{pq} = M(i_p, j_q)$. Each network N defines a rational map $\mathcal{E}_N \to \mathrm{Mat}_{k,m}$ given by M_N and called the *boundary measurement map* corresponding to N.

EXAMPLE 8.4. Consider the network shown in Figure 8.1. The corresponding boundary measurement matrix is a 2×2 matrix given by

$$\begin{pmatrix} \dfrac{w_3w_4w_5w_6w_{10}}{1+w_3w_7w_{10}w_{11}} & \dfrac{w_3w_5w_6w_8w_{11}}{1+w_3w_7w_{10}w_{11}} \\ \dfrac{w_1w_3w_4(w_2+w_6w_9w_{10})}{1+w_3w_7w_{10}w_{11}} & \dfrac{w_1w_3w_8w_{11}(w_2+w_6w_9w_{10})}{1+w_3w_7w_{10}w_{11}} \end{pmatrix}.$$

8.2. Poisson structures on the space of edge weights and induced Poisson structures on $\mathrm{Mat}_{k,m}$

8.2.1. A natural operation on networks is their *concatenation*, which consists, roughly speaking, in gluing some sinks/sources of one network to some of the sources/sinks of the other. We expect any Poisson structure associated with networks to behave naturally under concatenation. To obtain from two planar networks a new one by concatenation, one needs to select a segment from the boundary of each disk and identify these segments via a homeomorphism in such a way that every sink (resp. source) contained in the selected segment of the first network is glued to a source (resp. sink) of the second network. We can then erase the common piece of the boundary along which the gluing was performed and identify every pair of glued edges in the resulting network with a single edge of the same orientation and with the weight equal to the product of two weights assigned to the two edges that were glued.

As an illustration, let us review a particular but important case, in which sources and sinks of the network do not interlace. In this case, it is more convenient to view the network as located in a square rather than in a disk, with all sources located on the left side and sinks on the right side of the square. It will be also handy to label sources (resp. sinks) 1 to k (resp. 1 to m) going from the bottom to the top. This results in a different way of recording boundary measurements into a matrix. Namely, if M is the boundary measurements matrix we defined earlier, then now we associate with the network the matrix $A = MW_0$, where $W_0 = (\delta_{i,m+1-j})_{i,j=1}^m$ is the matrix of the longest permutation w_0.

We can now concatenate two networks of this kind, one with k sources and m sinks and another with m sources and l sinks, by gluing the sinks of the former to the sources of the latter. If A_1, A_2 are matrices associated with the two networks, then it is clear that the matrix associated with their concatenation is A_1A_2. Note that this "visualization" of the matrix multiplications is particularly relevant when one deals with factorization of matrices into products of elementary bidiagonal matrices that was mentioned in Section 1.2.4, in particular, in Example 1.2. Indeed, a $n \times n$ diagonal matrix $\mathrm{diag}(d_1, \dots, d_n)$ and elementary bidiagonal matrices

$$E_i^-(l) = \mathbf{1} + le_{i+1,i}, \qquad E_j^+(u) = \mathbf{1} + ue_{j,j+1} \tag{8.3}$$

correspond to planar networks shown in Figure 8.5 a, b and c, respectively; all weights not shown explicitly are equal to 1.

The construction of Poisson structures we are about to present is motivated by the way in which the network representation of the bidiagonal factorization reflect the *standard Poisson-Lie structure* (1.25) on SL_n.

Let us take another look at the explicit formulas for the standard Poisson-Lie structure on SL_2, which, when restricted to upper and lower Borel subgroups of

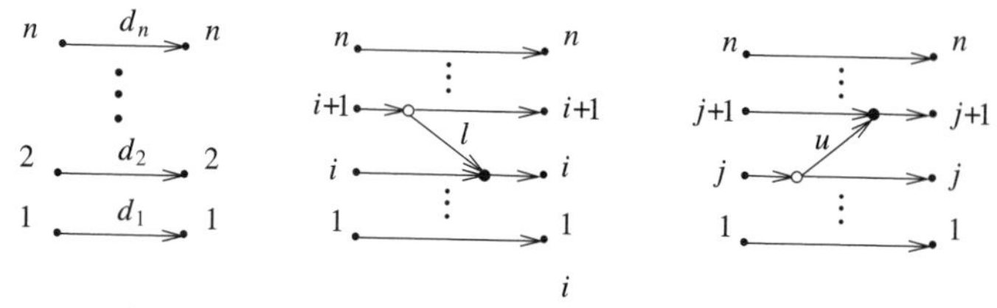

FIGURE 8.5. Three networks used in matrix factorization

$\mathcal{B}_\pm \subset SL_2$

$$\mathcal{B}_+ = \left\{ \begin{pmatrix} p & q \\ 0 & p^{-1} \end{pmatrix} \right\}, \qquad \mathcal{B}_- = \left\{ \begin{pmatrix} p & 0 \\ q & p^{-1} \end{pmatrix} \right\},$$

has an especially simple – log-canonical – form

$$\{p, q\} = \frac{1}{2} pq.$$

Recall now that Poisson embeddings $\rho_i : SL_2 \to SL_n$ $(i \in [1, n-1])$ that are used to characterize the standard Poisson-Lie structure on SL_n map SL_2 into SL_n as a diagonal 2×2 block occupying rows and columns i and $i+1$.

Note that the network that represents $\rho_i(\mathcal{B}_-)$ looks like the second network in the figure above with the weights p, p^{-1} and q attached to edges $(i-1) \to (i-1)$, $i \to i$ and $i \to (i-1)$ resp., while a network that represents $\rho_j(\mathcal{B}_+)$ looks like the third network in the figure above with the weights p, p^{-1} and q attached to edges $(j-1) \to (j-1)$, $j \to j$ and $(j-1) \to j$ resp. Concatenation of several networks $N_1, \cdots, N_r$, $r = n(n-1)$, of these two kinds, with appropriately chosen order and with each diagram having its own pair of nontrivial weights c_i, d_i, describes a generic element of SL_n.

As we explained in Section 1.2.4, the order is prescribed by a reduced word for the pair (w_0, w_0). For example, the network in Figure 8.6 corresponds to the reduced word $(-1\ -2\ -1\ -3\ -2\ -1 \ldots -(n-1) \ldots -2\ -1\ 1\ 2 \ldots (n-1) \ldots 1\ 2\ 3\ 1\ 2\ 1)$.

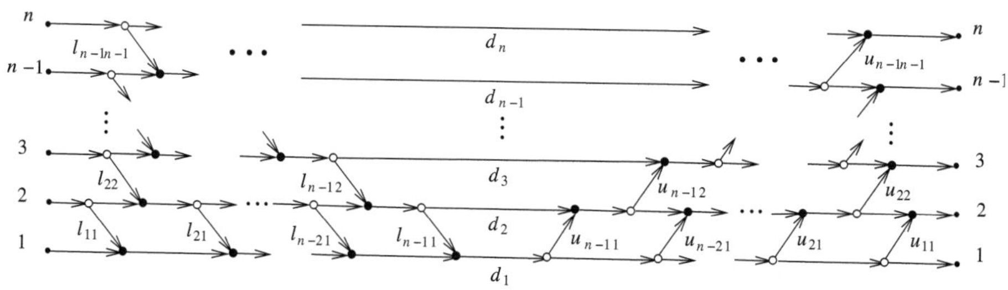

FIGURE 8.6. Generic planar network

On the other hand, due to the Poisson-Lie property, the Poisson structure on SL_n is then inherited from simple Poisson brackets for parameters c_i, d_i, which can be described completely in terms of networks: (i) the bracket is log-canonical; (ii) the only pairs of weights having nonzero brackets are those that correspond to two edges having a common source/sink; (iii) the constant entering the definition of a log-canonical bracket is $\pm\frac{1}{2}$ if the corresponding edges follow one another around the source/sink in the counterclockwise direction. The corresponding Poisson-Lie structure on GL_n is obtained by requiring the determinant to be a Casimir function.

This example motivates conditions we impose below on a natural Poisson structure associated with a 3-valent planar directed network.

8.2.2. Poisson structures on the space of edge weights. Let G be a directed planar graph in a disk as described in Section 8.1.1. A pair (v, e) is called a *flag* if v is an endpoint of e. To each internal vertex v of G we assign a 3-dimensional space $(\mathbb{R}\setminus 0)^3_v$ with coordinates x^1_v, x^2_v, x^3_v. We equip each $(\mathbb{R}\setminus 0)^3_v$ with a Poisson bracket $\{\cdot,\cdot\}_v$. It is convenient to assume that the flags involving v are labelled by the coordinates, as shown in Figure 8.7.

FIGURE 8.7. Edge labelling for $(\mathbb{R}\setminus 0)^3_v$

Besides, to each boundary vertex b_j of G we assign a 1-dimensional space $(\mathbb{R}\setminus 0)_j$ with the coordinate x^1_j (in accordance with the above convention, this coordinate labels the unique flag involving b_j). Define $\mathcal{R}$ to be the direct sum of all the above spaces; thus, the dimension of $\mathcal{R}$ equals twice the number of edges in G. Note that $\mathcal{R}$ is equipped with a Poisson bracket $\{\cdot,\cdot\}_{\mathcal{R}}$, which is defined as the direct sum of the brackets $\{\cdot,\cdot\}_v$; that is, $\{x, y\}_{\mathcal{R}} = 0$ whenever x and y are not defined on the same $(\mathbb{R}\setminus 0)^3_v$. We say that the bracket $\{\cdot,\cdot\}_{\mathcal{R}}$ is *universal* if each of $\{\cdot,\cdot\}_v$ depends only on the color of the vertex v.

Define the weights w_e by

$$w_e = x^i_v x^j_u, \tag{8.4}$$

provided the flag (v, e) is labelled by x^i_v and the flag (u, e) is labelled by x^j_u. In other words, the weight of an edge is defined as the product of the weights of the two flags involving this edge. Therefore, in this case the space of edge weights $\mathcal{E}_N$ coincides with the entire $(\mathbb{R}\setminus 0)^{|E|}$, and the weights define a *weight map* $w\colon (\mathbb{R}\setminus 0)^{2|E|} \to (\mathbb{R}\setminus 0)^{|E|}$. We require the pushforward of $\{\cdot,\cdot\}_{\mathcal{R}}$ to $(\mathbb{R}\setminus 0)^{|E|}$ by the weight map to be a well defined Poisson bracket; this can be regarded as an analog of the Poisson–Lie property for groups.

PROPOSITION 8.5. *Universal Poisson brackets $\{\cdot,\cdot\}_{\mathcal{R}}$ such that the weight map w is Poisson form a 6-parametric family defined by relations*

$$\{x^i_v, x^j_v\}_v = \alpha_{ij} x^i_v x^j_v, \quad i, j \in [1,3], i \neq j, \tag{8.5}$$

at each white vertex v and

$$\{x^i_v, x^j_v\}_v = \beta_{ij} x^i_v x^j_v, \quad i, j \in [1,3], i \neq j, \tag{8.6}$$

at each black vertex v.

PROOF. Indeed, let v be a white vertex, and let $e = (v, u)$ and $\bar{e} = (v, \bar{u})$ be the two outcoming edges. By definition, there exist $i, j, k, l \in [1, 3]$, $i \neq j$, such that $w_e = x_v^i x_u^k$, $w_{\bar{e}} = x_v^j x_{\bar{u}}^l$. Therefore,

$$\{w_e, w_{\bar{e}}\}_N = \{x_v^i x_u^k, x_v^j x_{\bar{u}}^l\}_{\mathcal{R}} = x_u^k x_{\bar{u}}^l \{x_v^i, x_v^j\}_v,$$

where $\{\cdot,\cdot\}_N$ stands for the pushforward of $\{\cdot,\cdot\}_{\mathcal{R}}$. Recall that the Poisson bracket in $(\mathbb{R}\setminus 0)_v^3$ depends only on x_v^1, x_v^2 and x_v^3. Hence the only possibility for the right hand side of the above relation to be a function of w_e and $w_{\bar{e}}$ occurs when $\{x_v^i, x_v^j\}_v = \alpha_{ij} x_v^i x_v^j$, as required.

Black vertices are treated in the same way. □

Let v be a white vertex. A *local gauge transformation* at v is a transformation $(\mathbb{R}\setminus 0)_v^3 \to (\mathbb{R}\setminus 0)_v^3$ defined by $(x_v^1, x_v^2, x_v^3) \mapsto (\bar{x}_v^1 = x_v^1 t_v, \bar{x}_v^2 = x_v^2 t_v^{-1}, \bar{x}_v^3 = x_v^3 t_v^{-1})$, where t_v is a Laurent monomial in x_v^1, x_v^2, x_v^3. A local gauge transformation at a black vertex is defined by the same formulas, with t_v replaced by t_v^{-1}.

A *global gauge transformation* $t\colon \mathcal{R} \to \mathcal{R}$ is defined by applying a local gauge transformation t_v at each vertex v. The composition map $w \circ t$ defines a network tN; the graph of tN coincides with the graph of N, and the weight w_e^t of an edge $e = (u, v)$ is given by $w_e^t = t_v w_e t_u^{-1}$. Therefore, the weights of the same path in N and tN coincide. It follows immediately that

$$M_{tN} \circ w \circ t = M_N \circ w, \tag{8.7}$$

provided both sides of the equality are well defined.

8.2.3. Induced Poisson structures on $\mathrm{Mat}_{k,m}$**.** Our next goal is to look at Poisson properties of the boundary measurement map. Fix an arbitrary partition $I \cup J = [1, n]$, $I \cap J = \varnothing$, and let $k = |I|$, $m = n - k = |J|$. Let $\mathrm{Net}_{I,J}$ stand for the set of all perfect planar networks in a disk with the sources b_i, $i \in I$, sinks b_j, $j \in J$, and edge weights w_e defined by (8.4). We assume that the space of edge weights $\mathcal{E}_N = (\mathbb{R}\setminus 0)^{|E|}$ is equipped with the Poisson bracket $\{\cdot,\cdot\}_N$ obtained as the pushforward of the 6-parametric family $\{\cdot,\cdot\}_{\mathcal{R}}$ described in Proposition 8.5.

THEOREM 8.6. *There exists a 2-parametric family of Poisson brackets on* $\mathrm{Mat}_{k,m}$ *with the following property: for any choice of parameters* α_{ij}, β_{ij} *in* (8.5), (8.6) *this family contains a unique Poisson bracket on* $\mathrm{Mat}_{k,m}$ *such that for any network* $N \in \mathrm{Net}_{I,J}$ *the map* $M_N\colon (\mathbb{R}\setminus 0)^{|E|} \to \mathrm{Mat}_{k,m}$ *is Poisson.*

PROOF. Relation (8.7) suggests that one may use global gauge transformations in order to decrease the number of parameters in the universal 6-parametric family described in Proposition 8.5. Indeed, for any white vertex v we consider a local gauge transformation $(x_v^1, x_v^2, x_v^3) \mapsto (\bar{x}_v^1, \bar{x}_v^2, \bar{x}_v^3)$ with $t_v = 1/x_v^1$. Evidently,

$$\{\bar{x}_v^2, \bar{x}_v^3\}_v = \alpha \bar{x}_v^2 \bar{x}_v^3, \quad \{\bar{x}_v^1, \bar{x}_v^2\}_v = \{\bar{x}_v^1, \bar{x}_v^3\}_v = 0 \tag{8.8}$$

with

$$\alpha = \alpha_{23} + \alpha_{13} - \alpha_{12}. \tag{8.9}$$

Similarly, for each black vertex v we consider a local gauge transformation $(x_v^1, x_v^2, x_v^3) \mapsto (\bar{x}_v^1, \bar{x}_v^2, \bar{x}_v^3)$ with $t_v = x_v^1$. Evidently,

$$\{\bar{x}_v^2, \bar{x}_v^3\}_v = \beta \bar{x}_v^2 \bar{x}_v^3, \quad \{\bar{x}_v^1, \bar{x}_v^2\}_v = \{\bar{x}_v^1, \bar{x}_v^3\}_v = 0 \tag{8.10}$$

with

$$\beta = \beta_{23} + \beta_{13} - \beta_{12}. \tag{8.11}$$

From now on we consider the 2-parametric family (8.8), (8.10) instead of the 6-parametric family (8.5), (8.6).

To define a Poisson bracket on $\mathrm{Mat}_{k,m}$, it suffices to calculate the bracket for any pair of matrix entries and to extend it further via bilinearity and the Leibniz identity. To do this we will need the following two auxiliary functions: for any $i, i' \in I$, $j, j' \in J$ define

$$s_=(i,j,i',j') = \begin{cases} 1 & \text{if } i \prec i' \prec j' \prec j, \\ -1 & \text{if } i' \prec i \prec j \prec j', \\ \frac{1}{2} & \text{if } i = i' \prec j' \prec j \text{ or } i \prec i' \prec j' = j, \\ -\frac{1}{2} & \text{if } i' = i \prec j \prec j' \text{ or } i' \prec i \prec j = j', \\ 0 & \text{otherwise,} \end{cases} \tag{8.12}$$

and

$$s_\times(i,j,i',j') = \begin{cases} 1 & \text{if } i' \prec i \prec j' \prec j, \\ -1 & \text{if } i \prec i' \prec j \prec j', \\ \frac{1}{2} & \text{if } i' = i \prec j' \prec j \text{ or } i' \prec i \prec j' = j, \\ -\frac{1}{2} & \text{if } i = i' \prec j \prec j' \text{ or } i \prec i' \prec j = j', \\ 0 & \text{otherwise.} \end{cases} \tag{8.13}$$

Note that both $s_=$ and $s_\times$ are skew-symmetric:

$$s_=(i,j,i',j') + s_=(i',j',i,j) = s_\times(i,j,i',j') + s_\times(i',j',i,j) = 0$$

for any $i, i' \in I$, $j, j' \in J$. Quadruples (i,j,i',j') such that at least one of $s_=(i,j,i',j')$ and $s_\times(i,j,i',j')$ is distinct from zero are shown in Figure 8.8. For a better visualization, pairs i,j and i',j' are joined by a directed edge; these edges should not be mistaken for edges of N.

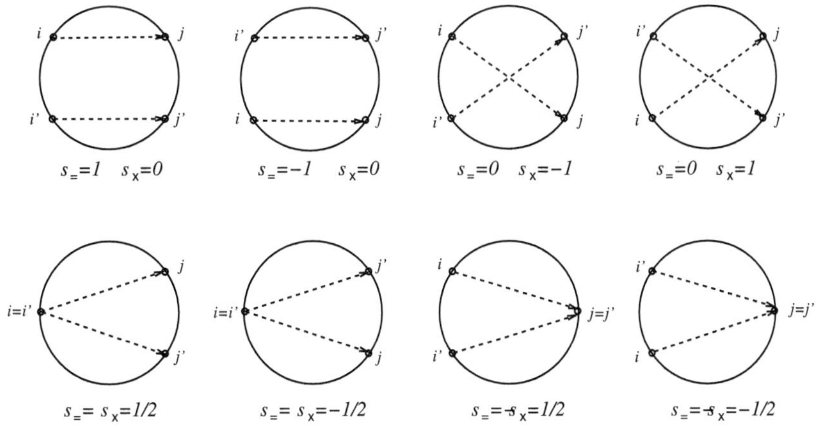

FIGURE 8.8. Nontrivial values of $s_=$ and $s_\times$

Theorem 8.6 is proved by presenting an explicit formula for the bracket on $\mathrm{Mat}_{k,m}$. □

THEOREM 8.7. *The 2-parametric family of Poisson brackets on* $\mathrm{Mat}_{k,m}$ *satisfying the conditions of Theorem 8.6 is given by*

$$\{M_{pq}, M_{\bar p\bar q}\}_{I,J} = (\alpha-\beta)s_=(i_p,j_q,i_{\bar p},j_{\bar q})M_{p\bar q}M_{\bar pq} + (\alpha+\beta)s_\times(i_p,j_q,i_{\bar p},j_{\bar q})M_{pq}M_{\bar p\bar q}, \tag{8.14}$$

where α *and* β *satisfy* (8.9), (8.11), $p,\bar p\in[1,k]$, $q,\bar q\in[1,m]$.

PROOF. First of all, let us check that that relations (8.14) indeed define a Poisson bracket on $\mathrm{Mat}_{k,m}$. Since bilinearity and the Leibnitz identity are built-in in the definition, and skew symmetry follows immediately from (8.12) and (8.13), it remains to check the Jacobi identity.

LEMMA 8.8. *The bracket* $\{\cdot,\cdot\}_{I,J}$ *satisfies the Jacobi identity.*

PROOF. The claim can be verified easily when at least one of the following five conditions holds true: $i=i'=i''$; $j=j'=j''$; $i=i'$ and $j=j'$; $i=i''$ and $j=j''$; $i'=i''$ and $j'=j''$. In what follows we assume that none of these conditions holds.

A simple computation shows that under this assumption, the Jacobi identity for $\{\cdot,\cdot\}_{I,J}$ is implied by the following three identities for the functions $s_=$ and $s_\times$: for any $i,i',i''\in I$, $j,j',j''\in J$,

$$\begin{aligned}
&s_\times(i',j',i'',j'')s_\times(i,j,i'',j'') + s_\times(i'',j'',i,j)s_\times(i',j',i,j)\\
&\quad + s_\times(i,j,i',j')s_\times(i'',j'',i',j') + s_\times(i',j',i'',j'')s_\times(i,j,i',j')\\
&\quad + s_\times(i'',j'',i,j)s_\times(i',j',i'',j'') + s_\times(i,j,i',j')s_\times(i'',j'',i,j) = 0,\\
&s_=(i',j',i'',j'')s_=(i,j,i'',j') + s_=(i'',j'',i,j)s_=(i',j',i,j'')\\
&\quad + s_=(i,j,i',j')s_=(i'',j'',i',j) = 0,\\
&s_=(i',j',i'',j'')(s_\times(i,j,i'',j') + s_\times(i,j,i',j'') - s_\times(i,j,i'',j'')\\
&\quad - s_\times(i,j,i,'j')) = 0.
\end{aligned}\tag{8.15}$$

The first identity in (8.15) is evident, since by (8.13), the first term is canceled by the fifth one, the second term is canceled by the sixth one, and the third term is canceled by the fourth one.

To prove the second identity, assume to the contrary that there exist $i,i',i''\in I$, $j,j',j''\in J$ such that the left hand side does not vanish. Consequently, at least one of the three terms in the left hand side does not vanish; without loss of generality we may assume that it is the first term.

Since $s_=(i',j',i'',j'')\neq 0$, we get either

$$i'\preceq i''\prec j''\preceq j', \tag{8.16}$$

or

$$i'\succeq i''\succ j''\succeq j'. \tag{8.17}$$

Assume that (8.16) holds, than we have five possibilities for j:

(i) $j''\prec j\prec j'$;
(ii) $j=j''$;
(iii) $i''\prec j\prec j''$;
(iv) $j'\prec j\prec i''$;
(v) $j=j'$.

In cases (i)-(iii) condition $s_=(i,j,i'',j') \neq 0$ implies $i'' \preceq i \prec j$. Therefore, in case (i) we get

$$s_=(i',j',i'',j'') = -s_=(i'',j'',i',j), \qquad s_=(i,j,i'',j') = s_=(i,j,i',j'),$$
$$s_=(i'',j'',i,j) = 0$$

provided $i'' \neq i$, and

$$s_=(i',j',i'',j'') = -s_=(i'',j'',i',j) = s_=(i',j',i,j'') = -s_=(i,j,i',j') = 1,$$
$$s_=(i,j,i'',j') = s_=(i'',j'',i,j) = -1/2$$

provided $i'' = i \neq i'$. In both situations the second identity in (8.15) follows immediately.

In case (ii) we get

$$s_=(i,j,i'',j') = s_=(i,j,i',j') = -s_=(i',j',i,j'') = -1, \qquad s_=(i'',j'',i,j) = 1/2,$$

and

$$s_=(i',j',i'',j'') = \begin{cases} 1 & \text{if } i' \neq i'', \\ 1/2 & \text{if } i' = i'', \end{cases} \qquad s_=(i'',j'',i',j) = \begin{cases} 1/2 & \text{if } i' \neq i'', \\ 0 & \text{if } i' = i''. \end{cases}$$

In both situations the second identity in (8.15) follows immediately.

In case (iii) we get

$$s_=(i',j',i'',j'') = s_=(i',j',i,j''), \qquad s_=(i,j,i'',j') = -s_=(i'',j'',i,j),$$
$$s_=(i'',j'',i',j) = 0$$

provided $i' \neq i''$, and

$$s_=(i'',j'',i,j) = -s_=(i,j,i'',j') = s_=(i',j',i,j'') = -s_=(i,j,i',j') = 1,$$
$$s_=(i',j',i'',j'') = s_=(i'',j'',i',j) = 1/2$$

provided $i' = i'' \neq i$. In both situations the second identity in (8.15) follows immediately.

Case (iv), in its turn, falls into three cases depending on the location of i; these three cases are parallel to the cases (i)-(iii) above and are treated in the same way.

Finally, in case (v) we have to distinguish two subcases: $i'' \prec i \prec i'$ and $i' \prec i \preceq i''$. In the first subcase we have

$$s_=(i',j',i'',j'') = -s_=(i'',j'',i',j), \qquad s_=(i,j,i'',j') = s_=(i,j,i',j'),$$
$$s_=(i'',j'',i,j) s_=(i',j',i,j'') = 0,$$

while in the second subcase,

$$s_=(i',j',i'',j'') = s_=(i',j',i,j'') = -s_=(i'',j'',i',j) = -1, \qquad s_=(i,j,i',j') = -1/2,$$

and

$$s_=(i'',j'',i,j) = \begin{cases} -1 & \text{if } i \neq i'', \\ -1/2 & \text{if } i = i'', \end{cases} \qquad s_=(i,j,i'',j') = \begin{cases} 1/2 & \text{if } i \neq i'', \\ 0 & \text{if } i = i''. \end{cases}$$

In both situations the second identity in (8.15) follows immediately.

If the points are ordered counterclockwise as prescribed by (8.17), the proof is very similar, with $\prec$ and $\preceq$ replaced by $\succ$ and $\succeq$, correspondingly.

To prove the third identity in (8.15), assume to the contrary that there exist $i, i', i'' \in I$, $j, j', j'' \in J$ such that the left hand side does not vanish. Consequently, $s_=(i',j',i'',j'') \neq 0$, and hence once again one of (8.16) and (8.17) holds.

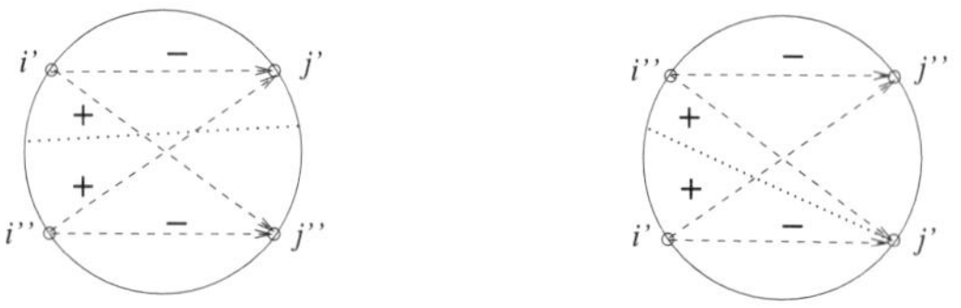

FIGURE 8.9. To the proof of Lemma 8.8

Denote by S the sum in the left hand side of the identity. If $i' = i''$ or $j' = j''$ then S vanishes due to the skew-symmetry of $s_\times$. The remaining case amounts to checking all possible ways to insert a chord ij in the configurations presented in Figure 8.9. The check itself in each case is trivial. For example, if the inserted chord is as shown by the dotted line in the left part of Figure 8.9, then the last two terms of S vanish, while the first two terms have opposite signs and absolute value 1. If the inserted chord is as shown by a dotted line in the right part of Figure 8.9, then one of the last two terms in S vanishes and one of the first two terms has absolute value 1. The two remaining terms have absolute value 1/2; besides, they have the same sign, which is opposite to the sign of the term with absolute value 1. Other cases are similar and left to the reader. □

To complete the proof of Theorem 8.7, it remains to check that

$$\{M_{pq}, M_{\bar{p}\bar{q}}\}_N = \{M_{pq}, M_{\bar{p}\bar{q}}\}_{I,J} \tag{8.18}$$

for any pair of matrix entries M_{pq} and $M_{\bar{p}\bar{q}}$. The proof is similar to the proof of Proposition 8.3 and relies on the induction on the number of inner vertices in $N \in \mathrm{Net}_{I,J}$.

Assume first that N does not have inner vertices, and hence each edge of N connects two boundary vertices. It is easy to see that in this case the Poisson bracket computed in $(\mathbb{R} \setminus 0)^{|E|}$ vanishes identically. Let us prove that the bracket given by (8.14) vanishes as well.

We start with the case when both (b_{i_p}, b_{j_q}) and $(b_{i_{\bar{p}}}, b_{j_{\bar{q}}})$ are edges in N. Then $p \neq \bar{p}$ and $q \neq \bar{q}$, since there is only one edge incident to each boundary vertex. Therefore $M_{p\bar{q}} = M_{\bar{p}q} = 0$, and the first term in the right hand side of (8.14) vanishes. Besides, since N is planar, $s_\times(i_p, j_q, i_{\bar{p}}, j_{\bar{q}}) = 0$, and the second term vanishes as well.

Next, let (b_{i_p}, b_{j_q}) be an edge of N, and $(b_{i_{\bar{p}}}, b_{j_{\bar{q}}})$ be a non-edge. Then $M_{pq} = M_{p\bar{q}} = 0$, and hence both terms in the right hand side of (8.14) vanish.

Finally, let both (b_{i_p}, b_{j_q}) and $(b_{i_{\bar{p}}}, b_{j_{\bar{q}}})$ be non-edges. Then $M_{pq} = M_{\bar{p}\bar{q}} = 0$, and the second term in the right hand side of (8.14) vanishes. The first term can be distinct from zero only if both $(b_{i_p}, b_{j_{\bar{q}}})$ and $(b_{i_{\bar{p}}}, b_{j_q})$ are edges in N. Once again we use planarity of N to see that in this case $s_=(i_p, j_q, i_{\bar{p}}, j_{\bar{q}}) = 0$, and hence the right hand side of (8.14) vanishes.

Now we may assume that N has r inner vertices, and that (8.18) is true for all networks with at most $r-1$ inner vertices and any number of boundary vertices. Consider the unique neighbor of b_{i_p} in N. If this neighbor is another boundary vertex then the same reasoning as above applies to show that $\{M_{pq}, M_{\bar{p}\bar{q}}\}_{I,J}$ vanishes identically for any choice of $i_{\bar p} \in I$, $j_q, j_{\bar q} \in J$, which agrees with the behavior of $\{\cdot,\cdot\}_N$.

Assume that the only neighbor of b_{i_p} is a white inner vertex v. Define

$$\widetilde{\mathcal{R}} = \left(\mathcal{R} \oplus (\mathbb{R}\setminus 0)_{i'_v} \oplus (\mathbb{R}\setminus 0)_{i''_v}\right) \ominus \left((\mathbb{R}\setminus 0)^3_v \oplus (\mathbb{R}\setminus 0)_{i_p}\right),$$

which corresponds to a network $\widetilde{N}$ obtained from N by deleting vertices b_{i_p}, v and the edge (b_{i_p}, v) from G and adding two new sources i'_v and i''_v so that $i_p - 1 \prec i'_v \prec i''_v \prec i_p + 1$, see Figure 8.3. Let $1, x', x''$ be the coordinates in $(\mathbb{R}\setminus 0)^3_v$ after the local gauge transformation at v, so that $\{x', x''\}_v = \alpha x' x''$. Then

$$M(i_p, j) = x'\widetilde{M}(i'_v, j) + x''\widetilde{M}(i''_v, j), \qquad M(i_{\bar p}, j) = \widetilde{M}(i_{\bar p}, j) \tag{8.19}$$

for any $j \in J$ and any $\bar p \neq p$. Since $\widetilde{N}$ has $r-1$ inner vertices, $\widetilde{M}(i,j)$ satisfy relations similar to (8.18) with N replaced by $\widetilde{N}$ and I replaced by $I \setminus i_p \cup i'_v \cup i''_v$. Relations (8.18) follow immediately from this fact and (8.19), provided $\bar p \neq p$. In the latter case we have

$$\begin{aligned}
\{M_{pq}, M_{p\bar q}\}_N &= \{x'\widetilde{M}(i'_v, j_q) + x''\widetilde{M}(i''_v, j_q), x'\widetilde{M}(i'_v, j_{\bar q}) + x''\widetilde{M}(i''_v, j_{\bar q})\}_N \\
&= (x')^2\{\widetilde{M}(i'_v, j_q), \widetilde{M}(i'_v, j_{\bar q})\}_{\widetilde{N}} + x''x'\{\widetilde{M}(i''_v, j_q), \widetilde{M}(i'_v, j_{\bar q})\}_{\widetilde{N}} \\
&\quad + \{x'', x'\}_v \widetilde{M}(i''_v, j_q)\widetilde{M}(i'_v, j_{\bar q}) + x'x''\{\widetilde{M}(i'_v, j_q), \widetilde{M}(i''_v, j_{\bar q})\}_{\widetilde{N}} \\
&\quad + \{x', x''\}_v \widetilde{M}(i'_v, j_q)\widetilde{M}(i''_v, j_{\bar q}) + (x'')^2\{\widetilde{M}(i''_v, j_q), \widetilde{M}(i''_v, j_{\bar q})\}_{\widetilde{N}}.
\end{aligned}$$

The first term in the right hand side of the expression above equals

$$((\alpha-\beta)s_=(i'_v, j_q, i'_v, j_{\bar q}) + (\alpha+\beta)s_\times(i'_v, j_q, i'_v, j_{\bar q}))\, x'\widetilde{M}(i'_v, j_q)x''\widetilde{M}(i'_v, j_{\bar q}).$$

Since $s_=(i'_v, j_q, i'_v, j_{\bar q}) = s_\times(i'_v, j_q, i'_v, j_{\bar q}) = \pm\frac{1}{2}$ (the sign is negative if $i_p \prec j_q \prec j_{\bar q}$ and positive if $i_p \prec j_{\bar q} \prec j_q$), the first term equals $\pm\alpha x'\widetilde{M}(i'_v, j_q)x''\widetilde{M}(i'_v, j_{\bar q})$.

Similarly, the second term equals

$$\begin{array}{ll}
-(\alpha-\beta)x'\widetilde{M}(i'_v, j_q)x''\widetilde{M}(i''_v, j_{\bar q}) & \text{if } i_p \prec j_q \prec j_{\bar q} \\
(\alpha+\beta)x'\widetilde{M}(i'_v, j_{\bar q})x''\widetilde{M}(i''_v, j_q) & \text{if } i_p \prec j_{\bar q} \prec j_q,
\end{array}$$

the fourth term equals

$$\begin{array}{ll}
-(\alpha+\beta)x'\widetilde{M}(i'_v, j_q)x''\widetilde{M}(i''_v, j_{\bar q}) & \text{if } i_p \prec j_q \prec j_{\bar q} \\
(\alpha-\beta)x'\widetilde{M}(i'_v, j_{\bar q})x''\widetilde{M}(i''_v, j_q) & \text{if } i_p \prec j_{\bar q} \prec j_q,
\end{array}$$

and the sixths term equals $\pm\alpha x''\widetilde{M}(i''_v, j_q)x''\widetilde{M}(i''_v, j_{\bar q})$ with the same sign rule as for the first term. We thus see that

$$\begin{aligned}
\{M_{pq}, M_{p\bar q}\}_N &= \pm\alpha(x'\widetilde{M}(i'_v, j_q) + x''\widetilde{M}(i''_v, j_q))(x'\widetilde{M}(i'_v, j_{\bar q}) + x''\widetilde{M}(i''_v, j_{\bar q})) \\
&= \pm\alpha M_{pq}M_{p\bar q} = \{M_{pq}, M_{p\bar q}\}_{I,J},
\end{aligned}$$

since $s_=(i_p, j_q, i_p, j_{\bar q}) = s_\times(i_p, j_q, i_p, j_{\bar q}) = \pm\frac{1}{2}$.

Assume now that the only neighbor of b_{i_p} is a black inner vertex u. Define

$$\widehat{\mathcal{R}} = \left(\mathcal{R} \oplus (\mathbb{R}\setminus 0)_{i_u} \oplus (\mathbb{R}\setminus 0)_{j_u}\right) \ominus \left((\mathbb{R}\setminus 0)^3_u \oplus (\mathbb{R}\setminus 0)_{i_p}\right),$$

which corresponds to a network $\widehat{N}$ obtained from N by deleting vertices b_{i_p}, u and the edge (b_{i_p}, u) from G and adding a new source i_u and a new sink j_u so that either $i_p - 1 \prec i_u \prec j_u \prec i_p + 1$, or $i_p - 1 \prec i_u \prec j_u \prec i_p + 1$, see Figure 8.4. Applying the local gauge transformation at u, we obtain new coordinates in $(\mathbb{R}\setminus 0)^3_u$: in the first case, $1, x', x''$ with $\{x', x''\}_u = \beta x'x''$, and in the second case, $1, x'', x'$ with $\{x', x''\}_u = -\beta x'x''$.

LEMMA 8.9. *Boundary measurements in the networks N and $\widehat{N}$ are related by*

$$M(i_p, j) = \frac{x'\widehat{M}(i_u, j)}{1 + x''\widehat{M}(i_u, j_u)},$$

$$M(i_{\bar{p}}, j) = \widehat{M}(i_{\bar{p}}, j) \pm \frac{x''\widehat{M}(i_{\bar{p}}, j_u)\widehat{M}(i_u, j)}{1 + x''\widehat{M}(i_u, j_u)}, \qquad \bar{p} \neq p;$$

in the second formula above, sign $+$ corresponds to the cases

$$i_p - 1 \prec j_u \prec i_u \prec i_p + 1 \preceq j \prec i_{\bar{p}} \qquad \text{or} \qquad i_{\bar{p}} \prec j \preceq i_p - 1 \prec i_u \prec j_u \prec i_p + 1,$$

and sign $-$ corresponds to the cases

$$i_p - 1 \prec i_u \prec j_u \prec i_p + 1 \preceq j \prec i_{\bar{p}} \qquad \text{or} \qquad i_{\bar{p}} \prec j \preceq i_p - 1 \prec j_u \prec i_u \prec i_p + 1.$$

PROOF. The first formula above was, in fact, already obtained in the proof of Lemma 8.3; one has to take into account that after the local gauge transformation at u we get $w_{e_+} = 1$, $w_{e_-} = x''$ and $w_{e_0} = x'$.

To get the second formula, we apply the same reasoning as in the proof of Lemma 8.3. The paths from $b_{i_{\bar{p}}}$ to b_j in N are classified according to the number of times they traverse the edge e_+. The total contribution of the paths not traversing e_+ at all to $M(i_{\bar{p}}, j)$ equals $\widehat{M}(i_{\bar{p}}, j)$. Each path P that traverses e_+ exactly once can be decomposed as $P = (P_1, e_-, e_+, P_2)$ so that $\widehat{P}_1 = (P_1, \hat{e}_-)$ is a path from $b_{i_{\bar{p}}}$ to b_{j_u} in $\widehat{N}$ and $\widehat{P}_2 = (\hat{e}_+, P_2)$ is a path from b_{i_u} to b_j in $\widehat{N}$. Define $\widehat{P} = (\widehat{P}_1, e, \widehat{P}_2)$, where e is the edge between j_u and i_u belonging to the convex hull of the boundary vertices of $\widehat{N}$ (see Figure 8.10). Clearly, $c(C_P) = c(C_{\widehat{P}})$.

Assume first that

$$i_{\bar{p}} \prec j \preceq i_p - 1 \prec j_u \prec i_u \prec i_p + 1,$$

which corresponds to the upper part of Figure 8.10. Then $c(C_{\widehat{P}_1}) = c_{\widehat{P}_1} + c_{i_u} + c'$, where $c_{\widehat{P}_1}$ is the contribution of all vertices of $\widehat{P}_1$, including $b_{i_{\bar{p}}}$ and b_{j_u}, c_{i_u} is the contribution of two consecutive edges of the convex hull of boundary vertices of $\widehat{N}$ calculated at b_{i_u}, and c' is the total contribution calculated at the vertices of the convex hull lying between b_{i_u} and $b_{i_{\bar{p}}}$. Similarly, $c(C_{\widehat{P}_2}) = c_{\widehat{P}_2} + c_{j_u} + c''$, where $c_{\widehat{P}_2}$ is the contribution of all vertices of $\widehat{P}_2$, including b_{i_u} and b_j, c_{j_u} is the contribution of two consecutive edges of the convex hull of boundary vertices of $\widehat{N}$ calculated at b_{j_u}, and c'' is the total contribution calculated at the vertices of the convex hull lying between b_j and b_{j_u}. Finally,

$$c(C_{\widehat{P}}) = c_{\widehat{P}_1} + c_{\widehat{P}_2} + c'' + c_{j_u} + c_{i_u} + c',$$

and so $c(C_{\widehat{P}}) = c(C_{\widehat{P}_1}) + c(C_{\widehat{P}_2})$. Therefore, in this case $w_P = -w_{P_1} w_{e_-} w_{e_+} w_{P_2}$, and the contribution of all such paths to $M(i_{\bar{p}}, j)$ equals $-x''\widehat{M}(i_{\bar{p}}, j_u)\widehat{M}(i_u, j)$. Each additional traversing of the edge e_+ results in multiplying this expression by $-x''\widehat{M}(i_u, j_u)$; the proof of this fact is similar to the proof of Lemma 8.3. Summing up we get the second formula with the $-$ sign, as desired.

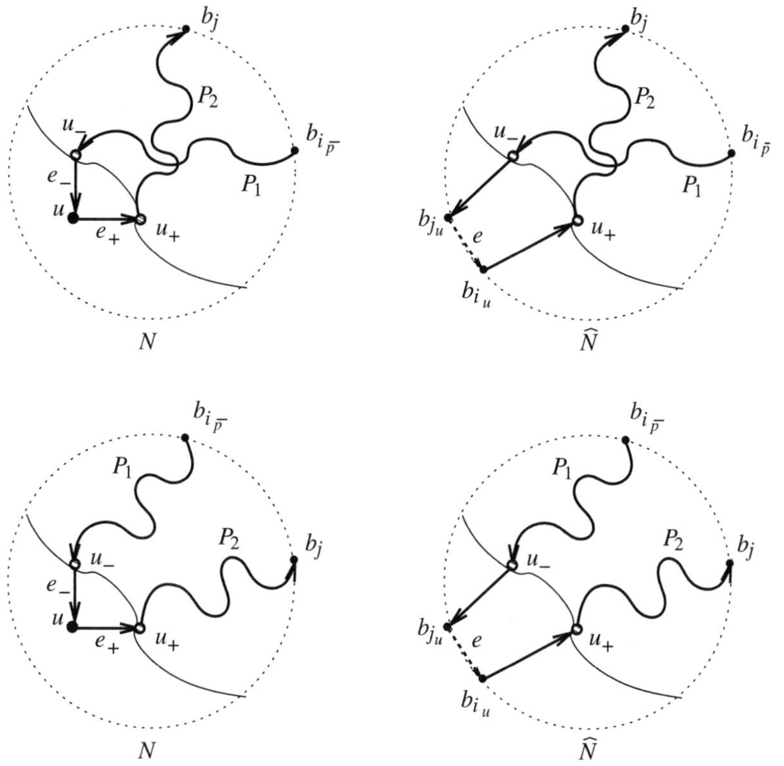

FIGURE 8.10. To the proof of Lemma 8.9

Assume now that

$$i_p - 1 \prec j_u \prec i_u \prec i_p + 1 \preceq j \prec i_{\bar p},$$

which corresponds to the lower part of Figure 8.10. Then

$$c(C_{\widehat{P}_1}) = c_{\widehat{P}_1} + c_{i_u} + c' + c_j + c''',$$

where $c_{\widehat{P}_1}$ and c_{i_u} are as in the previous case, c_j is the contribution of two consecutive edges of the convex hull of boundary vertices of $\widehat{N}$ calculated at b_j, c' is the total contribution calculated at the vertices of the convex hull lying between b_{i_u} and b_j, and c''' is the total contribution calculated at the vertices of the convex hull lying between b_j and $b_{i_{\bar p}}$. Similarly,

$$c(C_{\widehat{P}_2}) = c_{\widehat{P}_2} + c''' + c_{i_{\bar p}} + c'' + c_{j_u},$$

where $c_{\widehat{P}_2}$ and c_{j_u} are as in the previous case, $c_{i_{\bar p}}$ is the contribution of two consecutive edges of the convex hull of boundary vertices of $\widehat{N}$ calculated at $b_{i_{\bar p}}$, and c'' is the total contribution calculated at the vertices of the convex hull lying between $b_{i_{\bar p}}$ and b_{j_u}. Finally, $c(C_{\widehat{P}}) = c_{\widehat{P}_1} + c_{\widehat{P}_2} + c'''$, and so $c(C_{\widehat{P}}) = c(C_{\widehat{P}_1}) + c(C_{\widehat{P}_2}) + 1$, since $c' + c'' + c''' + c_{j_u} + c_{i_u} + c_j + c_{i_{\bar p}} = 1$. Therefore, in this case $w_P = w_{P_1} w_{e_-} w_{e_+} w_{P_2}$, and the contribution of all such paths to $M(i_{\bar p}, j)$ equals $x'' \widehat{M}(i_{\bar p}, j_u) \widehat{M}(i_u, j)$. Each additional traversing of e_+ once again results in multiplying this expression by $-x''\widehat{M}(i_u,j_u)$. Summing up we get the second formula with the $+$ sign, as desired.

To treat the remaining two cases one makes use of Remark 8.2 and applies the same reasoning. □

Since $\widehat{N}$ has $r-1$ inner vertices, $\widehat{M}(i,j)$ satisfy relations similar to (8.18) with N replaced by $\widehat{N}$, I replaced by $I\setminus i_p\cup i_v$ and J replaced by $J\cup j_v$. Relations (8.18) follow from this fact and Lemma 8.9 via simple though tedious computations.

For example, let $i_{\bar p}\prec j_{\bar q}\preceq i_p-1\prec j_u\prec i_u\prec i_p+1$. Then the left hand side of (8.18) is given by

$$\begin{aligned}
\{M_{pq},M_{\bar p\bar q}\}_N &= \left\{\frac{x'\widehat{M}(i_u,j_q)}{1+x''\widehat{M}(i_u,j_u)},\widehat{M}(i_{\bar p},j_{\bar q})-\frac{x''\widehat{M}(i_{\bar p},j_u)\widehat{M}(i_u,j_{\bar q})}{1+x''\widehat{M}(i_u,j_u)}\right\}_N\\
&=\frac{x'}{1+x''\widehat{M}(i_u,j_u)}\{\widehat{M}(i_u,j_q),\widehat{M}(i_{\bar p},j_{\bar q})\}_{\widehat{N}}\\
&-\frac{x'x''\widehat{M}(i_u,j_q)}{(1+x''\widehat{M}(i_u,j_u))^2}\{\widehat{M}(i_u,j_u),\widehat{M}(i_{\bar p},j_{\bar q})\}_{\widehat{N}}\\
&-\frac{\widehat{M}(i_u,j_q)\widehat{M}(i_{\bar p},j_u)\widehat{M}(i_u,j_{\bar q})}{(1+x''\widehat{M}(i_u,j_u))^2}\{x',x''\}_u\\
&-\frac{x'x''\widehat{M}(i_u,j_{\bar q})}{(1+x''\widehat{M}(i_u,j_u))^2}\{\widehat{M}(i_u,j_q),\widehat{M}(i_{\bar p},j_u)\}_{\widehat{N}}\\
&-\frac{x'x''\widehat{M}(i_{\bar p},j_u)}{(1+x''\widehat{M}(i_u,j_u))^2}\{\widehat{M}(i_u,j_q),\widehat{M}(i_u,j_{\bar q})\}_{\widehat{N}}\\
&+\frac{x'(x'')^2\widehat{M}(i_u,j_{\bar q})\widehat{M}(i_u,j_q)}{(1+x''\widehat{M}(i_u,j_u))^3}\{\widehat{M}(i_u,j_u),\widehat{M}(i_{\bar p},j_u)\}_{\widehat{N}}\\
&+\frac{x'(x'')^2\widehat{M}(i_u,j_q)\widehat{M}(i_{\bar p},j_u)}{(1+x''\widehat{M}(i_u,j_u))^3}\{\widehat{M}(i_u,j_u),\widehat{M}(i_u,j_{\bar q})\}_{\widehat{N}}\\
&+\frac{x'(x'')^2\widehat{M}(i_{\bar p},j_u)\widehat{M}(i_u,j_{\bar q})}{(1+x''\widehat{M}(i_u,j_u))^3}\{\widehat{M}(i_u,j_q),\widehat{M}(i_u,j_u)\}_{\widehat{N}}\\
&+\frac{x''\widehat{M}(i_{\bar p},j_u)\widehat{M}(i_u,j_{\bar q})\widehat{M}(i_u,j_q),\widehat{M}(i_u,j_u)}{(1+x''\widehat{M}(i_u,j_u))^3}\{x',x''\}_u,
\end{aligned}$$

and the right hand side is given by

$$\begin{aligned}
&(\alpha-\beta)s_=(i_p,j_q,i_{\bar p},j_{\bar q})\frac{x'\widehat{M}(i_u,j_{\bar q})}{1+x''\widehat{M}(i_u,j_u)}\left(\widehat{M}(i_{\bar p},j_q)-\frac{x''\widehat{M}(i_{\bar p},j_u)\widehat{M}(i_u,j_q)}{1+x''\widehat{M}(i_u,j_u)}\right)\\
&\quad+(\alpha+\beta)s_\times(i_p,j_q,i_{\bar p},j_{\bar q})\frac{x'\widehat{M}(i_u,j_q)}{1+x''\widehat{M}(i_u,j_u)}\left(\widehat{M}(i_{\bar p},j_{\bar q})-\frac{x''\widehat{M}(i_{\bar p},j_u)\widehat{M}(i_u,j_{\bar q})}{1+x''\widehat{M}(i_u,j_u)}\right).
\end{aligned}$$

Treating $\widehat{M}(\cdot,\cdot)$, $\alpha-\beta$ and $\alpha+\beta$ as independent variables and equating coefficients of the same monomials in the above two expressions we arrive to the following identities:

$$s_=(i_u, j_q, i_{\bar p}, j_{\bar q}) = s_=(i_p, j_q, i_{\bar p}, j_{\bar q}), \qquad s_\times(i_u, j_q, i_{\bar p}, j_{\bar q}) = s_\times(i_p, j_q, i_{\bar p}, j_{\bar q}),$$
$$s_=(i_u, j_u, i_{\bar p}, j_{\bar q}) + s_=(i_u, j_q, i_{\bar p}, j_u) + s_=(i_u, j_q, i_u, j_{\bar q}) - 1/2 = s_=(i_p, j_q, i_{\bar p}, j_{\bar q}),$$
$$s_\times(i_u, j_u, i_{\bar p}, j_{\bar q}) + s_\times(i_u, j_q, i_{\bar p}, j_u) + s_\times(i_u, j_q, i_u, j_{\bar q}) + 1/2 = s_\times(i_p, j_q, i_{\bar p}, j_{\bar q}),$$
$$s_=(i_u, j_u, i_{\bar p}, j_u) + s_=(i_u, j_u, i_u, j_{\bar q}) + s_=(i_u, j_q, i_u, j_u) - 1/2 = 0,$$
$$s_\times(i_u, j_u, i_{\bar p}, j_u) + s_\times(i_u, j_u, i_u, j_{\bar q}) + s_\times(i_u, j_q, i_u, j_u) + 1/2 = 0.$$

The latter can be checked easily by considering separately the following cases: $i_p \prec j_q \prec i_{\bar p}$; $i_{\bar p} \prec j_q \prec j_{\bar q}$; $j_q = j_{\bar q}$; $j_{\bar q} \prec j_q \prec i_p$. In each one of these cases all the functions involved in the above identities take constant values. □

8.3. Grassmannian boundary measurement map and induced Poisson structures on $G_k(n)$

8.3.1. Recovering the Sklyanin bracket on Mat_k**.** Let us take a closer look at the 2-parameter family of Poisson brackets obtained in Theorem 8.7 in the case when vertices $b_1, \dots, b_k$ on the boundary of the disk are sources and vertices $b_{k+1}, \dots, b_n$ are sinks, that is, when $I = [1, k]$ and $J = [k+1, n]$. To simplify notation, in this situation we will write $\{\cdot,\cdot\}_{k,m}$ and $\mathrm{Net}_{k,m}$ instead of $\{\cdot,\cdot\}_{[1,k],[k+1,n]}$ and $\mathrm{Net}_{[1,k],[k+1,n]}$. Therefore, formula (8.14) can be re-written as

$$(8.20)\quad 2\{M_{ij}, M_{\bar\imath\bar\jmath}\}_{k,m} = (\alpha - \beta)\left(\mathrm{sign}(\bar\imath - i) - \mathrm{sign}(\bar\jmath - j)\right) M_{i\bar\jmath} M_{\bar\imath j} + (\alpha + \beta)\left(\mathrm{sign}(\bar\imath - i) + \mathrm{sign}(\bar\jmath - j)\right) M_{ij} M_{\bar\imath\bar\jmath},$$

where M_{ij} corresponds to the boundary measurement between b_i and b_{j+k}. The first term in the equation above coincides (up to a multiple) with formula (4.7) for a Poisson bracket on the open cell $G_k^0(n)$ in the Grassmannian $G_k(n)$ viewed as a Poisson homogeneous space of the group GL_n equipped with the standard Poisson-Lie structure. This suggests that it makes sense to investigate Poisson properties of the boundary measurement map viewed as a map into $G_k(n)$, which will be the goal of this section.

First, however, we will go back to the example, considered in Sect. 8.2.1, where we associated with a network $N \in \mathrm{Net}_{k,k}$ a matrix $A_N = M_N W_0 \in \mathrm{Mat}_k$. Written in terms of matrix entries A_{ij} of A, bracket (8.20) becomes

$$(8.21)\quad 2\{A_{ij}, A_{\bar\imath\bar\jmath}\}_{k,m} = (\alpha - \beta)\left(\mathrm{sign}(\bar\imath - i) + \mathrm{sign}(\bar\jmath - j)\right) A_{i\bar\jmath} A_{\bar\imath j} + (\alpha + \beta)\left(\mathrm{sign}(\bar\imath - i) - \mathrm{sign}(\bar\jmath - j)\right) A_{ij} A_{\bar\imath\bar\jmath}.$$

If A_{N_1}, A_{N_2} are matrices that correspond to networks $N_1, N_2 \in \mathrm{Net}_{k,k}$ then their product $A_{N_1} A_{N_2}$ corresponds to the concatenation of N_1 and N_2. This fact combined with Theorem 8.6 implies that the bracket (8.21) possesses the Poisson-Lie property.

In fact, (8.21) is the Sklyanin bracket (4.6) on Mat_k associated with a deformation of the standard R-matrix (1.23). Indeed, it is known that if R_0 is the standard R-matrix, S is any linear operator on the set of diagonal matrices that is skew-symmetric w.r.t. the trace-form, and π_0 is the natural projection onto a subspace of diagonal matrices, then for any scalar c_1, c_2 the linear combination $c_1 R_0 + c_2 S \pi_0$ satisfies the MCYBE and thus gives rise to a Sklyanin Poisson-Lie bracket.

Define S by

$$S(e_{jj}) = \sum_{i=1}^{k} \operatorname{sign}(j-i) e_{ii}, \quad j = 1, \dots, k,$$

and put

$$R_{\alpha,\beta} = (\alpha - \beta) R_0 + (\alpha + \beta) S \pi_0. \tag{8.22}$$

Substituting coordinate functions A_{ij}, $A_{i'j'}$ into the expression (4.6) for the Sklyanin Poisson-Lie bracket associated with the R-matrix $R_{\alpha,\beta}$, we recover equation (8.21).

To summarize, we obtained

THEOREM 8.10. *For any $N \in \mathrm{Net}_{k,k}$ and any choice of parameters α_{ij}, β_{ij} in* (8.5), (8.6) *the map $A_N : (\mathbb{R} \setminus 0)^{|E|} \to \mathrm{Mat}_k$ is Poisson with respect to the Sklyanin bracket* (4.6) *associated with the R-matrix $R_{\alpha,\beta}$, where α and β satisfy relations* (8.9) *and* (8.11).

8.3.2. Induced Poisson structures on $G_k(n)$. Let N be a network with the sources b_i, $i \in I$ and sinks b_j, $j \in J$. We are going to interpret the boundary measurement map as a map into the Grassmannian $G_k(n)$. To this end, we extend M_N to a $k \times n$ matrix $\bar{X}_N$ as follows:

(i) the $k \times k$ submatrix of $\bar{X}_N$ formed by k columns indexed by I is the identity matrix $\mathbf{1}_k$;

(ii) for $p \in [1, k]$ and $j = j_q \in J$, the (p, j)-entry of $\bar{X}_N$ is $m_{pj}^I = (-1)^{s(p,j)} M_{pq}$, where $s(p, j)$ is the number of elements in I lying strictly between $\min\{i_p, j\}$ and $\max\{i_p, j\}$ in the linear ordering; note that the sign is selected in such a way that the minor $(\bar{X}_N)_{[1,k]}^{I(i_p \to j)}$ coincides with M_{pq}.

We will view $\bar{X}_N$ as a matrix representative of an element $X_N \in G_k(n)$. The corresponding rational map $X_N \colon \mathcal{E}_N \to G_k(n)$ is called the *Grassmannian boundary measurement map*. For example, the network presented in Figure 8.11 defines a map of $(\mathbb{R} \setminus 0)^8$ to $G_2(4)$ given by the matrix

$$\begin{pmatrix} 1 & \dfrac{w_1 w_4 w_6}{1 + w_2 w_4 w_5 w_7} & 0 & -\dfrac{w_1 w_3 w_4 w_5 w_7}{1 + w_2 w_4 w_5 w_7} \\[2ex] 0 & \dfrac{w_2 w_4 w_5 w_6 w_8}{1 + w_2 w_4 w_5 w_7} & 1 & \dfrac{w_3 w_5 w_8}{1 + w_2 w_4 w_5 w_7} \end{pmatrix}.$$

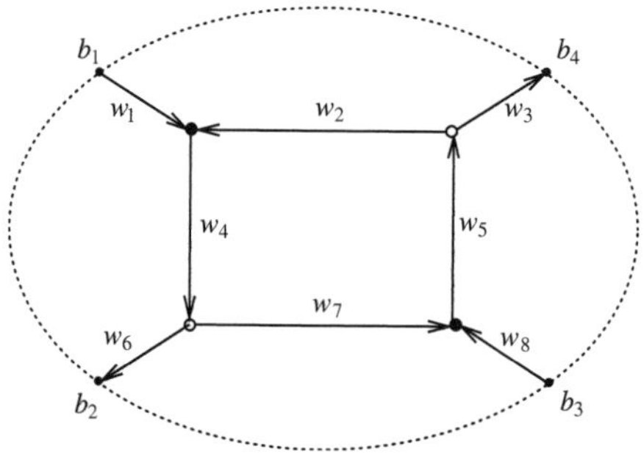

FIGURE 8.11. To the definition of the Grassmannian boundary measurement map

Clearly, X_N belongs to the cell $G_k^I(n) = \{X \in G_k(n) \ : \ x_I \neq 0\}$; here, as before, x_I is the corresponding Plücker coordinate. Therefore, we can regard matrix entries m_{pj}^I as coordinate functions on $G_k^I(n)$ and rewrite (8.14) as follows:

(8.23) $$\{m_{pj}^I, m_{\bar p\bar j}^I\}_{I,J} = (\alpha-\beta)s_=(i_p, j, i_{\bar p}, \bar j)m_{p\bar j}^I m_{\bar p j}^I + (\alpha+\beta)s_\times(i_p, j, i_{\bar p}, \bar j)m_{pj}^I m_{\bar p\bar j}^I.$$

REMARK 8.11. If $I = [1,k]$, then the 2-parametric family (8.23) of Poisson brackets on $G_k^I(n) = G_k^0(n)$ is defined by (8.20). A computation identical to the one presented in Sect. 4.2.1 shows that (8.20) is the pushforward to $G_k^0(n)$ of the Sklyanin Poisson-Lie bracket (4.6) on Mat_n associated with the R-matrix $R_{\alpha,\beta}$.

The following result says that the families $\{\cdot,\cdot\}_{I,J}$ on different cells $G_k^I(n)$ can be glued together to form the unique 2-parametric family of Poisson brackets on $G_k(n)$ that makes all maps X_N Poisson.

THEOREM 8.12. (i) *For any choice of parameters α and β there exists a unique Poisson bracket $\mathcal{P}_{\alpha,\beta}$ on $G_k(n)$ such that for any network N with k sources, $n-k$ sinks and weights defined by* (8.4), *the map X_N: $(\mathbb{R}\setminus 0)^{|E|} \to G_k(n)$ is Poisson provided the parameters α_{ij} and β_{ij} defining the bracket $\{\cdot,\cdot\}_N$ on $(\mathbb{R}\setminus 0)^{|E|}$ satisfy relations* (8.9) *and* (8.11).

(ii) *For any $I \subset [1,n]$, $|I| = k$, and $J = [1,n]\setminus I$, the restriction of $\mathcal{P}_{\alpha,\beta}$ to the cell $G_k^I(n)$ coincides with the bracket $\{\cdot,\cdot\}_{I,J}$ given by* (8.23) *in coordinates m_{pj}^I.*

PROOF. Both statements follow immediately from Theorems 8.6 and 8.7, the fact that $G_k^0(n)$ is an open dense subset in $G_k(n)$, and the following proposition.

PROPOSITION 8.13. *For any $i_p, i_{\bar p} \in I$ and any $j, \bar j \in J$,*

(8.24) $$\{m_{pj}^I, m_{\bar p\bar j}^I\}_{k,n-k} = \{m_{pj}^I, m_{\bar p\bar j}^I\}_{I,J}.$$

PROOF. It is convenient to rewrite (8.23) as

$$\{m_{pj}^I, m_{\bar p\bar j}^I\}_{I,J} = \{m_{pj}^I, m_{\bar p\bar j}^I\}_{I,J}^1 + \{m_{pj}^I, m_{\bar p\bar j}^I\}_{I,J}^2$$

and to treat the brackets $\{\cdot,\cdot\}_{I,J}^1$ and $\{\cdot,\cdot\}_{I,J}^2$ separately for a suitable choice of parameters α and β.

Let $I = [1,k]$ and denote $m_{ij}^{[1,k]}$ simply by m_{ij}. Then two independent brackets are given by

(8.25) $$\{m_{ij}, m_{\bar\imath\bar\jmath}\}_{k,n-k}^1 = (\mathrm{sign}(\bar\imath - i) - \mathrm{sign}(\bar\jmath - j))\, m_{i\bar\jmath} m_{\bar\imath j}.$$

(for $\alpha - \beta = 2$) that coincides (up to a multiple) with the Poisson homogeneous bracket we defined in (4.7) on $G_k^0(n)$ and

(8.26) $$\{m_{ij}, m_{\bar\imath\bar\jmath}\}_{k,n-k}^2 = (\mathrm{sign}(\bar\imath - i) + \mathrm{sign}(\bar\jmath - j))\, m_{ij} m_{\bar\imath\bar\jmath}$$

(for $\alpha + \beta = 2$).

Let us start with the first of the two brackets above. For a generic element X in $G_k(n)$ represented by a matrix $\bar X = \bar X_N$, consider any two minors, $\bar X_\mu^\nu$ and $\bar X_{\mu'}^{\nu'}$, with row sets $\mu = (\mu_1, \dots, \mu_l)$, $\mu' = (\mu_1', \dots, \mu_{l'}')$, $\mu, \mu' \subseteq [1,k]$ and column sets $\nu = (\nu_1, \dots, \nu_l)$, $\nu' = (\nu_1', \dots, \nu_{l'}')$, $\nu, \nu' \subseteq [k+1, n]$. Using considerations similar to those in the proof of Lemma 4.7, we obtain from (8.25)

$$(8.27)\quad \{\bar X^\nu_\mu, \bar X^{\nu'}_{\mu'}\}^1_{k,n-k} = \sum_{p=1}^{l}\sum_{q=1}^{l'}\Big(\operatorname{sign}(\mu'_p-\mu_q)\bar X^\nu_{\mu(\mu_p\to\mu'_q)}\bar X^{\nu'}_{\mu'(\mu'_q\to\mu_p)} - \operatorname{sign}(\nu'_p-\nu_q)\bar X^{\nu(\nu_p\to\nu'_q)}_\mu \bar X^{\nu'(\nu'_q\to\nu_p)}_{\mu'}\Big).$$

Plücker coordinates x_I (for $I\subset[1,n]$, $|I|=k$) and minors $\bar X^\nu_\mu$ are related via

$$\bar X^\nu_\mu = (-1)^{l(l-1)/2}\frac{x_{([1,k]\setminus\mu)\cup\{\nu\}}}{x_{[1,k]}}.$$

(The difference between this formula and the one used in Sect. 4.2.1 is due to signs we assigned to entries of $\bar X_N$.) Denote

$$a_I = \frac{x_I}{x_{[1,k]}} = (-1)^{l(l-1)/2}\bar X^{I\setminus[1,k]}_{[1,k]\setminus I},$$

where $l=|[1,k]\setminus I|$. Then (8.27) gives rise to Poisson relations

$$\begin{aligned}\{a_I,a_{I'}\}^1_{k,n-k} &= \textstyle\sum_{i\in I\cap[1,k]}\sum_{i'\in I'\cap[1,k]}\operatorname{sign}(i-i')a_{I(i\to i')}a_{I'(i'\to i)}\\ &\quad+\textstyle\sum_{i\in I\setminus[1,k]}\sum_{i'\in I'\setminus[1,k]}\operatorname{sign}(i-i')a_{I(i\to i')}a_{I'(i'\to i)}\\ &=\textstyle\sum_{i\in I}\sum_{i'\in I'}\varepsilon_{ii'}a_{I(i\to i')}a_{I'(i'\to i)},\end{aligned}$$

where

$$\varepsilon_{ii'} = \begin{cases} 0 & \text{if}\quad i\le k<\bar\imath \quad\text{or}\quad \bar\imath\le k<i,\\ \operatorname{sign}(i-i') & \text{otherwise.}\end{cases}$$

Let us fix a k-element index set $I=\{1\le i_1<\ldots<i_k\le n\}$ and use (8.28) to compute, for any $p,\bar p\in[1,k]$ and $j,\bar\jmath\notin I$, the bracket $\{m^I_{pj},m^I_{\bar p\bar\jmath}\}^1_{k,n-k}$. Taking into account that

$$m^I_{pj} = \frac{x_{I(i_p\to j)}}{x_I}\quad\text{for any } p\in[1,k],\ j\in J,$$

we get

$$\begin{aligned}\{m^I_{pj},m^I_{\bar p\bar\jmath}\}^1_{k,n-k} &= \frac{1}{a_I^2}\Big(\{a_{I(i_p\to j)},a_{I(i_{\bar p}\to\bar\jmath)}\}^1_{k,n-k} - \frac{a_{I(i_p\to j)}}{a_I}\{a_I,a_{I(i_{\bar p}\to\bar\jmath)}\}^1_{k,n-k}\\ &\qquad \frac{a_{I(i_{\bar p}\to\bar\jmath)}}{a_I}\{a_{I(i_p\to j)},a_I\}^1_{k,n-k}\Big)\\ &= \frac{1}{a_I^2}\big((\varepsilon_{j\bar\jmath}+\varepsilon_{i_{\bar p}i_p})a_{I(i_p\to\bar\jmath)}a_{I(i_{\bar p}\to j)} + (\varepsilon_{ji_p}+\varepsilon_{i_{\bar p}\bar\jmath})a_Ia_{I(i_p\to j,i_{\bar p}\to\bar\jmath)}\\ &\qquad -(\varepsilon_{ji_p}+\varepsilon_{i_{\bar p}\bar\jmath})a_{I(i_p\to j)}a_{I(i_{\bar p}\to\bar\jmath)}\big)\\ &= \frac{1}{a_I^2}(\varepsilon_{j\bar\jmath}+\varepsilon_{i_{\bar p}i_p}-\varepsilon_{ji_p}-\varepsilon_{i_{\bar p}\bar\jmath})a_{I(i_p\to\bar\jmath)}a_{I(i_{\bar p}\to j)}\\ &= (\varepsilon_{j\bar\jmath}+\varepsilon_{i_{\bar p}i_p}-\varepsilon_{ji_p}-\varepsilon_{i_{\bar p}\bar\jmath})m^I_{p\bar\jmath}m^I_{\bar pj},\end{aligned}$$

where in the third step we have used the short Plücker relation.

Relation (8.24) for the first bracket follows from

LEMMA 8.14. *For any $i_p,i_{\bar p}\in I$ and any $j,\bar\jmath\notin I$,*

$$(8.28)\qquad \varepsilon_{j\bar\jmath}+\varepsilon_{i_{\bar p}i_p}-\varepsilon_{ji_p}-\varepsilon_{i_{\bar p}\bar\jmath} = 2s_=(i_p,j,i_{\bar p},\bar\jmath).$$

PROOF. Denote the left hand side of (8.28) by $\varepsilon^1(i_p, j, \bar{\jmath}, i_{\bar{p}})$. Let us prove that this expression depends only on the counterclockwise order of the numbers i_p, j, $\bar{\jmath}$, $i_{\bar{p}}$ and does not depend on the numbers themselves.

Assume first that all four numbers are distinct. In this case $\varepsilon^1(i_p, j, \bar{\jmath}, i_{\bar{p}})$ is invariant with respect to the cyclic shift of the variables, hence it suffices to verify the identity $\varepsilon^1(i_p, j, \bar{\jmath}, i_{\bar{p}}) = \varepsilon^1(i_p, j-1 \mod n, \bar{\jmath}, i_{\bar{p}})$ provided the counterclockwise orders of i_p, j, $\bar{\jmath}$, $i_{\bar{p}}$ and i_p, $j-1 \mod n$, $\bar{\jmath}$, $i_{\bar{p}}$ coincide. This is trivial unless $j=1$ or $j=k$, since all four summands retain their values. In the remaining cases the second and the fourth summands retain their values, and we have to check the identities $\varepsilon_{1\bar{\jmath}} - \varepsilon_{1i_p} = \varepsilon_{n\bar{\jmath}} - \varepsilon_{ni_p}$ and $\varepsilon_{k\bar{\jmath}} - \varepsilon_{ki_p} = \varepsilon_{k-1,\bar{\jmath}} - \varepsilon_{k-1,i_p}$. The first of them follows from the identity $\varepsilon_{ni} = \varepsilon_{1i} + 1$ for $i \neq 1, n$, and the second one, from the identity $\varepsilon_{k-1,i} = \varepsilon_{ki} + 1$ for $i \neq k-1, k$.

If $j = \bar{\jmath}$ or $i_p = i_{\bar{p}}$ (other coincidences are impossible since $i_p, i_{\bar{p}} \in I$, $j, \bar{\jmath} \notin I$), ε^1 degenerates to a function of three variables, which is again invariant with respect to cyclic shifts, therefore all the above argument applies as well.

To obtain (8.28) it remains to check that $\varepsilon^1(i_p, 1, k, i_{\bar{p}}) = 2s_=(i_p, 1, i_{\bar{p}}, k)$, which can be done separately in all the cases mentioned in (8.12). □

Next, let us turn to the Poisson bracket (8.26). For any two monomials in matrix entries of $\bar{X}_N$, $\mathbf{x} = \prod_{p=1}^{l} \bar{X}_{\mu_p\nu_p}$, and $\mathbf{x}' = \prod_{q=1}^{l'} \bar{X}_{\mu'_q\nu'_q}$, we have

$$\{\mathbf{x}, \mathbf{x}'\}^2_{k,n-k} = \mathbf{x}\,\mathbf{x}' \sum_{p=1}^{l}\sum_{q=1}^{l'} \left(\operatorname{sign}(\mu'_p - \mu_q) + \operatorname{sign}(\nu'_p - \nu_q)\right).$$

Observe that the double sum above is invariant under any permutation of indices within sets $\mu = (\mu_1, \dots, \mu_l)$, $\mu' = (\mu'_1, \dots, \mu'_{l'})$, $\nu = (\nu_1, \dots, \nu_l)$, $\nu' = (\nu'_1, \dots, \nu'_{l'})$. This means that for minors $\bar{X}^{\nu}_{\mu}, \bar{X}^{\nu'}_{\mu'}$ we have

$$\{\bar{X}^{\nu}_{\mu}, \bar{X}^{\nu'}_{\mu'}\}^2_{k,n-k} = \bar{X}^{\nu}_{\mu}\bar{X}^{\nu'}_{\mu'} \sum_{p=1}^{l}\sum_{q=1}^{l'} \left(\operatorname{sign}(\mu'_p - \mu_q) + \operatorname{sign}(\nu'_p - \nu_q)\right),$$

and the resulting Poisson relations for functions a_I are

$$\begin{aligned}\{a_I, a_{I'}\}^2_{k,n-k} &= a_I a_{I'} \left(\sum_{i\in[1,k]\setminus I}\ \sum_{i'\in[1,k]\setminus I'} \operatorname{sign}(i-i') \right.\\ &\qquad \left. + \sum_{i\in I\setminus[1,k]}\ \sum_{i'\in I'\setminus[1,k]} \operatorname{sign}(i-i') \right)\\ &= a_I a_{I'} \sum_{i\in I}\sum_{i'\in I'} \varepsilon_{ii'},\end{aligned}$$

where $\varepsilon_{ii'}$ has the same meaning as above. These relations imply

$$\{m^I_{i_p j}, m^I_{i_{\bar{p}}\bar{\jmath}}\}^2_{k,n-k} = m^I_{i_p j} m^I_{i_{\bar{p}}\bar{\jmath}} \left(\varepsilon_{i_p i_{\bar{p}}} - \varepsilon_{i_p\bar{\jmath}} - \varepsilon_{q i_{\bar{p}}} + \varepsilon_{j\bar{\jmath}}\right).$$

Relation (8.24) for the second bracket follows now from

LEMMA 8.15. *For any $i_p, i_{\bar{p}} \in I$ and any $j, \bar{\jmath} \notin I$,*

$$\varepsilon_{i_p i_{\bar{p}}} - \varepsilon_{i_p\bar{\jmath}} - \varepsilon_{q i_{\bar{p}}} + \varepsilon_{j\bar{\jmath}} = 2s_\times(i_p, j, i_{\bar{p}}, \bar{\jmath}). \tag{8.29}$$

PROOF. The proof of (8.29) is similar to the proof of (8.28). □

This proves Proposition 8.13, and hence Theorem 8.12. □

□

8.3.3. GL_n-action on the Grassmannian via networks. In this subsection we interpret the natural action of GL_n on $G_k(n)$ in terms of planar networks.

First, note that *any* element of GL_n can be represented by a planar network built by concatenation from building blocks (elementary networks) described in Fig. 8.5. To see this, one needs to observe that an elementary transposition matrix $S_i = \mathbf{1} - e_{ii} - e_{i+1,i+1} + e_{i,i+1} + e_{i+1,i}$ can be factored as

$$S_i = (\mathbf{1} - 2e_{i+1,i+1})E^-_{i+1}(-1)E^+_{i+1}(1)E^-_{i+1}(-1),$$

which implies that any permutation matrix can be represented via concatenation of elementary networks. Consequently, any elementary triangular matrices $\mathbf{1} + ue_{ij}$, $\mathbf{1} + le_{ji}$, $i < j$, can be factored as

$$\mathbf{1} + ue_{ij} = W_{ij}E^+_{i+1}(u)W^{-1}_{ij}, \qquad \mathbf{1} + le_{ji} = W^{-1}_{ij}E^-_{i+1}(l)W_{ij},$$

where W_{ij} is the permutation matrix that corresponds to the permutation

$$(1)\cdots(i)(i+1\dots j)(j+1)\cdots(n).$$

The claim then follows from the Bruhat decomposition and constructions presented in Section 8.2.1.

Consider now a network $N \in \mathrm{Net}_{I,J}$ and a network $N(A)$ representing an element $A \in GL_n$ as explained above. We will concatenate N and $N(A)$ according to the following rule: in $N(A)$ reverse directions of all horizontal paths $i \to i$ for $i \in I$, changing every edge weight w involved to w^{-1}, and then glue the left boundary of $N(A)$ to the boundary of N in such a way that boundary vertices with the same label are glued to each other. Denote the resulting net by $N \circ N(A)$. Let $\bar X_N$ and $\bar X_{N\circ N(A)}$ be the signed boundary measurement matrices constructed according to the recipe outlined at the beginning of Section 8.3.2.

LEMMA 8.16. *Matrices $\bar X_{N\circ N(A)}$ and $\bar X_N A$ are representatives of the same element in $G_k(n)$.*

PROOF. To check that $\bar X_{N\circ N(A)}$ coincides with the result of the natural action of GL_n on $G_k(n)$ induced by the right multiplication, it suffices to consider the case when A is a diagonal or an elementary bidiagonal matrix, that is when $N(A)$ is one of the elementary networks in Fig. 8.5. If $N(A)$ is the first diagram in Fig. 8.5, then the boundary measurements in $N' = N\circ N(A)$ are given by $M'(i,j) = d_i^{-1}M(i,j)d_j$, $i \in I, j \in J$, where $M(i,j)$ are the boundary measurements in N. This is clearly consistent with the natural GL_n action.

Now let $A = E^-_i(l)$ (the case $A = E^+_i(u)$ can be treated similarly). Then

$$M'(i_p,j) = \begin{cases} M(i_p,j) + \delta_{ii_p} lM(i_{p-1},j) & \text{if } i, i-1 \in I \\ M(i_p,j) + \delta_{i-1,j} lM(i_p,j+1) & \text{if } i, i-1 \in J \\ M(i_p,j) + \delta_{ii_p}\delta_{i-1,j} l & \text{if } i \in I,\ i-1 \in J \\ (1-\delta_{i_p,i-1})M(i_p,j) + \dfrac{M(i-1,j)}{1+lM(i-1,i)}(\delta_{i_p,i-1} \pm (1-\delta_{i_p,i-1})lM(i_p,i)) & \\ & \text{if } i \in J,\ i-1 \in I \end{cases} \tag{8.30}$$

(in the last line above we used Lemma 8.9).

Recall that for $j \in J$, an entry $\bar{X}_{pj}$ of $\bar{X}_N$ coincides with $M(i_p, j)$ up to the sign $(-1)^{s(p,j)}$. By the construction in Section 8.3.2, if $i, i-1 \in I$, then $(-1)^{s(p,j)} = -(-1)^{s(p-1,j)}$ for p such that $i_p = i$ and for all $j \in J$. Then the first line of (8.30) shows that $\bar{X}_{N'} = E_p^-(-l)\bar{X}_N E_i^-(l)$, where the first elementary matrix is $k \times k$ and the second one is $n \times n$.

If $i, i-1 \in J$, then $(-1)^{s(p,i-1)} = (-1)^{s(p,i)}$ for all $p \in [1, k]$, and the second line of (8.30) results in $\bar{X}_{N'} = \bar{X}_N E_i^-(l)$. The latter equality is also clearly valid for the case $i \in I$, $i-1 \in J$.

Finally, if $i \in J$, $i-1 \in I$, let $r \in [1, k]$ be the index such that $i-1 = i_r$, and let $\bar{X}_N^{(i)}$ denote the ith column of $\bar{X}_N$. Then the $k \times k$ submatrix of $\bar{X}_N E_i^-(l)$ formed by the columns indexed by I is $C = \mathbf{1}_k + l\bar{X}_N^{(i)} e_r^T$, where $e_r = (0, \dots, 0, 1, 0, \dots, 0)$ with the only nonzero element in the rth position. Therefore,

$$\tilde{X} = C^{-1}\bar{X}_N E_i^-(l) = \left(\mathbf{1}_k - \frac{1}{1 + l\bar{X}_{ri}} l\bar{X}_N^{(i)} e_r^T\right) \bar{X}_N E_i^-(l)$$

is the representative of $[\bar{X}_N E_i^-(l)] \in G_k(n)$ that has an identity matrix as its $k \times k$ submatrix formed by columns indexed by I. We need to show that $\tilde{X} = \bar{X}_{N'}$. First note that $\bar{X}_{ri} = M(i_r, i) = M(i-1, i)$ and so, for all $j \in J$,

$$\tilde{X}_{rj} = \frac{\bar{X}_{rj}}{1 + l\bar{X}_{ri}} = \frac{(-1)^{s(r,j)} M(i-1, j)}{1 + lM(i-1, i)} = (-1)^{s(r,j)} M'(i-1, j) = (\bar{X}_{N'})_{rj}.$$

If $p \neq r$, then

$$\begin{aligned} \tilde{X}_{pj} &= \bar{X}_{pj} - \frac{l\bar{X}_{rj}\bar{X}_{pi}}{1 + l\bar{X}_{ri}} \\ &= (-1)^{s(p,j)} \left(M(i_p, j) - (-1)^{s(p,i)+s(r,j)-s(p,j)} \frac{lM(i_p, i)M(i-1, j)}{1 + lM(i-1, i)} \right). \end{aligned}$$

Thus, to see that $\tilde{X}_{pj} = (\bar{X}_{N'})_{rj}$, it is enough to check that the sign assignment that was used in Lemma 8.9 is consistent with the formula $-(-1)^{s(p,i)+s(r,j)-s(p,j)}$. This can be done by direct inspection and is left to the reader as an exercise. □

Now that we have established that the natural right action of GL_n on $G_k(n)$ can be realized via the operation on networks described above, Theorems 8.7, 8.10 and 8.12 immediately imply the following

THEOREM 8.17. *For any choice of parameters α, β, the Grassmannian $G_k(n)$, equipped with the bracket $\mathcal{P}_{\alpha,\beta}$ is a Poisson homogeneous space for GL_n equipped with the "matching" Poisson-Lie bracket* (8.21).

8.4. Face weights

In Section 4.2.2 we constructed the cluster algebra $\mathcal{A}_{G_k^0(n)}$ on the open cell $G_k^0(n)$ in the Grassmannian $G_k(n)$. By construction, $\mathcal{A}_{G_k^0(n)}$ is compatible with the Sklyanin Poisson structure on $G_k(n)$ induced by the standard Sklyanin Poisson–Lie structure on SL_n.

The main goal of this section is to use networks in order to show that, in fact, every Poisson structure in the 2-parameter family $\mathcal{P}_{\alpha,\beta}$ described in Theorem 8.12 is compatible with $\mathcal{A}_{G_k^0(n)}$.

8.4.1. Face weights and boundary measurements. Let $N = (G, w)$ be a perfect planar network in a disk. Graph G divides the disk into a finite number of connected components called *faces*. The boundary of each face consists of edges of G and, possibly, of several arcs bounding the disk. A face is called *bounded* if its boundary contains only edges of G and *unbounded* otherwise. In this Section we additionally require that each edge of G belongs to a path from a source to a sink. This is a technical condition that ensures that the two faces separated by an edge are distinct. Clearly, the edges that violate this condition do not influence the boundary measurement map and may be eliminated from the graph.

Given a face f, we define its *face weight* y_f as the Laurent monomial in the edge weights w_e, $e \in E$, given by

$$y_f = \prod_{e \in \partial f} w_e^{\gamma_e}, \tag{8.31}$$

where $\gamma_e = 1$ if the direction of e is compatible with the counterclockwise orientation of the boundary ∂f and $\gamma_e = -1$ otherwise. For example, the face weights for the network shown in Figure 8.11 are $w_1 w_2^{-1} w_3$, $w_3^{-1} w_5^{-1} w_8^{-1}$, $w_6 w_7^{-1} w_8$, $w_1^{-1} w_4^{-1} w_6^{-1}$ for four unbounded faces and $w_2 w_4 w_5 w_7$ for the only bounded face.

Similarly to the space of edge weights $\mathcal{E}_N$, we can define the *space of face weights* $\mathcal{F}_N$; a point of this space is the graph G as above with the faces weighted by real numbers obtained by specializing the variables $x_1, \dots, x_d$ in the expressions for w_e to nonzero values and subsequent computation via (8.31) (recall that by definition, w_e do not vanish on $\mathcal{E}_N$). By the Euler formula, the number of faces of G equals $|E| - |V| + 1$, and so $\mathcal{F}_N$ is a semialgebraic subset in $\mathbb{R}^{|E|-|V|+1}$. In particular, the product of the face weights over all faces equals 1 identically, since each edge enters the boundaries of exactly two faces in opposite directions.

In general, the edge weights can not be restored from the face weights. However, the following proposition holds true.

LEMMA 8.18. *The weight w_P of an arbitrary path P between a source b_i and a sink b_j is a monomial in the face weights.*

PROOF. Assume first that all the edges in P are distinct. Extend P to a cycle C_P by adding the arc on the boundary of the disk between b_j and b_i in the counterclockwise direction. Then the product of the face weights over all faces lying inside C_P equals the product of the edge weights over the edges of P, which is w_P. Similarly, for any simple cycle C in G, its weight w_C equals the product of the face weights over all faces lying inside C. It remains to use (8.1) and to care that the cycles that are split off are simple. The latter can be guaranteed by the *loop erasure* procedure that consists in traversing P from b_i to b_j and splitting off the cycle that occurs when the first time the current edge of P coincides with the edge traversed earlier. □

It follows from Lemma 8.18 that one can define boundary measurement maps $M_N^{\mathcal{F}}\colon \mathcal{F}_N \to \mathrm{Mat}_{k,m}$ and $X_N^{\mathcal{F}}\colon \mathcal{F}_N \to G_k(n)$ so that

$$M_N^{\mathcal{F}} \circ y = M_N, \qquad X_N^{\mathcal{F}} \circ y = X_N,$$

where $y\colon \mathcal{E}_N \to \mathcal{F}_N$ is given by (8.31).

Our next goal is to write down the 2-parametric family of Poisson structures induced on $\mathcal{F}_N$ by the map y. Let $N = (G, w)$ be a perfect planar network. In this section it will be convenient to assume that boundary vertices are colored in *gray*.

Define the *directed dual network* $N^* = (G^*, w^*)$ as follows. Vertices of G^* are the faces of N. Edges of G^* correspond to the edges of N with endpoints of different colors; note that there might be several edges between the same pair of vertices in G^*. An edge e^* of G^* corresponding to e is directed in such a way that the white endpoint of e (if it exists) lies to the left of e^* and the black endpoint of e (if it exists) lies to the right of e. The weight $w^*(e^*)$ equals $\alpha - \beta$ if the endpoints of e are white and black, α if the endpoints of e are white and gray and $-\beta$ if the endpoints of e are black and gray. An example of a planar network and its directed dual network is given in Fig. 8.12.

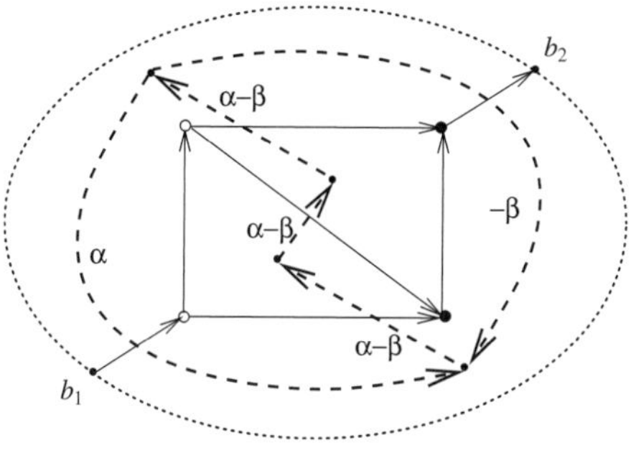

FIGURE 8.12. Planar network and its directed dual

LEMMA 8.19. *The 2-parametric family* $\{\cdot,\cdot\}_{\mathcal{F}_N}$ *is given by*

$$\{y_f, y_{f'}\}_{\mathcal{F}_N} = \left(\sum_{e^*: f \to f'} w^*(e^*) - \sum_{e^*: f' \to f} w^*(e^*) \right) y_f y_{f'}.$$

PROOF. Let $e = (u, v)$ be a directed edge. We say that the flag (u, e) is *positive*, and the flag (v, e) is *negative*. The color of a flag is defined as the color of the vertex participating in the flag.

Let f and f' be two faces of N. We say that a flag (v, e) is *common* to f and f' if both v and e belong to $\partial f \cap \partial f'$. Clearly, the bracket $\{y_f, y_{f'}\}_{\mathcal{F}_N}$ can be calculated as the sum of the contributions of all flags common to f and f'.

Assume that (v, e) is a positive white flag common to f and f', see Fig. 8.13. Then $y_f = \dfrac{x_v^3}{x_v^2} \bar{y}_f$ and $y_{f'} = x_v^1 x_v^2 \bar{y}_{f'}$, where x_v^i are the weights of flags involving v and $\{x_v^i, \bar{y}_f\}_{\mathcal{R}} = \{x_v^i, \bar{y}_{f'}\}_{\mathcal{R}} = 0$, see Section 8.2.2. Therefore, by (8.5), the contribution of (v, e) equals $(\alpha_{12} - \alpha_{13} - \alpha_{23}) y_f y_{f'}$, which by (8.9) equals $-\alpha y_f y_{f'}$.

Assume now that (v, e) is a negative white flag common to f and f', see Fig. 8.13. In this case $y_f = \dfrac{1}{x_v^1 x_v^3} \bar{y}_f$ and $y_{f'} = x_v^1 x_v^2 \bar{y}_{f'}$, so the contribution of (v, e) equals $(\alpha_{13} + \alpha_{23} - \alpha_{12}) y_f y_{f'} = \alpha y_f y_{f'}$.

In a similar way one proves that the contribution of a positive black flag common to f and f' equals $-\beta y_f y_{f'}$, and the contribution of a negative black flag common to f and f' equals $\beta y_f y_{f'}$. Finally, the contributions of positive and negative gray flags are clearly equal to zero.

The statement of the lemma now follows from the definition of the directed dual network. □

FIGURE 8.13. Contribution of a white common flag: a) positive flag; b) negative flag

8.4.2. Compatibility theorem. The main result of this section is the following theorem.

THEOREM 8.20. *Any Poisson structure in the two-parameter family $\mathcal{P}_{\alpha,\beta}$ is compatible with the standard cluster algebra structure on the Grassmannian $G_k(n)$.*

PROOF. By Theorem 4.5, it is enough to choose an initial extended cluster and to compare the coefficient matrix of $\mathcal{P}_{\alpha,\beta}$ in the basis τ with the exchange matrix for this cluster.

For the initial extended cluster we take the cluster f_{ij}, $i \in [1,k]$, $j \in [1,m]$, defined by (4.24). The corresponding exchange matrix (more exactly, its diagram $\Gamma(\widetilde{B})$) is presented on Fig. 4.2. Recall that f_{ij} are cluster variables for $i \in [1,k-1]$, $j \in [2,m]$ and stable variables otherwise.

To find the coefficient matrix of $\mathcal{P}_{\alpha,\beta}$ we define a special network $N(k,m)$ with k sources and m sinks. The graph of $N(k,m)$ has $km+1$ faces. Each bounded face is a hexagon; all bounded faces together form a $(k-1)\times(m-1)$ parallelogram on the hexagonal lattice. Edges of the hexagons are directed North, South-East and South-West. Each vertex of degree 2 on the left boundary of this parallelogram is connected to a source, and each vertex of degree 2 on the upper boundary of this parallelogram is connected to a sink. The remaining vertices of degree 2 on the lower and the right boundaries of the parallelogram are eliminated: two edges $u_1 \to v \to u_2$ are replaced by one edge $u_1 \to u_2$ and the intermediate vertex v is deleted. The sources are labelled counterclockwise from 1 to k, sinks from $k+1$ to n. The faces are labelled by pairs (ij) such that $i \in [1,k]$, $j \in [k+1,n]$. The unbounded faces are labelled $(1j)$ and (in) except for one face, which is not labelled at all. The network $N(k,m)$ is defined by assigning a face weight y_{ij} to each labelled face (ij). Consequently, the directed dual network $N^*(k,m)$ forms the dual triangular lattice. All edges of $N^*(k,m)$ incident to the vertices (ij), $i \in [2,k]$, $j \in [k+1,n-1]$ are of weight $\alpha-\beta$.

For an example of the construction for $k=3$, $m=4$ see Fig. 8.14. The edge weights of the dual network that are not shown explicitly are equal to $\alpha-\beta$.

As the first step of the proof, we express the cluster variables f_{ij} via the face weights of $N(k,m)$.

LEMMA 8.21. *For any $i \in [1,k]$, $j \in [1,m]$, one has*

$$f_{ij} = \pm \prod_{p=1}^{i} \prod_{q=j+k}^{n} y_{pq}^{1+\min\{i-p,q-j-k\}}. \tag{8.32}$$

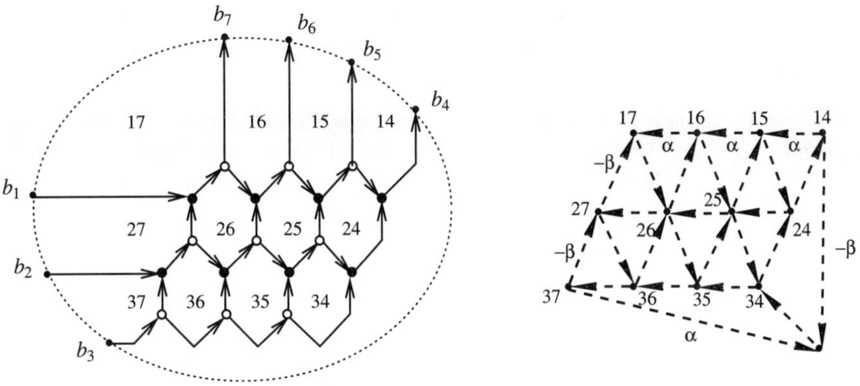

FIGURE 8.14. The graph of $N(3,4)$ and its directed dual $N^*(3,4)$

PROOF. Since the graph of $N(k,m)$ is acyclic, one can use the Lindström lemma to calculate the minors of the boundary measurement matrix. Assume first that $i+j \le n+1-k$. By (4.13) and the Lindström lemma, f_{ij} equals to the sum of the products of the path weights for all i-tuples of nonintersecting paths between the sources $b_1, b_2, \dots, b_i$ and the sinks $b_{j+k}, b_{j+k+1}, \dots, b_{j+k+i-1}$, each product being taken with a certain sign. Note that there exists a unique path from b_1 to $b_{i+j+k-1}$ in $N(k,m)$. After this path is chosen, there remains a unique path from b_2 to $b_{i+j+k-2}$, and so on. Moreover, a path from b_1 to any sink b_p with $p < i+j+k-1$ cuts off the sink $b_{i+j+k-1}$. Therefore, there exists exactly one i-tuple of paths as required. Relation (8.32) for $i+j \le n+1-k$ now follows from the proof of Lemma 8.18.

The case $i+j \ge n+1-k$ is similar and relays on the uniqueness of an $(n+1-k-j)$-tuple of nonintersecting paths between the sources $b_{i+j+k-n}, b_{i+j+k-n+1}, \dots, b_i$ and the sinks $b_{j+k}, b_{j+k+1}, \dots, b_n$. Here we start with the unique path from b_i to b_n, then choose the unique remaining path between b_{i-1} and b_{n-1}, and so on. □

The next step is the calculation of τ-coordinates for the initial extended cluster (4.24) via face weights.

LEMMA 8.22. *The τ-coordinates corresponding to the cluster variables of the initial extended cluster* (4.24) *are given by*

$$\tau_{ij} = \pm y_{i+1,j+k-1}, \qquad i \in [1,k-1],\ j \in [2,m]. \tag{8.33}$$

PROOF. Combining the definition of τ-coordinates via (4.1) and the description of the exchange matrix in the basis $\{f_{ij}\}$ provided by Fig. 4.2, we conclude that

$$\tau_{ij} = \frac{f_{i+1,j-1} f_{i,j+1} f_{i-1,j}}{f_{i,j-1} f_{i+1,j} f_{i-1,j+1}} = \frac{\dfrac{f_{i+1,j-1}}{f_{ij}} \cdot \dfrac{f_{ij}}{f_{i-1,j+1}}}{\dfrac{f_{i+1,j}}{f_{i,j+1}} \cdot \dfrac{f_{i,j-1}}{f_{i-1,j}}}, \qquad i \in [1,k-1],\ j \in [2,m];$$

here we assume that $f_{ij} = 1$ if $i = 0$ or $j = m+1$. Next, by (8.32),

$$\frac{f_{ij}}{f_{i-1,j+1}} = \pm \prod_{p=1}^{i} \prod_{q=j+k}^{n} y_{pq}$$

for $i \in [1, k-1]$, $j \in [2, m]$, and the result follows. $\square$

REMARK 8.23. The operation that expresses τ-coordinates in terms of face weights is an analog of the twist mentioned in Section 1.2.

The expressions for the τ-coordinates that correspond to the stable variables of the initial extended cluster (4.24) are somewhat cumbersome. Recall that by (4.1), each stable variable enters the expression for the corresponding τ-coordinate with some integer exponent $\varkappa_{ij}$. Let us denote

$$\tau^*_{ij} = \tau_{ij} f_{ij}^{-\varkappa_{ij}}$$

for $i = k$, $j \in [1, m]$ and $i \in [1, k]$, $j = 1$.

LEMMA 8.24. *The τ-coordinates corresponding to the stable variables of the initial extended cluster* (4.24) *are given by*

$$\tau^*_{ij} = \begin{cases} \pm \prod_{p=1}^{k-1} \prod_{q=j+k}^{\min\{n,j+k-p-1\}} y_{pq} & \text{for } i = k,\ j \in [2, m-1], \\ \pm \prod_{q=k+2}^{n} \prod_{p=i}^{\min\{n,i-q+k+2\}} y_{pq} & \text{for } i \in [2, k-1],\ j = 1, \\ \pm \prod_{p=1}^{k-1} y_{pn} & \text{for } i = k,\ j = m, \\ \pm \prod_{q=k+2}^{n} y_{1q} & \text{for } i = 1,\ j = 1, \\ \pm \prod_{p=1}^{k} \prod_{q=k+1}^{n} y_{pq}^{-\min\{k-p,q-k-1\}} & \text{for } i = k,\ j = 1. \end{cases} \tag{8.34}$$

PROOF. It suffices to note that

$$\tau^*_{ij} = \begin{cases} f_{k-1,j}/f_{k-1,j+1} & \text{for } i = k,\ j \in [2, m-1], \\ f_{i,2}/f_{i-1,2} & \text{for } i \in [2, k-1],\ j = 1, \\ f_{k-1,m} & \text{for } i = k,\ j = m, \\ f_{12} & \text{for } i = 1,\ j = 1, \\ 1/f_{k-1,2} & \text{for } i = k,\ j = 1, \end{cases}$$

and to apply Lemma 8.21. $\square$

To conclude the proof of the theorem we build the network $\Gamma_{\alpha,\beta}$ representing the family of Poisson brackets $\mathcal{P}_{\alpha,\beta}$ in the basis τ. The vertices of $\Gamma_{\alpha,\beta}$ are τ-coordinates τ_{ij}, $i \in [1, k]$, $j \in [1, m]$. An edge from τ_{ij} to τ_{pq} with weight c means that $\{\tau_{ij}, \tau_{pq}\}_{\mathcal{F}_{N(k,m)}} = c\tau_{ij}\tau_{pq}$.

It follows immediately from Lemmas 8.19 and 8.22 that the induced subnetwork $\Gamma^0_{\alpha,\beta}$ of $\Gamma_{\alpha,\beta}$ spanned by the vertices τ_{ij}, $i \in [1, k-1]$, $j \in [2, m]$ is isomorphic to the subnetwork $N^*_0(k, m)$ of $N^*(k, m)$ spanned by the vertices (pq), $p \in [2, k]$, $q \in [k+1, n-1]$. We thus get an isomorphism between $\Gamma^0_{\alpha,\beta}$ and the corresponding induced subgraph of $\Gamma(\widetilde{B})$, see Fig. 8.15. Under this isomorphism each edge of weight $\alpha - \beta$ is mapped to an edge of weight 1.

For the remaining vertices of $\Gamma_{\alpha,\beta}$, that is, τ_{ij}, $i \in [1, k]$, $j = 1$ or $i = k$, $j \in [1, m]$, we have to check the edges connecting them to the vertices of $\Gamma^0_{\alpha,\beta}$.

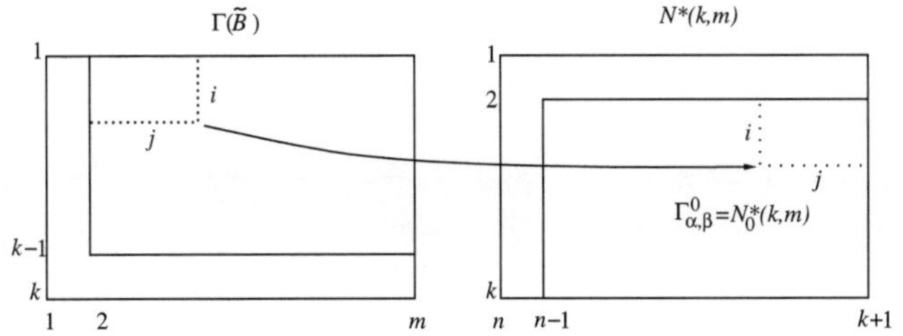

FIGURE 8.15. The isomorphisms between induced subgraphs of $\Gamma_{\alpha,\beta}$ and $\Gamma(\widetilde{B})$

Consider for example the case $i = k$, $j \in [2, m-1]$. Then, by Lemma 8.24, the bracket $\{\tau^*_{kj}, \tau_{pq}\}_{\mathcal{F}_{N(k,m)}}$ for $\tau_{pq} \in \Gamma^0_{\alpha,\beta}$ is defined by the edges incident to the vertex subset of $N^*(k,m)$ corresponding to the factors in the right hand side of the first formula in (8.34), see Fig. 8.16. Clearly, any vertex $(p,q) \in N^*_0(k,m)$ other than $(k, j+k)$ and $(k, j+k-1)$ is connected to this subset by an even number of edges (more exactly, 0, 2, 4 or 6), all of them of weight $\alpha - \beta$. Since exactly half of the edges are directed to (pq), the bracket between y_{pq} and τ^*_{kj} vanishes by Lemma 8.19. The remaining two edges connecting the contracted set to the vertices $(k, j+k)$ and $(k, j+k-1)$ correspond to the two edges connecting the vertex (k,j) to $(k-1,j)$ and $(k-1, j+1)$ in $\Gamma(\widetilde{B})$, see Fig. 8.16.

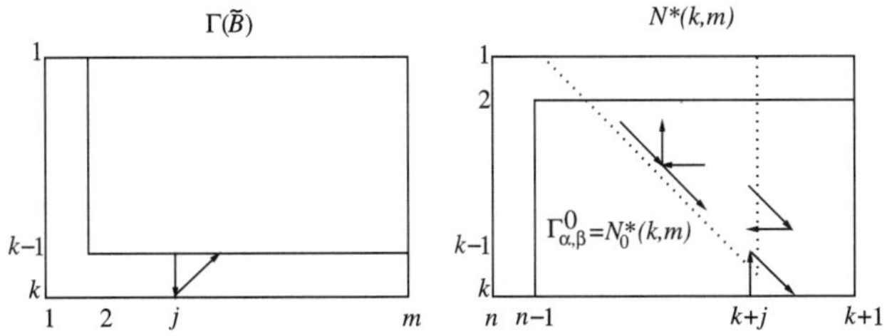

FIGURE 8.16. The isomorphisms in the neighborhood of τ_{kj}

Finally, factor f_{kj} commutes with all $\tau_{pq} \in \Gamma^0_{\alpha,\beta}$. Indeed, given an edge between $\tau_{pq} \in \Gamma^0_{\alpha,\beta}$ and y_{rs}, define its degree as the exponent of y_{rs} in expression (8.32) for f_{kj}. We have to prove that the sum of the degrees of all edges entering τ_{pq} equals the sum of the degrees of all edges leaving τ_{pq}. This fact can be proved by analyzing all possible configurations of edges, see Fig. 8.17 presenting these configurations up to reflection.

Other cases listed in Lemma 8.24 are treated in the same way. □

8.5. Summary

- We introduce a notion of a perfect planar network in a disk and associate with every such network a matrix of boundary measurements. Each boundary measurement is a rational function in the weights of the edges (edge weights) and admits a subtraction-free rational expression, see Proposition 8.3.

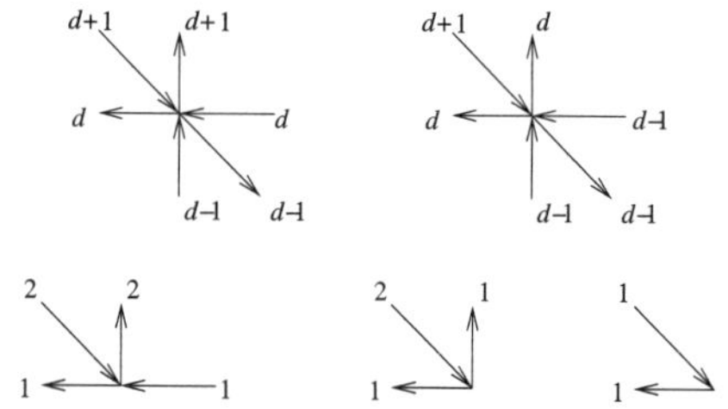

FIGURE 8.17. Edges incident to τ_{pq} and their degrees

- We construct a family of *universal* Poisson brackets on the space of edge weights of a given network, such that every bracket in the family respects the natural operation of gluing networks. Furthermore, we establish that the family of universal brackets induces a two-parameter family of Poisson brackets on boundary measurement matrices, see Theorem 8.6. This family depends on a mutual location of sources and sinks, but not on the network itself.
- We provide an explicit description of this family in Theorem 8.7. In particular, if n sources and n sinks are not intermixed along the boundary of the disk, we recover a 2-parametric family of R-matrices and the corresponding R-matrix brackets on GL_n, see Theorem 8.10.
- The boundary measurement map is extended to a map into the Grassmannian $G_k(n)$ defined by a network with k sources and $n-k$ sinks. The Poisson family on boundary measurement matrices allows us to equip the Grassmannian with a two-parameter family of Poisson brackets $\mathcal{P}_{\alpha,\beta}$ in such a way, that for any choice of a universal Poisson bracket on the edge weights there is a unique member of $\mathcal{P}_{\alpha,\beta}$ that makes the extended boundary measurement map Poisson, see Theorem 8.12. This family depends only on the number of sources and sinks and does not depend on its mutual location.
- We give an interpretation of the natural GL_n action on $G_k(n)$ in terms of networks and establish that every member of the Poisson family $\mathcal{P}_{\alpha,\beta}$ on $G_k(n)$ (i) makes $G_k(n)$ into a Poisson homogeneous space of GL_n equipped with the above described R-matrix bracket, see Theorem 8.17, and (ii) is compatible with the cluster algebra structure on the Grassmannian that was discussed in Chapter 4, see Theorem 8.20.

Bibliographical notes

Planar networks in a disk (in particular, perfect planar networks) and boundary measurements in such networks were introduced and studied in [**Po**]. The exposition in this Chapter follows [**GSV5**].

8.1. The fact that $c_l(C)$ does not depend on l follows from Theorem 1 in [**GrSh**]. Proposition 8.3 was first proved in [**Po**] and extended in [**Ta**]. Our proof follows [**GSV5**].

8.3. The Grassmannian boundary measurement map was introduced and studied in [**Po**]. The fact that $c_1R_0 + c_2S\pi_0$ satisfies the MCYBE and thus gives rise to a Sklyanin Poisson–Lie bracket can be found in [**ReST**]. Formula (8.22) in [**GSV5**] contains a superfluous factor 1/2.

CHAPTER 9

Perfect planar networks in an annulus and rational loops in Grassmannians

In this Chapter, we continue the line of investigation started in Chapter 8 and build a parallel theory for directed weighted networks in an annulus (or, equivalently, on a cylinder). First, we have to modify the definition of the boundary measurement map, whose image now consists of rational valued matrix functions of an auxiliary parameter λ associated with the notion of a *cut*. We then show that the analogue of Postnikov's construction leads to a map into the space of loops in the Grassmannian. Universal Poisson brackets for networks in an annulus are defined in exactly the same way as in the case of a disk. We show that they induce a two-parameter family of Poisson brackets on rational-valued boundary measurement matrices. In particular, when sources and sinks belong to distinct circles bounding the annulus, one of the generators of these family coincides with the Sklyanin R-matrix bracket associated with the trigonometric solution of the classical Yang-Baxter equation in $sl(n)$. Moreover, we prove that the two-parameter family of Poisson brackets can be further pushed forward to the space of loops in the Grassmannian. In proving the latter, we depart from the approach of Chapter 8 where a similar result is obtained via a more or less straightforward calculation. Such an approach would have been too cumbersome in our current setting. Instead, we find a way to utilize *face weights* and their behavior under *path-reversal maps*.

9.1. Perfect planar networks and boundary measurements

9.1.1. Networks, cuts, paths and weights. Let $G = (V, E)$ be a directed planar graph drawn inside an annulus with the vertex set V and the edge set E. Exactly n of its vertices are located on the boundary circles of the annulus and are called *boundary vertices*; $n_1 \geq 0$ of them lie on the outer circle, and $n_2 = n - n_1 \geq 0$ on the inner circle. The graph is considered up to an isotopy relative to the boundary (with fixed boundary vertices).

Each boundary vertex is marked as a source or a sink. Sources, sinks and black or white internal vertices are defined exactly in the same way as in the case of planar networks in a disk. That is, sources are vertices of degree 1 with one outcoming edge, sinks are vertices of degree 1 with one incoming edge, and all internal vertices have degree 3 and either have exactly one incoming edge (white vertices), or exactly one outcoming edge (black vertices).

A *cut* ρ is an oriented non-selfintersecting curve starting at a *base point* on the inner circle and ending at a base point on the outer circle considered up to an isotopy relative to the boundary (with fixed endpoints). We assume that the base points of the cut are distinct from the boundary vertices of G. For an arbitrary oriented curve γ with endpoints not lying on the cut ρ we denote by $\mathrm{ind}(\gamma)$ the algebraic intersection number of γ and ρ. Recall that each transversal intersection point of γ and ρ contributes to this number 1 if the oriented tangents to γ and ρ at this point form a positively oriented basis, and -1 otherwise. Non-transversal intersection points are treated in a similar way.

Let $x_1, \dots, x_d$ be independent variables. A *perfect planar network in an annulus* $N = (G, \rho, w)$ is obtained from a graph G equipped with a cut ρ as above by assigning a weight $w_e \in \mathbb{Z}(x_1, \dots, x_d)$ to each edge $e \in E$. In this chapter we occasionally write "network" instead of "perfect planar network in an annulus". Each network defines a rational map $w : \mathbb{R}^d \to \mathbb{R}^{|E|}$; the *space of edge weights* $\mathcal{E}_N$ is defined as the intersection of the image of w with $(\mathbb{R} \setminus 0)^{|E|}$. In other words, a point in $\mathcal{E}_N$ is a graph G as above with edges weighted by nonzero real numbers obtained by specializing the variables $x_1, \dots, x_d$ in the expressions for w_e.

Paths and cycles in N are defined exactly in the same way as in the case of perfect planar networks in a disk. In what follows we assume without loss of generality that N is drawn in such a way that all its edges and the cut are smooth curves. Moreover, any simple path in N is a piecewise-smooth curve with no cusps, at any boundary vertex of N the edge and the circle intersect transversally, and the same holds for the cut at both of its base points. Given a path P between a source b' and a sink b'', we define a closed piecewise-smooth curve C_P in the following way: if both b' and b'' belong to the same boundary circle, C_P is obtained by adding to P the path between b'' and b' that goes counterclockwise along the boundary of the corresponding circle. Otherwise, if b' and b'' belong to distinct circles, C_P is obtained by adding to P the path that starts at b'', goes counterclockwise along the corresponding circle to the base point of the cut, follows the cut to the other base point and then goes counterclockwise along the other circle up to b'. Clearly, the obtained curve C_P does not have cusps, so its concordance number $c(C_P)$ can be defined in a straightforward manner via polygonal approximation. Finally the weight of P is defined as

$$w_P = w_P(\lambda) = (-1)^{c(C_P)-1} \lambda^{\operatorname{ind}(P)} \prod_{e \in P} w_e, \tag{9.1}$$

where λ is an auxiliary independent variable. Occasionally, it will be convenient to assume that the internal vertices of G do not lie on the cut and to rewrite the above formula as

$$w_P = (-1)^{c(C_P)-1} \prod_{e \in P} \bar{w}_e, \tag{9.2}$$

where $\bar{w}_e = w_e \lambda^{\operatorname{ind}(e)}$ are *modified edge weights*. Observe that the weight of a path is a relative isotopy invariant, while modified edge weights are not. The weight of an arbitrary cycle in N is defined in the same way via the concordance number of the cycle. As in the case of the disk, if P can be decomposed into a path P' and a cycle C^0, relation (8.1) holds true.

An example of a perfect planar network in an annulus is shown in Fig. 9.1 on the left. It has two sources, b on the outer circle and b'' on the inner circle, and one sink b' on the inner circle. Each edge e_i is labelled by its weight. The cut is shown by the dashed line. The same network is shown in Fig. 9.1 on the right; it differs from the original picture by an isotopic deformation of the cut.

Consider the path $P_1 = (e_1, e_2, e_3, e_4)$ from b to b'. Its algebraic intersection number with the cut equals 0 (in the left picture, two intersection points contribute 1

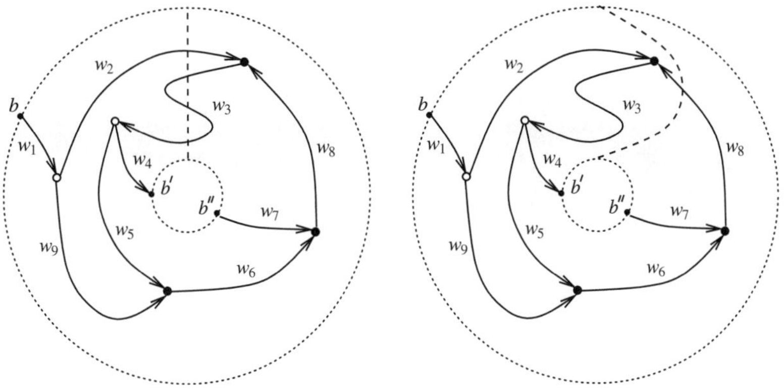

FIGURE 9.1. A perfect planar network in an annulus

each, and two other intersection points contribute -1 each; in the right picture there are no intersection points). The concordance number of the corresponding closed curve C_{P_1} equals 1. Therefore, (9.1) gives $w_{P_1} = w_1w_2w_3w_4$. On the other hand, the modified weights for the left picture are given by $\bar{w}_1 = w_1$, $\bar{w}_2 = \lambda w_2$, $\bar{w}_3 = \lambda^{-1}w_3$, $\bar{w}_4 = w_4$, and hence, computation via (9.2) gives $w_{P_1} = \bar{w}_1\bar{w}_2\bar{w}_3\bar{w}_4 = w_1w_2w_3w_4$. Finally, the modified weights for the relevant edges in the right picture coincide with the original weights, and we again get the same result.

Consider the path $P_2 = (e_7, e_8, e_3, e_4)$ from b'' to b'. Its algebraic intersection number with the cut equals -1, and the concordance number of the corresponding closed curve C_{P_2} equals 1. Therefore, $w_{P_2} = \lambda^{-1}w_3w_4w_7w_8$. The same result can be obtained by using modified weights.

Finally, consider the path $P_3 = (e_1, e_2, e_3, e_5, e_6, e_8, e_3, e_4)$ from b to b'. Clearly, $w_{P_3} = -\lambda^{-1}w_1w_2w_3^2w_4w_5w_6w_8$. The path P_3 can be decomposed into the path P_1 as above and a cycle $C^0 = (e_3, e_5, e_6, e_8)$ with the weights $w_{P_1} = w_1w_2w_3w_4$ and $w_{C^0} = \lambda^{-1}w_3w_5w_6w_8$, hence relation (8.1) yields the same expression for w_{P_3} as before.

Let us see how moving a base point of the cut affects the weights of paths. Let $N = (G, \rho, w)$ and $N' = (G, \rho', w)$ be two networks with the same graph and the same weights, and assume that the cuts ρ and ρ' are not isotopic. More exactly, let us start moving a base point of the cut in the counterclockwise direction. Assume that b is the first boundary vertex in the counterclockwise direction from the base point of ρ that is being moved. Clearly, nothing is changed while the base point and b do not interchange. Let ρ' be the cut obtained after the interchange, and assume that no other interchanges occurred. Then the relation between the weight w_P of a path P in N and its weight w'_P in N' is given by the following proposition.

PROPOSITION 9.1. *For N and N' as above,*

$$w'_P(\lambda) = ((-1)^{\alpha(P)}\lambda)^{\beta(b,P)}w_P,$$

where $\alpha(P)$ *equals* 0 *if the endpoints of* P *lie on the same circle and* 1 *otherwise, and*

$$\beta(b,P)=\begin{cases}1 & \text{if } b \text{ is the sink of } P,\\ -1 & \text{if } b \text{ is the source of } P,\\ 0 & \text{otherwise.}\end{cases}$$

PROOF. The proof is straightforward. □

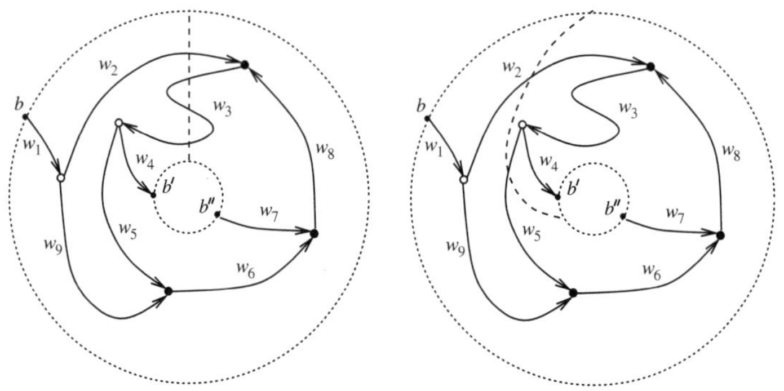

FIGURE 9.2. Moving the base point of the cut

For example, consider the networks N and N' shown in Fig. 9.2. The base point of the cut lying on the inner circle interchanges positions with the sink b'. The path P_1 from the previous example goes from the outer circle to the inner circle, so $\alpha(P_1)=1$; besides, b' is the sink of P_1, so $\beta(b',P_1)=1$. Therefore, by Proposition 9.1, $w'_{P_1}=-\lambda w_{P_1}=-\lambda w_1w_2w_3w_4$, which coincides with the value obtained via (9.2). The path P_2 from the previous example starts and ends at the inner circle, so $\alpha(P_2)=0$. Besides, $\beta(b',P_2)=1$, and hence, by Proposition 9.1, $w'_{P_2}=\lambda w_{P_2}=w_1w_2w_3w_4$; once again, this coincides with the value obtained via (9.2).

9.1.2. Boundary measurements. Given a perfect planar network in an annulus as above, we label its boundary vertices $b_1,\dots,b_n$ in the following way. The boundary vertices lying on the outer circle are labelled $b_1,\dots,b_{n_1}$ in the counterclockwise order starting from the first vertex that follows after the base point of the cut. The boundary vertices lying on the inner circle are labelled $b_{n_1+1},\dots,b_n$ in the clockwise order starting from the first vertex that follows after the base point of the cut. For example, for the network N in Fig. 9.2, the boundary vertices are labelled as $b_1=b$, $b_2=b''$, $b_3=b'$, while for the network N', one has $b_1=b$, $b_2=b'$, $b_3=b''$.

The number of sources lying on the outer circle is denoted by k_1, and the corresponding set of indices, by $I_1\subseteq[1,n_1]$; the set of the remaining $m_1=n_1-k_1$ indices is denoted by J_1. Similarly, the number of sources lying on the inner circle is denoted by k_2, and the corresponding set of indices, by $I_2\subseteq[n_1+1,n]$; the set of the remaining $m_2=n_2-k_2$ indices is denoted by J_2. Finally, we denote

$I = I_1 \cup I_2$ and $J = J_1 \cup J_2$; the cardinalities of I and J are denoted $k = k_1 + k_2$ and $m = m_1 + m_2$.

Given a source b_i, $i \in I$, and a sink b_j, $j \in J$, we define the *boundary measurement* $M(i,j)$ as the sum of the weights of all paths starting at b_i and ending at b_j. Assume first that the weights of the paths are calculated via (9.2). The boundary measurement thus defined is a formal infinite series in variables $\bar{w}_e$, $e \in E$. The following analog of Proposition 8.3 holds true in $\mathbb{Z}[[\bar{w}_e, e \in E]]$.

PROPOSITION 9.2. *Let N be a perfect planar network in an annulus, then each boundary measurement in N is a rational function in the modified weights $\bar{w}_e$, $e \in E$.*

PROOF. The proof by induction on the number of internal vertices literally follows the proof of Proposition 8.3. The only changes are that modified weights are used instead of original weights and that the counterclockwise cyclic order $\prec$ is replaced by the cyclic order $\mod n$ induced by the labeling. □

Taking into account the definition of the modified weights, we immediately get the following corollary.

COROLLARY 9.3. *Let N be a perfect planar network in an annulus, then each boundary measurement in N is a rational function in the parameter λ and the weights w_e, $e \in E$.*

For example, the boundary measurement $M(1,2)$ in the network N' shown in Fig. 9.2 equals

$$\frac{w_1 w_3 w_4 (w_6 w_8 w_9 - \lambda w_2)}{1 + \lambda^{-1} w_3 w_5 w_6 w_8}.$$

Boundary measurements can be organized into a $k \times m$ *boundary measurement matrix* M_N exactly as in the case of planar networks in the disk: let $I = \{i_1 < i_2 < \cdots < i_k\}$ and $J = \{j_1 < j_2 < \cdots < j_m\}$, then $M_N = (M_{pq})$, $p \in [1,k]$, $q \in [1,m]$, where $M_{pq} = M(i_p, j_q)$. Let $\mathrm{Rat}_{k,m}$ stand for the space of real rational $k \times m$ matrix functions in one variable. Then each network N defines a map $\mathcal{E}_N \to \mathrm{Rat}_{k,m}$ given by M_N and called the *boundary measurement map* corresponding to N.

The boundary measurement matrix has a block structure

$$M_N = \begin{pmatrix} M_1 & M_2 \\ M_3 & M_4 \end{pmatrix},$$

where M_1 is $k_1 \times m_1$ and M_4 is $k_2 \times m_2$. Moving a base point of the cut changes the weights of paths as described in Proposition 9.1, which affects M_N in the following way. Define

$$\Lambda_+ = \begin{pmatrix} 0 & 1 & \dots & 0 \\ \vdots & \vdots & \ddots & \vdots \\ 0 & 0 & \dots & 1 \\ \lambda^{-1} & 0 & \dots & 0 \end{pmatrix}, \qquad \Lambda_- = \begin{pmatrix} 0 & 1 & \dots & 0 \\ \vdots & \vdots & \ddots & \vdots \\ 0 & 0 & \dots & 1 \\ -\lambda^{-1} & 0 & \dots & 0 \end{pmatrix},$$

then interchanging the base point of the cut with b_1 implies transformation

$$\begin{pmatrix} M_1 & M_2 \\ M_3 & M_4 \end{pmatrix} \mapsto \begin{pmatrix} \Lambda_+ M_1 & \Lambda_- M_2 \\ M_3 & M_4 \end{pmatrix},$$

if b_1 is a source and

$$\begin{pmatrix} M_1 & M_2 \\ M_3 & M_4 \end{pmatrix} \mapsto \begin{pmatrix} M_1\Lambda_+^{-1} & M_2 \\ M_3\Lambda_-^{-1} & M_4 \end{pmatrix},$$

if b_1 is a sink, while interchanging the base point of the cut with b_n yields

$$\begin{pmatrix} M_1 & M_2 \\ M_3 & M_4 \end{pmatrix} \mapsto \begin{pmatrix} M_1 & M_2 \\ \Lambda_+^T M_3 & \Lambda_-^T M_4 \end{pmatrix}.$$

if b_n is a source and

$$\begin{pmatrix} M_1 & M_2 \\ M_3 & M_4 \end{pmatrix} \mapsto \begin{pmatrix} M_1 & M_2(\Lambda_+^{-1})^T \\ M_3 & M_4(\Lambda_-^{-1})^T \end{pmatrix},$$

if b_1 is a sink. Note that Λ_+ and Λ_- are $k_1 \times k_1$ in the first case, $m_1 \times m_1$ in the second case, $k_2 \times k_2$ in the third case and $m_2 \times m_2$ in the fourth case.

9.1.3. The space of face and trail weights. Let $N = (G, \rho, w)$ be a perfect network. Consider the $\mathbb{Z}$-module $\mathbb{Z}^E$ generated by the edges of G. Clearly, points of $\mathcal{E}_N$ can be identified with the elements of $\mathrm{Hom}(\mathbb{Z}^E, \mathbb{R}^*)$ via $w(\sum n_i e_i) = \prod w_{e_i}^{n_i}$, where $\mathbb{R}^*$ is the abelian multiplicative group $\mathbb{R} \setminus 0$. Further, consider the $\mathbb{Z}$-module $\mathbb{Z}^V$ generated by the vertices of G and its $\mathbb{Z}$-submodule $\mathbb{Z}^{V_0}$ generated by the internal vertices. An arbitrary element $\varphi \in \mathrm{Hom}(\mathbb{Z}^V, \mathbb{R}^*)$ acts on $\mathcal{E}_N$ as follows: if $e = (u, v)$ then

$$w_e \mapsto w_e \frac{\varphi(v)}{\varphi(u)}.$$

Therefore, the weight of a path between the boundary vertices b_i and b_j is multiplied by $\varphi(b_j)/\varphi(b_i)$. It follows that the *gauge group* $\mathfrak{G}$, which preserves the weights of all paths between boundary vertices, consists of all $\varphi \in \mathrm{Hom}(\mathbb{Z}^V, \mathbb{R}^*)$ such that $\varphi(b) = 1$ for any boundary vertex b, and can be identified with $\mathrm{Hom}(\mathbb{Z}^{V_0}, \mathbb{R}^*)$. Thus, the boundary measurement map $\mathcal{E}_N \to \mathrm{Rat}_{k,m}$ factors through the quotient space $\mathcal{F}_N = \mathcal{E}_N/\mathfrak{G}$ as follows: $M_N = M_N^{\mathcal{F}} \circ y$, where $y : \mathcal{E}_N \to \mathcal{F}_N$ is the projection and $M_N^{\mathcal{F}}$ is a map $\mathcal{F}_N \to \mathrm{Rat}_{k,m}$. The space $\mathcal{F}_N$ is called the *space of face and trail weights* for the following reasons.

First, by considering the cochain complex

$$0 \to \mathfrak{G} \to \mathcal{E}_N \to 0$$

with the coboundary operator $\delta : \mathfrak{G} \to \mathcal{E}_N$ defined by $\delta(\varphi)(e) = \varphi(v)/\varphi(u)$ for $e = (u, v)$, we can identify $\mathcal{F}_N$ with the first relative cohomology $H^1(G, \partial G; \mathbb{R}^*)$ of the complex, where ∂G is the set of all boundary vertices of G.

Second, consider a slightly more general situation, when the annulus is replaced by an arbitrary Riemann surface Σ with the boundary $\partial\Sigma$, and G is embedded into Σ in such a way that all vertices of degree 1 belong to $\partial\Sigma$ (boundary vertices). Then the exact sequence of relative cohomology with coefficients in $\mathbb{R}^*$ gives

$$\begin{aligned} 0 \to H^0(G \cup \partial\Sigma, \partial\Sigma) \to & H^0(G \cup \partial\Sigma) \to H^0(\partial\Sigma) \to \\ & H^1(G \cup \partial\Sigma, \partial\Sigma) \to H^1(G \cup \partial\Sigma) \to H^1(\partial\Sigma) \to 0. \end{aligned}$$

Evidently, $H^1(G, \partial G) \simeq H^1(G \cup \partial\Sigma, \partial\Sigma)$. Next, $H^0(G \cup \partial\Sigma, \partial\Sigma) = 0$, since each connected component of G is connected to at least one connected component of $\partial\Sigma$, and hence

$$H^1(G, \partial G) \simeq H^1(G \cup \partial\Sigma)/H^1(\partial\Sigma) \oplus H^0(\partial\Sigma)/H^0(G \cup \partial\Sigma) = \mathcal{F}_N^f \oplus \mathcal{F}_N^t.$$

If the genus of Σ equals zero, the space $\mathcal{F}_N^f$ can be described as follows. Graph G divides Σ into a finite number of connected components called *faces*. The boundary of each face consists of edges of G and, possibly, of several arcs of $\partial\Sigma$. A face is called *bounded* if its boundary contains only edges of G and *unbounded* otherwise.

Given a face f, we define its *face weight* y_f as the function on $\mathcal{E}_N$ that assigns to the edge weights w_e, $e \in E$, the value

$$y_f = \prod_{e\in\partial f} w_e^{\gamma_e}, \tag{9.3}$$

where $\gamma_e = 1$ if the direction of e is compatible with the counterclockwise orientation of the boundary ∂f and $\gamma_e = -1$ otherwise. It follows immediately from the definition that face weights are invariant under the gauge group action, and hence are functions on $\mathcal{F}_N^f$, and, moreover, form a basis in the space of such functions.

Consider now the space $\mathcal{F}_N^t$. If Σ is a disk, then $\mathcal{F}_N^t = 0$, and hence $\mathcal{F}_N$ and $\mathcal{F}_N^f$ coincide. This case was studied in the previous Chapter, and the space $\mathcal{F}_N$ was called there the *space of face weights*. If Σ is an annulus, there are two possible cases. Indeed, $\dim H^0(\partial\Sigma) = 2$. The dimension of $H^0(G \cup \partial\Sigma)$ is either 2 or 1, depending on the existence of a trail connecting the components of $\partial\Sigma$. Here a *trail* is a sequence $(v_1, \dots, v_{k+1})$ of vertices such that either (v_i, v_{i+1}) or (v_{i+1}, v_i) is an edge in G for all $i \in [1, k]$ an the endpoints v_1 and v_{k+1} are boundary vertices of G. Given a trail t, the *trail weight* y_t is defined as

$$y_t = \prod_{i=1}^{k} w(v_i, v_{i+1}),$$

where

$$w(v_i, v_{i+1}) = \begin{cases} w_e & \text{if } e = (v_i, v_{i+1}) \in E, \\ w_e^{-1} & \text{if } e = (v_{i+1}, v_i) \in E. \end{cases}$$

Clearly, the trail weights are invariant under the action of the gauge group.

If G does not contain a trail connecting the inner and the outer circles, then $\dim H^0(G \cup \partial\Sigma) = 2$, and hence $\mathcal{F}_N^t = 0$. Otherwise, $\dim H^1(G \cup \partial\Sigma) = 1$, and hence $\dim \mathcal{F}_N^t = 1$. The functions on $\mathcal{F}_N^t$ are generated by the weight of any connecting trail.

9.2. Poisson properties of the boundary measurement map

9.2.1. Induced Poisson structures on $\mathrm{Rat}_{k,m}$. Fix an arbitrary pair of partitions $I_1 \cup J_1 = [1, n_1]$, $I_1 \cap J_1 = \varnothing$, $I_2 \cup J_2 = [n_1 + 1, n]$, $I_2 \cap J_2 = \varnothing$, and denote $k = |I_1| + |I_2|$, $m = n - k = |J_1| + |J_2|$. Let $\mathrm{Net}_{I_1,J_1,I_2,J_2}$ stand for the set of all perfect planar networks in an annulus with the sources b_i, $i \in I_1$ and sinks b_j, $j \in J_1$, on the outer circle, sources b_i, $i \in I_2$ and sinks b_j, $j \in J_2$, on the inner circle, and edge weights w_e defined by (8.4). We assume that the space of edge weights $\mathcal{E}_N = (\mathbb{R} \setminus 0)^{|E|}$ is equipped with the Poisson bracket $\{\cdot,\cdot\}_N$ obtained as the pushforward of the 6-parametric family $\{\cdot,\cdot\}_R$ described in Proposition 8.5.

THEOREM 9.4. *There exists a 2-parametric family of Poisson brackets on* $\mathrm{Rat}_{k,m}$*, denoted* $\{\cdot,\cdot\}_{I_1,J_1,I_2,J_2}$*, with the following property: for any choice of parameters* α_{ij}*,* β_{ij} *in* (8.5), (8.6) *this family contains a unique Poisson bracket on* $\mathrm{Rat}_{k,m}$ *such that for any network* $N \in \mathrm{Net}_{I_1,J_1,I_2,J_2}$ *the map* $M_N\colon (\mathbb{R} \setminus 0)^{|E|} \to \mathrm{Rat}_{k,m}$ *is Poisson.*

PROOF. First of all, we use the factorization $M_N = M_N^{\mathcal{F}} \circ y$ to decrease the number of parameters.

LEMMA 9.5. *The 6-parametric family $\{\cdot,\cdot\}_N$ induces a 2-parametric family of Poisson brackets $\{\cdot,\cdot\}_{\mathcal{F}_N}$ on $\mathcal{F}_N$ with parameters α and β given by*

$$\alpha = \alpha_{23} + \alpha_{13} - \alpha_{12}, \qquad \beta = \beta_{23} + \beta_{13} - \beta_{12}. \tag{9.4}$$

PROOF. The proof literally follows the proof of Lemma 8.19. □

By the above lemma, it suffices to consider the 2-parametric family (8.8), (8.10) instead of the 6-parametric family (8.5), (8.6).

The rest of the proof consists of two major steps. First, we compute the induced Poisson bracket on the image of the boundary measurement map. More exactly, we show that the bracket $\{\cdot,\cdot\}_N$ of any pair of pullbacks of coordinate functions on the image can be expressed in terms of pullbacks of other coordinate functions, and that for fixed I_1, J_1, I_2, J_2 these expressions do not depend on the network $N \in \mathrm{Net}_{I_1,J_1,I_2,J_2}$. Second, we prove that any rational matrix function belongs to the image of the boundary measurement map (for a sufficiently large $N \in \mathrm{Net}_{I_1,J_1,I_2,J_2}$), and therefore $\{\cdot,\cdot\}_N$ induces $\{\cdot,\cdot\}_{I_1,J_1,I_2,J_2}$ on $\mathrm{Rat}_{k,m}$. This approach allows us to circumvent technical difficulties one encounters when attempting to prove an analog of Lemma 8.8.

To compute the induced Poisson bracket on the image of the boundary measurement map, we consider coordinate functions $\mathrm{val}_t\colon f \mapsto f(t)$ that assign to any $f \in \mathrm{Rat}_{1,1}$ its value at point t. Given a pair of two matrix entries, it suffices to calculate the bracket between arbitrary pair of functions val_t and val_s defined on two copies of $\mathrm{Rat}_{1,1}$ representing these entries. Since the pullback of val_t is the corresponding component of $M_N(t)$, we have to deal with expressions of the form $\{M_{pq}(t), M_{\bar p\bar q}(s)\}_N$.

To avoid overcomplicated formulas we use the same trick as in the proof of Proposition 8.13 and consider separately two particular representatives of the family (8.8), (8.10): 1) $\alpha = -\beta = 1$, and 2) $\alpha = \beta = 1$. Any member of the family can be represented as a linear combination of the above two.

Denote by $\{\cdot,\cdot\}_N^1$ the member of the 2-parametric family (8.8), (8.10) corresponding to the case $\alpha = -\beta = 1$ (cp. to the bracket (8.25) in the proof of Proposition 8.13). Besides, define $\sigma_=(i,j,i',j') = \mathrm{sign}(i'-i) - \mathrm{sign}(j'-j)$; clearly, $\sigma_=(i,j,i',j')$ is closely related to $s_=(i,j,i',j')$ defined in (8.12). The bracket induced by $\{\cdot,\cdot\}_N^1$ on the image of the boundary measurement map is completely described by the following statement.

PROPOSITION 9.6. (i) *Let $i_p, i_{\bar p} \in [1, n_1]$ and $1 \le \max\{i_p, i_{\bar p}\} < j_{\bar q} < j_q \le n$, then*

$$\{M_{pq}(t), M_{\bar p\bar q}(s)\}_N^1 = \sigma_=(i_p, j_q, i_{\bar p}, j_{\bar q}) M_{p\bar q}(s) M_{\bar p q}(t) - \frac{2}{t-s}\Phi_{pq}^{\bar p\bar q}(t,s), \tag{9.5}$$

where

$$\Phi_{pq}^{\bar p\bar q}(t,s) = \begin{cases} \big(M_{p\bar q}(t) - M_{p\bar q}(s)\big)\big(sM_{\bar p q}(t) - tM_{\bar p q}(s)\big), & j_{\bar q} < j_q \le n_1, \\ sM_{\bar p q}(t)\big(M_{p\bar q}(t) - M_{p\bar q}(s)\big), & j_{\bar q} \le n_1 < j_q, \\ s\big(M_{p\bar q}(t) M_{\bar p q}(s) - M_{p\bar q}(s) M_{\bar p q}(t)\big), & n_1 < j_{\bar q} < j_q. \end{cases}$$

(ii) *Let $j_{\bar q}, j_q \in [n_1+1, n]$ and $1 \le i_p < i_{\bar p} < \min\{j_{\bar q}, j_q\} \le n$, then*

$$\{M_{pq}(t), M_{\bar p\bar q}(s)\}_N^1 = \sigma_=(i_p, j_q, i_{\bar p}, j_{\bar q}) M_{p\bar q}(t) M_{\bar p q}(s) - \frac{2}{t-s}\Psi_{pq}^{\bar p\bar q}(t,s), \tag{9.6}$$

where

$$\Psi_{pq}^{\bar p\bar q}(t,s)=\begin{cases} t\big(M_{p\bar q}(t)M_{\bar pq}(s)-M_{p\bar q}(s)M_{\bar pq}(t)\big), & i_p<i_{\bar p}\le n_1,\\ -tM_{p\bar q}(t)\big(M_{\bar pq}(t)-M_{\bar pq}(s)\big), & i_p\le n_1<i_{\bar p},\\ -\big(tM_{p\bar q}(t)-sM_{p\bar q}(s)\big)\big(M_{\bar pq}(t)-M_{\bar pq}(s)\big), & n_1<i_p<i_{\bar p}.\end{cases}$$

(iii) *Let* $1\le i_p=i_{\bar p}<j_q=j_{\bar q}\le n$, *then*

(9.7)

$$\{M_{pq}(t),M_{pq}(s)\}_N^1 =\begin{cases} -\frac{2}{t-s}\big(M_{pq}(t)-M_{pq}(s)\big)\big(sM_{pq}(t)-tM_{pq}(s)\big), & i_p<j_q\le n_1,\\ 0, & i_p\le n_1<j_q.\end{cases}$$

(iv) *Let* $1\le i_p<\min\{i_{\bar p},j_q,j_{\bar q}\}$, *then*

(9.8)

$$\{M_{pq}(t),M_{\bar p\bar q}(s)\}_N^1 =\begin{cases} \frac{2t}{t-s}\big(M_{p\bar q}(t)-M_{p\bar q}(s)\big)\big(M_{\bar pq}(t)-M_{\bar pq}(s)\big), & j_{\bar q}<i_{\bar p}<j_q\le n_1,\\ -\frac{2t}{t-s}\big(M_{p\bar q}(t)M_{\bar pq}(t)-M_{p\bar q}(s)M_{\bar pq}(s)\big), & j_q\le n_1<i_{\bar p}<j_{\bar q},\\ 0, & j_{\bar q}\le n_1<i_{\bar p}<j_q.\end{cases}$$

PROOF. Let us first make sure that relations (9.5)–(9.8) indeed allow to compute $\{M_{pq}(t),M_{\bar p\bar q}(s)\}_N^1$ for any $p,\bar p\in[1,k]$ and $q,\bar q\in[1,m]$. This is done by employing the following three techniques:

– moving a base point of the cut;
– reversing the direction of the cut;
– reversing the orientation of boundary circles.

The first of the above techniques has been described in detail in Section 9.1.1. For example, let $1\le j_q<i_p<i_{\bar p}<j_{\bar q}\le n_1$. This case is not covered explicitly by relations (9.5)–(9.8). Consider the network N' obtained from N by moving the base point of the cut on the outer circle counterclockwise and interchanging it with j_q. In this new network one has $1\le i_{p'}<i_{\bar p'}<j_{\bar q'}<j'_q=n_1$, so the conditions of Proposition 9.6(i) are satisfied and (9.5) yields

$$\{M_{p'q'}(t),M_{\bar p'\bar q'}(s)\}_{N'}^1=2M_{p'\bar q'}(s)M_{\bar p'q'}(t) -\frac{2}{t-s}\big(M_{p'\bar q'}(t)-M_{p'\bar q'}(s)\big)\big(sM_{\bar p'q'}(t)-tM_{\bar p'q'}(s)\big).$$

By Lemma 9.1,

$$M_{p'q'}(t)=tM_{pq}(t),\quad M_{\bar p'\bar q'}(t)=M_{\bar p\bar q}(t),\quad M_{p'\bar q'}(t)=M_{p\bar q}(t),\quad M_{\bar p'q'}(t)=tM_{\bar p'q'}(t).$$

Finally, $\{\cdot,\cdot\}_N=\{\cdot,\cdot\}_{N'}$ for any pair of edge weights, so we get

$$\{M_{pq}(t),M_{\bar p\bar q}(s)\}_N^1=2M_{p\bar q}(s)M_{\bar pq}(t)-\frac{2s}{t-s}\big(M_{p\bar q}(t)-M_{p\bar q}(s)\big)\big(M_{\bar pq}(t)-M_{\bar pq}(s)\big).$$

Reversing the direction of the cut transforms the initial network N to a new network N'; the graph G remains the same, while the labeling of its boundary vertices is changed. Namely, the $n'_1=n_2$ boundary vertices lying on the inner circle are labelled $b_1,\dots,b_{n'_1}$ in the clockwise order starting from the first vertex that follows after the base point of the cut. The boundary vertices lying on the outer circle are labelled $b_{n'_1+1},\dots,b_n$ in the counterclockwise order starting from the first vertex that follows after the base point of the cut. The transformation $N\mapsto N'$ is better visualized if the network is drawn on a cylinder, instead of an annulus. The boundary circles of a cylinder are identical, and reversing the direction of the cut simply interchanges them. Clearly, the boundary measurements in N and N' are related by $M_{r's'}(t) = M_{rs}(1/t)$ for any $i_r \in I$, $j_s \in J$. Besides, $\{\cdot,\cdot\}_N = \{\cdot,\cdot\}_{N'}$ for any pair of edge weights. Therefore, an expression for $\{M_{pq}(t), M_{\bar p\bar q}(s)\}^1_N$ via $M_{p\bar q}(t)$, $M_{p\bar q}(s)$, $M_{\bar pq}(t)$, $M_{\bar pq}(s)$ is transformed to the expression for $\{M_{p'q'}(t), M_{\bar p'\bar q'}(s)\}^1_{N'}$ via $M_{p'\bar q'}(t)$, $M_{p'\bar q'}(s)$, $M_{\bar p'q'}(t)$, $M_{\bar p'q'}(s)$ by the substitution $t \mapsto 1/t$ and $s \mapsto 1/s$ in coefficients. For example, let $1 \le i_{\bar p} < j_q < j_{\bar q} \le n_1 < i_p \le n$. This case is not covered explicitly by relations (9.5)–(9.8). Consider the network N' obtained from N by reversing the direction of the cut. In this new network one has $1 \le i_{p'} \le n'_1 < i_{\bar p'} < j_q < j_{\bar q'} \le n$, so the conditions of Proposition 9.6(ii) are satisfied and (9.6) yields

$$\{M_{p'q'}(t), M_{\bar p'\bar q'}(s)\}^1_{N'} = -2M_{p'\bar q'}(t)M_{\bar p'q'}(s) + \frac{2t}{t-s}M_{p'\bar q'}(t)\big(M_{\bar p'q'}(t) - M_{\bar p'q'}(s)\big).$$

Applying the above described rule one gets

$$\begin{aligned}\{M_{pq}(t), M_{\bar p\bar q}(s)\}^1_N &= -2M_{p\bar q}(t)M_{\bar pq}(s) + \frac{2t^{-1}}{t^{-1}-s^{-1}}M_{p\bar q}(t)\big(M_{\bar pq}(t) - M_{\bar pq}(s)\big)\\ &= -2M_{p\bar q}(t)M_{\bar pq}(s) - \frac{2s}{t-s}M_{p\bar q}(t)\big(M_{\bar pq}(t) - M_{\bar pq}(s)\big).\end{aligned}$$

Finally, reversing the orientation of boundary circles also retains the graph G and changes the labeling of its boundary vertices. Namely, the n_1 boundary vertices of N' lying on the outer circle are labelled $b_1, \dots, b_{n_1}$ in the clockwise order starting from the first vertex that follows after the base point of the cut. The boundary vertices lying on the inner circle are labelled $b_{n_1+1}, \dots, b_n$ in the counterclockwise order starting from the first vertex that follows after the base point of the cut. The transformation $N \mapsto N'$ may be visualized as a mirror reflection. Clearly, the boundary measurements in N and N' are related by $M_{r's'}(t) = M_{rs}(1/t)$ for any $i_r \in I$, $j_s \in J$. Besides, $\{\cdot,\cdot\}_N = -\{\cdot,\cdot\}_{N'}$ for any pair of edge weights. Therefore, the transformation of the expressions for the brackets differs from the one for the case of cut reversal by factor -1. For example, let $1 \le j_q < i_p < i_{\bar p} < j_{\bar q} \le n_1$. This case is not covered explicitly by relations (9.5)–(9.8). Consider the network N' obtained from N by reversing the orientation of boundary circles. In this new network one has $1 \le i_{p'} < j_{\bar q'} < i_{\bar p'} < j_q \le n_1$, so the conditions of Proposition 9.6(iv) are satisfied and (9.8) yields

$$\{M_{p'q'}(t), M_{\bar p'\bar q'}(s)\}^1_{N'} = \frac{2t}{t-s}\big(M_{p'\bar q'}(t) - M_{p'\bar q'}(s)\big)\big(M_{\bar p'q'}(t) - M_{\bar p'q'}(s)\big).$$

Applying the above described rule one gets

$$\begin{aligned}\{M_{pq}(t), M_{\bar p\bar q}(s)\}^1_N &= -\frac{2t^{-1}}{t^{-1}-s^{-1}}\big(M_{p\bar q}(t) - M_{p\bar q}(s)\big)\big(M_{\bar pq}(t) - M_{\bar pq}(s)\big)\\ &= \frac{2s}{t-s}\big(M_{p\bar q}(t) - M_{p\bar q}(s)\big)\big(M_{\bar pq}(t) - M_{\bar pq}(s)\big).\end{aligned}$$

Elementary, though tedious, consideration of all possible cases reveals that indeed any quadruple $(i_p, j_q, i_{\bar p}, j_{\bar q})$ can be reduced by the above three transformations to one of the quadruples mentioned in the statement of Proposition 9.6.

It is worth to note that cases (i) and (ii) are not independent. First, they both apply if $1 \le i_p < i_{\bar p} \le n_1 < j_{\bar q} < j_q \le n$; the expressions prescribed by (9.5) and (9.6) are distinct, but yield the same result:

$$2M_{p\bar q}(s)M_{\bar pq}(t) - \frac{2s}{t-s}\left(M_{p\bar q}(t)M_{\bar pq}(s) - M_{p\bar q}(s)M_{\bar pq}(t)\right)$$
$$= 2M_{p\bar q}(t)M_{\bar pq}(s) - \frac{2t}{t-s}\left(M_{p\bar q}(t)M_{\bar pq}(s) - M_{p\bar q}(s)M_{\bar pq}(t)\right).$$

Besides, the expression for $n_1 < i_p < i_{\bar p} < j_{\bar q} < j_q \le n$ in case (ii) can be obtained from the expressions for $1 \le i_p < i_{\bar p} < j_{\bar q} < j_q \le n_1$ in case (i) by reversing the direction of the cut. However, we think that the above presentation, though redundant, better emphasizes the underlying symmetries of the obtained expressions.

The proof of relations (9.5)–(9.8) is similar to the proof of Theorem 8.7 and is based on the induction on the number of internal vertices in N. The key ingredient of the proof is the following straightforward analog of Lemma 8.9.

Let $\widehat N$ be the network obtained from N by splitting a black neighbor of a boundary vertex (see Figure 8.4). We may assume without loss of generality that the cut ρ in N does not intersect the edge e_0, and hence ρ remains a valid cut in $\widehat N$.

LEMMA 9.7. *Boundary measurements in the networks N and $\widehat N$ are related by*

$$M(i_p, j) = \frac{w_{e_0}w_{e_+}\widehat M(i_u, j)}{1 + w_{e_-}w_{e_+}\widehat M(i_u, j_u)},$$

$$M(i_{\bar p}, j) = \widehat M(i_{\bar p}, j) \pm \frac{w_{e_-}w_{e_+}\widehat M(i_{\bar p}, j_u)\widehat M(i_u, j)}{1 + w_{e_-}w_{e_+}\widehat M(i_u, j_u)}, \qquad \bar p \ne p;$$

in the second formula above, sign + corresponds to the cases

$$i_p - 1 \prec j_u \prec i_u \prec i_p + 1 \preceq j \prec i_{\bar p} \qquad or \qquad i_{\bar p} \prec j \preceq i_p - 1 \prec i_u \prec j_u \prec i_p + 1,$$

and sign − corresponds to the cases

$$i_p - 1 \prec i_u \prec j_u \prec i_p + 1 \preceq j \prec i_{\bar p} \qquad or \qquad i_{\bar p} \prec j \preceq i_p - 1 \prec j_u \prec i_u \prec i_p + 1,$$

where $\prec$ is the cyclic order mod n.

We leave the details of the proof to the interested reader. □

Denote by $\{\cdot,\cdot\}^2_N$ the member of the 2-parametric family (8.8), (8.10) corresponding to the case $\alpha = \beta = 1$ (cp. to the bracket (8.26) in the proof of Proposition 8.13). Besides, define $\sigma_\times(i, j, i', j') = \operatorname{sign}(i' - i) + \operatorname{sign}(j' - j)$; clearly, $\sigma_\times(i, j, i', j')$ is closely related to $s_\times(i, j, i', j')$ defined in (8.12). The bracket induced by $\{\cdot,\cdot\}^2_N$ on the image of the boundary measurement map is completely described by the following statement.

PROPOSITION 9.8. (i) *Let $1 \le \max\{i_p, i_{\bar p}\} < j_{\bar q} < j_q \le n$, then*

$$\{M_{pq}(t), M_{\bar p\bar q}(s)\}^2_N = \sigma_\times(i_p, j_q, i_{\bar p}, j_{\bar q})M_{pq}(t)M_{\bar p\bar q}(s) - 2\Gamma^{\bar p\bar q}_{pq}(t, s), \tag{9.9}$$

where

$$\Gamma^{\bar p\bar q}_{pq}(t, s) = \begin{cases} 0, & j_q \le n_1, \\ -sM_{pq}(t)M'_{\bar p\bar q}(s), & j_{\bar q} \le n_1 < j_q, \\ tM'_{pq}(t)M_{\bar p\bar q}(s) - sM_{pq}(t)M'_{\bar p\bar q}(s), & \max\{i_p, i_{\bar p}\} \le n_1 < j_{\bar q}, \\ tM'_{pq}(t)M_{\bar p\bar q}(s), & i_p \le n_1 < i_{\bar p}, \end{cases}$$

and M'_{pq}, $M'_{\bar p\bar q}$ *are the derivatives of* M_{pq} *and* $M_{\bar p\bar q}$.

(ii) *Let* $1 \le i_p < j_{\bar q} < i_{\bar p} < j_q \le n$ *and either* $j_{\bar q} \le n_1 < i_{\bar p}$ *or* $j_q \le n_1$, *then*

$$\{M_{pq}(t), M_{\bar p\bar q}(s)\}^2_N = 0. \tag{9.10}$$

(iii) *Let* $1 \le i_p < j_q < i_{\bar p} < j_{\bar q} \le n$ *and either* $j_q \le n_1 < i_{\bar p}$ *or* $j_{\bar q} \le n_1$, *then*

$$\{M_{pq}(t), M_{\bar p\bar q}(s)\}^2_N = 0. \tag{9.11}$$

PROOF. The proof is similar to the proof of Proposition 9.6. We leave the details to the interested reader. □

REMARK 9.9. It is worth to mention that bracket (8.14) can be considered as a particular case of (9.5)–(9.8) (for $\alpha = -\beta = 1$) or (9.9)–(9.11) (for $\alpha = \beta = 1$). To see this it suffices to consider only networks without edges that intersect the cut ρ, and to cut the annulus along ρ in order to get a disk.

9.2.2. Realization theorem. To conclude the proof of Theorem 9.4 we need the following statement. We say that $F \in \mathrm{Rat}_{k,m}$ is *represented by* a network N if F belongs to the image of M_N.

THEOREM 9.10. *For any* $F \in \mathrm{Rat}_{k,m}$ *there exists a network* $N \in \mathrm{Net}_{I_1,J_1,I_2,J_2}$ *such that* F *is represented by* N.

PROOF. We preface the proof by the following simple observation concerning perfect planar networks in a disk.

LEMMA 9.11. *Let* $n = 4$, $I = \{1,2\}$, $J = \{3,4\}$. *There exists a network* $N_{\mathrm{id}} \in \mathrm{Net}_{I,J}$ *such that the* 2×2 *identity matrix is represented by* N_{id}.

PROOF. The proof is furnished by the network depicted in Fig. 9.3. The corresponding boundary measurement matrix is given by

$$\begin{pmatrix} w_1w_8\big(w_3w_{11}(w_2+w_6w_9w_{10})+w_6w_7w_9\big) & w_1w_3w_4(w_2+w_6w_9w_{10}) \\ w_1w_6w_8(w_7+w_3w_{10}w_{11}) & w_3w_4w_5w_6w_{10} \end{pmatrix},$$

which yields the identity matrix for $w_5 = w_{10} = -1$ and $w_i = 1$ for $i \ne 5, 10$.

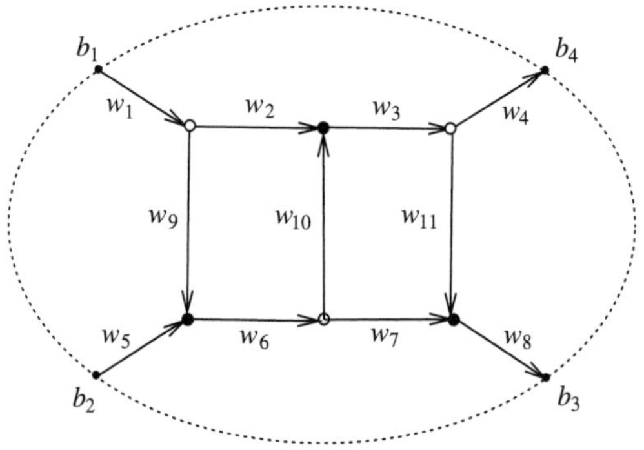

FIGURE 9.3. Network N_{id}

□

Effectively, Lemma 9.11 says that the planarity restriction can be omitted in the proof of Theorem 9.10. Indeed, if $F \in \mathrm{Rat}_{k,m}$ is represented by a nonplanar perfect network in an annulus, one can turn it to a planar perfect network in annulus by replacing each intersection by a copy of N_{id}.

In what follows we make use of the concatenation of planar networks in an annulus. Similarly to the case of networks in a disk (see Section 8.2.1), the most important particular case of concatenation arises when the sources and the sinks are separated, that is, all sources lie on the outer circle, and all sinks lie on the inner circle. We can concatenate two networks of this kind, one with k sources and m sinks and another with m sources and l sinks, by gluing the sinks of the former to the sources of the latter. More exactly, we glue together the inner circle of the former network and the outer circle of the latter in such a way that the corresponding base points of the cuts are identified, and the ith sink of the former network is identified with the $(m+1-i)$th source of the latter. The erasure of the common boundary and the identification of edges are performed exactly as in the case of a disk.

Let us start with representing any rational function $F \in \mathrm{Rat}_{1,1}$ by a network with the only source on the outer circle and the only sink on the inner circle.

LEMMA 9.12. *Any rational function* $F \in \mathrm{Rat}_{1,1}$ *can be represented by a network* $N \in \mathrm{Net}_{1,\varnothing,\varnothing,2}$.

PROOF. First, if networks $N_1, N_2 \in \mathrm{Net}_{1,\varnothing,\varnothing,2}$ represent functions F_1 and F_2 respectively, their concatenation $N_1 \circ N_2 \in \mathrm{Net}_{1,\varnothing,\varnothing,2}$ represents F_1F_2.

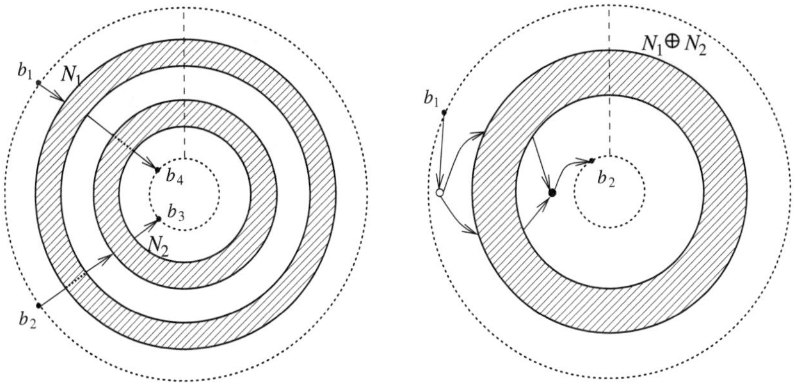

FIGURE 9.4. The direct sum of two networks (left) and a network representing the sum of two functions (right)

Second, define the *direct sum* $N_1 \oplus N_2 \in \mathrm{Net}_{[1,2],\varnothing,\varnothing,[3,4]}$ as shown in the left part of Fig. 9.4. The shadowed annuli contain networks N_1 and N_2. The intersections of the dashed parts of additional edges with the edges of N_1 and N_2 are resolved with the help of N_{id} (not shown). Note that the direct sum operation is not commutative. Clearly, $N_1 \oplus N_2$ represents the 2×2 matrix $\begin{pmatrix} 0 & F_1 \\ F_2 & 0 \end{pmatrix}$. The direct sum of networks is used to represent the sum $F_1 + F_2$ as shown in the right part of Fig. 9.4.

Third, if $N \in \mathrm{Net}_{1,\varnothing,\varnothing,2}$ represents F, the network shown in Fig. 9.5 represents $F/(1+F)$, and, with a simple adjustment of weights, can also be used to represent $-F/(1+F)$. Taking the direct sum with the trivial network representing 1, we get a representation for $1/(1+F)$.

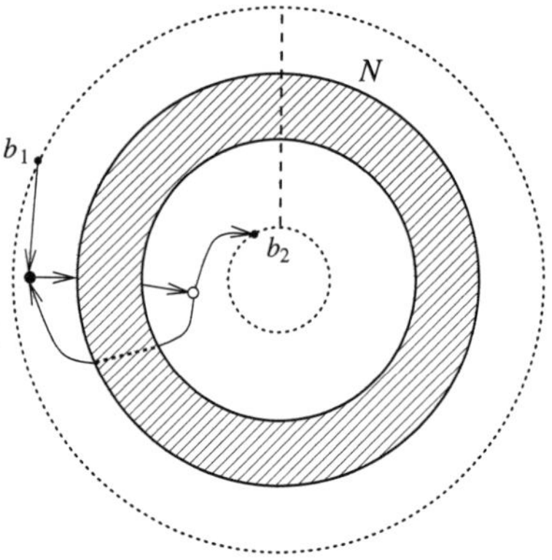

FIGURE 9.5. Representing $F/(1+F)$

Finally, functions $a\lambda^k$ for any integer k can be represented by networks in $\mathrm{Net}_{1,\varnothing,\varnothing,2}$. The cases $k=2$ and $k=-2$ are shown in Fig. 9.6. Other values of k are obtained in the same way.

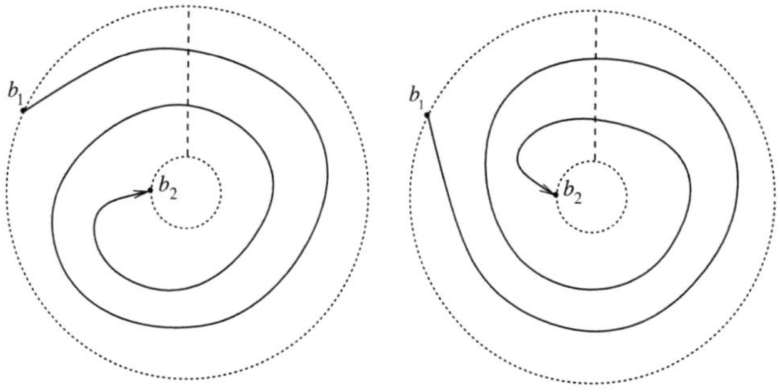

FIGURE 9.6. Representing $a\lambda^k$ for $k=2$ (left) and $k=-2$ (right)

We now have all the ingredients for the proof of the lemma. Any rational function F can be represented as $F(\lambda)=\sum_{i=0}^{r} a_i\lambda^{d+i}/Q(\lambda)$, where d is an integer and Q is a polynomial satisfying $Q(0)=1$. Therefore, it suffices to represent each of the summands, and to use the direct sum construction. Each summand, in its turn, is represented by the concatenation of a network representing $a_i\lambda^{d+i}$ with a network representing $1/Q = 1/(1+(Q-1))$. The latter network is obtained as explained above from a network representing $Q-1=\sum_{j=1}^{p} b_j\lambda^j$ via the direct sum construction. □

To get an analog of Lemma 9.12 for networks with the only source and the only sink on the outer circle, one has to use once again Lemma 9.11.

LEMMA 9.13. *Any rational function* $F \in \mathrm{Rat}_{1,1}$ *can be represented by a network* $N \in \mathrm{Net}_{1,2,\varnothing,\varnothing}$.

PROOF. Such a representation is obtained from the one constructed in the proof of Lemma 9.12 by replacing the edge incident to the sink with a new edge sharing the same tail. The arising intersections, if any, are resolved with the help of N_{id}. For example, representation of $a(1+b\lambda)^{-1}$ obtained this way is shown in Fig. 9.7 on the right. It makes use of the network N_{id} described in Lemma 9.11; the latter is shown in thin lines inside a dashed circle. Note that the network on the left, which represents $a(1+b\lambda)^{-1}$ in $\mathrm{Net}_{1,\varnothing,\varnothing,2}$, is not the one built in the proof of Lemma 9.12.

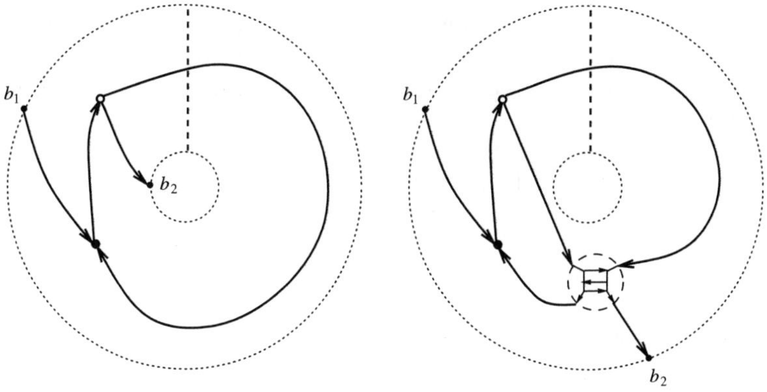

FIGURE 9.7. Representing $a(1+b\lambda)^{-1}$ by a network in $\mathrm{Net}_{1,\varnothing,\varnothing,2}$ (left) and $\mathrm{Net}_{1,2,\varnothing,\varnothing}$ (right)

□

Representation of rational functions by networks in $\mathrm{Net}_{2,1,\varnothing,\varnothing}$ and $\mathrm{Net}_{\varnothing,1,2,\varnothing}$ is obtained in a similar way. In the latter case one has to replace also the edge incident to the source with a new one sharing the head.

The next step is to prove Theorem 9.10 in the case when all sources lie on the outer circle and all sinks lie on the inner circle.

LEMMA 9.14. *For any rational matrix* $F \in \mathrm{Rat}_{k,m}$ *there exists a network* $N \in \mathrm{Net}_{[1,k],\varnothing,\varnothing,[k+1,k+m]}$ *such that* F *is represented by* N.

PROOF. First of all, we represent F as $F = A\widetilde{F}B$, where $A = \{a_{ij}\}$ is the $k \times km$ constant matrix given by

$$a_{ij} = \begin{cases} 1, & \text{if } (k-i)m < j \le (k-i+1)m, \\ 0, & \text{otherwise,} \end{cases}$$

$\widetilde{F}$ is the $km \times km$ diagonal matrix

$$\widetilde{F} = \mathrm{diag}\{F_{km}, F_{k,m-1}, \dots, F_{k1}, F_{k-1,m}, \dots, F_{11}\},$$

and B is the $km \times m$ constant matrix

$$B = \begin{pmatrix} W_0 \\ \vdots \\ W_0 \end{pmatrix}$$

with $W_0 = (\delta_{i,m+1-j})_{i,j=1}^m$. Similarly to the case of networks in a disk (see Section 8.2.1), the concatenation of networks representing matrices F_1 and F_2 produces a network representing $F_1 W_0 F_2$. Therefore, in order to get F as above, we have to represent matrices A, $W_0\widetilde{F}$ and $W_0 B$.

The first representation is achieved trivially as the disjoint union of networks representing the $1 \times m$ matrix $(1\ 1\ \dots 1)$; in fact, since A is constant, it can be represented by a network in a disk. The second representation is obtained as the direct sum of networks representing each of F_{ij}. Finally, the third representation can be also achieved by a network in a disk, via a repeated use of the network N_{id}.

Observe that in order to represent a $k \times m$ matrix we have to use intermediate matrices of a larger size. $\square$

To complete the proof of Theorem 9.10 we rely on Lemma 9.14 and use the same idea of replacing edges incident to boundary vertices as in the proof of Lemma 9.13.

So, Theorem 9.10 has been proved, and hence the proof of Theorem 9.4 is completed. $\square$

As we already mentioned in the proof, the 2-parametric family of Poisson brackets on $\mathrm{Rat}_{k,m}$ induced by $(\alpha-\beta)\{\cdot,\cdot\}_N^1 + (\alpha+\beta)\{\cdot,\cdot\}_N^2$, where $\{\cdot,\cdot\}_n^1$ and $\{\cdot,\cdot\}_n^2$ are described in Propositions 9.6 and 9.8, respectively, is denoted $\{\cdot,\cdot\}_{I_1,J_1,I_2,J_2}$.

9.2.3. Recovering the trigonometric R-matrix bracket on $\mathrm{Rat}_{k,k}$**.** As an application of results obtained in Section 9.2.1, consider the set $N_{[1,k],\varnothing,\varnothing,[k+1,2k]}$ of perfect networks with k sources on the outer circle and k sinks on the inner circle. Clearly, in this case the boundary measurement map takes $\mathcal{E}_N$ to $\mathrm{Rat}_{k,k}$. Just as we did in Section 8.2.1 and in the proof of Theorem 9.10, we can replace M_N with $A_N = M_N W_0$ and observe that the concatenation N of networks $N_1, N_2 \in N_{[1,k],\varnothing,\varnothing,[k+1,2k]}$ leads to $A_N = A_{N_1} A_{N_2}$. We would like to take a closer look at the bracket $\{\cdot,\cdot\}_N^1$ in this case.

First, recall that the space $\mathrm{Rat}_{k,k}$ can be equipped with an R-matrix (Sklyanin) Poisson bracket

$$\{A(t), A(s)\} = [R(t,s), A(s) \otimes A(t)]\,, \tag{9.12}$$

where the left-hand side should be understood as

$$\{A(t), A(s)\}_{p\bar{p}}^{q\bar{q}} = \{a_{pq}(t), a_{\bar{p}\bar{q}}(s)\}$$

and the R-matrix $R(t,s)$ is an operator acting in $\mathbb{R}^{2k} \otimes \mathbb{R}^{2k}$ that depends on parameters t, s and solves the classical Yang-Baxter equation. Of interest to us is a bracket (9.12) that corresponds to the so-called *trigonometric R-matrix*

$$R(t,s) = \frac{t+s}{s-t}\sum_{k=1}^{n} E_{kk} \otimes E_{kk} \;+\; \frac{2}{s-t}\sum_{1 \le l < m \le 2k} (tE_{lm} \otimes E_{ml} + sE_{ml} \otimes E_{lm})\,. \tag{9.13}$$

Bracket (9.12), (9.13) can be re-written in terms of matrix entries of $A(t)$ as follows (we only list non-zero brackets): for $p < \bar p$ and $q < \bar q$,

$$\{a_{pq}(t), a_{\bar p\bar q}(s)\} = 2\frac{ta_{p\bar q}(s)a_{\bar pq}(t) - sa_{p\bar q}(t)a_{\bar pq}(s)}{t-s}. \tag{9.14}$$

$$\{a_{p\bar q}(t), a_{\bar pq}(s)\} = 2t\frac{a_{pq}(s)a_{\bar p\bar q}(t) - a_{pq}(t)a_{\bar p\bar q}(s)}{t-s}. \tag{9.15}$$

$$\{a_{pq}(t), a_{p\bar q}(s)\} = \frac{(t+s)a_{p\bar q}(s)a_{pq}(t) - 2sa_{p\bar q}(t)a_{pq}(s)}{t-s}. \tag{9.16}$$

$$\{a_{pq}(t), a_{\bar pq}(s)\} = \frac{2ta_{pq}(s)a_{\bar pq}(t) - (t+s)a_{pq}(t)a_{\bar pq}(s)}{t-s}. \tag{9.17}$$

It is now straightforward to check that for $N \in N_{[1,k],\varnothing,\varnothing,[k+1,2k]}$, the Poisson algebra satisfied by the entries of A_N coincides with that of the Sklyanin bracket (9.12), (9.13). More exactly, relations (9.14) and (9.16) are equivalent to (9.5) with $\Phi_{pq}^{\bar p\bar q}(t,s)$ calculated according to the third case, while (9.15) and (9.17) are equivalent to (9.6) with $\Psi_{pq}^{\bar p\bar q}(t,s)$ calculated according to the first case. Finally, the brackets that vanish identically, correspond exactly to the situations listed in the second case in (9.7) and in the third case in (9.8).

To summarize, we obtained the following statement.

THEOREM 9.15. *For any $N \in N_{[1,k],\varnothing,\varnothing,[k+1,2k]}$ and any choice of parameters α_{ij}, β_{ij} in* (8.5), (8.6) *such that $\alpha = 1$ and $\beta = -1$ in* (8.9), (8.11), *the map $A_N : (\mathbb{R}\setminus 0)^{|E|} \to \mathrm{Rat}_{k,k}$ is Poisson with respect to the Sklyanin bracket* (9.12) *associated with the R-matrix* (9.13).

REMARK 9.16. Equations (9.14)–(9.17) can be also used to define a Poisson bracket in the "rectangular" case of Rat_{k_1,k_2}. In this case, a concise description (9.12) of the bracket should be modified as follows:

$$\{A(t), A(s)\} = R_{k_1}(t,s)\,(A(t)\otimes A(s)) - (A(t)\otimes A(s))\,R_{k_2}(t,s),$$

where $R_{k_i}(t,s)$ denotes the R-matrix (9.13) acting in $\mathbb{R}^{k_i}\otimes\mathbb{R}^{k_i}$.

9.3. Poisson properties of the Grassmannian boundary measurement map

9.3.1. Grassmannian boundary measurement map and path reversal. Let $N \in \mathrm{Net}_{I_1,J_1,I_2,J_2}$ be a perfect planar network in an annulus. Similarly to Section 8.3.2, we are going to provide a Grassmannian interpretation of the boundary measurement map defined by N. To this end, we extend the boundary measurement matrix M_N to a $k\times n$ matrix $\bar X_N = (m^I_{pj})$ exactly in the same way as it was done in Section 8.3.2, that is, by inserting a $k\times k$ identity submatrix in the columns indexed by $I = I_1 \cup I_2$ and setting $m^I_{pj} = (-1)^{s(p,j)}M_{pq}$ for $p \in [1,k]$ and $j = j_q \in J$, where $s(p,j)$ is the number of elements in I lying strictly between $\min\{i_p, j\}$ and $\max\{i_p, j\}$ in the linear ordering; recall that the sign is selected in such a way that the minor $(\bar X_N)^{I(i_p\to j)}_{[1,k]}$ coincides with M_{pq}.

We will view $\bar X_N$ as a matrix representative of an element X_N in the space $LG_k(n)$ of rational functions $X\colon \mathbb{R}\to G_k(n)$. The latter space is called the *space of Grassmannian loops*, and the corresponding rational map $X_N\colon \mathcal{E}_N \to LG_k(n)$ is called the *Grassmannian boundary measurement map*.

Given a network N and a simple path P from a source b_i to a sink b_j in N, we define the *reversal* of P as follows: for every edge $e \in P$, change its direction and replace its weight w_e by $1/w_e$; equivalently, the modified weight $\bar{w}_e$ is replaced by $1/\bar{w}_e$. Clearly, after the reversal of P all vertices preserve their color.

Denote by N^P the network obtained from N by the reversal of P, and by R^P the corresponding *path reversal map* $\mathcal{E}_N \to \mathcal{E}_{N^P}$. Besides, put $t^P = 1$ if both endpoints of P belong to the same boundary circle, and $t^P = -1$ otherwise. Define two maps from $LG_k(n)$ to itself: S_1 is the identity map, while S_{-1} takes any $X(t) \in LG_k(n)$ to $X(-t)$. Our next goal is to prove that the path reversal map for paths not intersecting the cut commutes with the Grassmannian boundary measurement map up to S_{t^P}.

THEOREM 9.17. *Let P be a simple path from a source b_i to a sink b_j in N such that $M(i,j)$ does not vanish identically and P does not intersect the cut. Then*

$$S_{t^P} \circ X_N = X_{N^P} \circ R^P.$$

PROOF. Let I be the index set of the sources in N. The statement of the Theorem is equivalent to the equality $S_{t^P}(x_K)x_I^P = x_K^P$ for any subset $K \subset [1,n]$ of size k. Here and in what follows the superscript P means that the corresponding value is related to the network N^P. The signs of the elements m_{pj}^I are chosen in such a way that $x_I^P = M^P(j,i)$, so we have to prove

$$S_{t^P}(x_K)M^P(j,i) = x_K^P. \tag{9.18}$$

The proof of (9.18) is similar to the proof of Proposition 8.3 and relation (8.18) and relies on the induction on the number of inner vertices in N.

Let us start with the case when N does not have inner vertices. In this case it suffices to prove (9.18) with $K = I(i_r \to l)$ for all edges $e = (b_{i_r}, b_l)$. Assume first that the intersection index of each edge with the cut ρ equals ± 1 or 0; consequently, after a suitable isotopy, each edge either intersects ρ exactly once, or does not intersect it at all. Let $e^* = (b_i, b_j)$ be the edge to be reversed, and hence $\operatorname{ind}(e^*) = 0$. If both b_i and b_j belong to the same boundary circle, then exactly the following two cases are prohibited:

$$\min\{i_r, l\} < \min\{i, j\} < \max\{i_r, l\} < \max\{i, j\},$$
$$\min\{i, j\} < \min\{i_r, l\} < \max\{i, j\} < \max\{i_r, l\}.$$

Consequently, reversing e^* does not change $(-1)^{s(r,l)}$, which corresponds to the map S_1. If b_i and b_j belong to distinct boundary circles, then the above two cases are prohibited whenever e does not intersect the cut. If e intersects the cut then the above two cases are the only possibilities. Consequently, reversing e^* does not change $(-1)^{s(r,l)}$ for the edges not intersecting the cut and reverses it for the edges intersecting the cut, which corresponds to the map S_{-1}.

It remains to lift the restriction on the intersection index of edges with ρ. If there exists an edge e' such that $|\operatorname{ind}(e')| > 1$ then the endpoints of e' belong to distinct boundary circles, and for any other edge with the endpoints on distinct boundary circles, the intersection index with ρ does not vanish. Consequently, only edges with the endpoints on the same boundary circle can be reversed, and the above reasoning applies, which leads to S_1.

Let now N have inner vertices, and assume that the first inner vertex v on P is white. Denote by e and e' the first two edges of P, and by e'' the third edge incident to v. In what follows we assume without loss of generality that the cut in N does not intersect e. To find $M^P(j,i)$ consider the network $\widehat{N^P}$ that is related to N^P exactly in the same way as the network $\widehat{N}$ defined as in the proof of Proposition 8.3 is related to N. Similarly to the first relation in Lemma 9.7, we find

$$M^P(j,i) = \frac{w_e^P w_{e'}^P \widehat{M^P}(j,j_v)}{1 + w_{e'}^P w_{e''}^P \widehat{M^P}(i_v,j_v)}.$$

Taking into account that $w_e^P = 1/w_e$, $w_{e'}^P = 1/w_{e'}$, $w_{e''}^P = w_{e''}$, we finally get

$$M^P(j,i) = \frac{\widehat{M^P}(j,j_v)}{w_e w_{e'} + w_e w_{e''} \widehat{M^P}(i_v,j_v)}. \tag{9.19}$$

To find x_K^P we proceed as follows.

LEMMA 9.18. *Let the first inner vertex v of P be white, then*

$$x_K^P = \begin{cases} \dfrac{w_{e'}(\widehat{x^P})_{K\cup i_v} + w_{e''}(\widehat{x^P})_{K\cup j_v}}{w_{e'} + w_{e''}\widehat{M^P}(i_v,j_v)} & \text{if } i \notin K \\[3ex] \dfrac{(\widehat{x^P})_{K(i\to j_v)\cup i_v}}{w_e w_{e'} + w_e w_{e''}\widehat{M^P}(i_v,j_v)} & \text{if } i \in K. \end{cases}$$

PROOF. The proof utilizes explicit formulas (similar to those provided by Lemma 9.7) that relate boundary measurements in the networks N^P and $\widehat{N^P}$. What is important, the sign $\pm$ in the second formula in Lemma 9.7 and the sign $(-1)^{s(p,j)}$ defined at the beginning of this Section interplay in such a way that any submatrix of $\bar{X}_{N^P}$ is the sum of the corresponding submatrix of $\bar{X}_{\widehat{N^P}}$ and a submatrix of the rank 1 matrix that is equal to the tensor product of the ith column of $\bar{X}_{N^P}$ and the j_vth row of $\bar{X}_{\widehat{N^P}}$. □

To find x_K, create a new network $\widetilde{N}$ by deleting b_i and the edge e from G, splitting v into 2 sources $b_{i'_v}, b_{i''_v}$ (so that either $i-1 < i'_v < i''_v < i+1$ or $i-1 < i''_v < i'_v < i+1$) and replacing the edges $e' = (v,v')$ and $e'' = (v,v'')$ by $(b_{i'_v}, v')$ and $(b_{i''_v}, v'')$, respectively (see Figure 8.3).

LEMMA 9.19. *Let the first inner vertex v of P be white, then*

$$x_K = \begin{cases} w_e w_{e'}\widetilde{x}_{K\cup i''_v} + w_e w_{e''}\widetilde{x}_{K\cup i'_v} & \text{if } i \notin K \\[2ex] \widetilde{x}_{K(i\to i'_v)\cup i''_v} & \text{if } i \in K. \end{cases}$$

PROOF. The proof is a straightforward computation. □

By (9.19) and Lemmas 9.18 and 9.19, relation (9.18) boils down to

$$S_{t^P}(w_{e'}\widetilde{x}_{K\cup i''_v} + w_{e''}\widetilde{x}_{K\cup i'_v})\widehat{M^P}(j,j_v) = w_{e'}(\widehat{x^P})_{K\cup i_v} + w_{e''}(\widehat{x^P})_{K\cup j_v}$$

for $i \notin K$ and

$$S_{t^P}(\widetilde{x}_{K(i\to i'_v)\cup i''_v})\widehat{M^P}(j,j_v) = (\widehat{x^P})_{K(i\to j_v)\cup i_v}$$

for $i \in K$. To prove these two equalities, we identify $b_{i'_v}$ with b_{j_v} and $b_{i''_v}$ with b_{i_v}. Under this identification we have $\widetilde{N}^{\widetilde{P}} = \widehat{N^P}$, where $\widetilde{P}$ is the path from $b_{i'_v}$ to b_j in $\widetilde{N}$ induced by P. Observe that $\widetilde{N}$ has less inner vertices than N, and that the index set of the sources in $\widetilde{N}$ is $I(i \to i'_v) \cup i''_v$. Therefore, by the induction hypothesis,

$$S_{t^{\widetilde{P}}}(\widetilde{x}_{\widetilde{K}})\widetilde{x}^{\widetilde{P}}_{I(i\to i'_v)\cup i''_v} = \widetilde{x}^{\widetilde{P}}_{\widetilde{K}} \tag{9.20}$$

for any $\widetilde{K}$ of size $k+1$. Besides, $\widetilde{x}^{\widetilde{P}}_{I(i\to i'_v)\cup i''_v} = \widetilde{M}^{\widetilde{P}}(j, i'_v) = \widehat{M^P}(j, j_v)$ and $t^P = t^{\widetilde{P}}$. Therefore, using (9.20) for $\widetilde{K} = K \cup i'_v = K \cup j_v$, $\widetilde{K} = K \cup i''_v = K \cup i_v$ and $\widetilde{K} = K(i \to i'_v) \cup i''_v = K(i \to j_v) \cup i_v$ we get both equalities above.

Assume now that the first inner vertex v on P is black. Denote by e and e' the first two edges of P, and by e'' the third edge incident to v. Once again we assume without loss of generality that the cut does not intersect e. To find $M^P(j,i)$, consider the network $\widetilde{N^P}$ similar to the one defined immediately before Lemma 9.19, the difference being that the two new boundary vertices j'_v and j''_v are sinks rather than sources. Clearly,

$$M^P(j,i) = \frac{1}{w_e w_{e'}}\left(\widetilde{M^P}(j, j'_v) + w_{e''}w_{e'}\widetilde{M^P}(j, j''_v)\right). \tag{9.21}$$

To find x^P_K we proceed as follows.

LEMMA 9.20. *Let the first inner vertex v of P be black, then*

$$x^P_K = \begin{cases} (\widetilde{x^P})_K & \text{if } i \notin K \\ \dfrac{1}{w_e w_{e'}}\left((\widetilde{x^P})_{K(i\to j'_v)} + w_{e''}w_{e'}(\widetilde{x^P})_{K(i\to j''_v)}\right) & \text{if } i \in K. \end{cases}$$

PROOF. The proof is a straightforward computation. □

To find x_K we consider the network $\widehat{N}$ defined as in the proof of Proposition 8.3.

LEMMA 9.21. *Let the first inner vertex v of P be black, then*

$$x_K = \begin{cases} \dfrac{w_e w_{e'}\widehat{x}_K}{1 + w_{e''}w_{e'}\widehat{M}(i_v, j_v)} & \text{if } i \notin K \\ \dfrac{\widehat{x}_{K(i\to i_v)} + w_{e''}w_{e'}\widehat{x}_{K(i\to j_v)}}{1 + w_{e''}w_{e'}\widehat{M}(i_v, j_v)} & \text{if } i \in K. \end{cases}$$

PROOF. The proof is similar to the proof of Lemma 9.18. □

By (9.21) and Lemmas 9.20 and 9.21, relation (9.18) boils down to

$$S_{t^P}(\widehat{x}_K)\widetilde{M^P}(j, j'_v)\left(1 + w_{e''}w_{e'}\frac{\widetilde{M^P}(j, j''_v)}{\widetilde{M^P}(j, j'_v)}\right) = (\widetilde{x^P})_K S_{t^P}\left(1 + w_{e''}w_{e'}\widehat{M}(i_v, j_v)\right)$$

for $i \notin K$ and

$$S_{t^P}\left(\widehat{x}_{K(i\to i_v)} + w_{e''}w_{e'}\widehat{x}_{K(i\to j_v)}\right)\widetilde{M^P}(j, j'_v)\left(1 + w_{e''}w_{e'}\frac{\widetilde{M^P}(j, j''_v)}{\widetilde{M^P}(j, j'_v)}\right)$$
$$= \left((\widetilde{x^P})_{K(i\to j'_v)} + w_{e''}w_{e'}(\widetilde{x^P})_{K(i\to j''_v)}\right) S_{t^P}\left(1 + w_{e''}w_{e'}\widehat{M}(i_v, j_v)\right)$$

for $i \in K$. To prove these two equalities, we identify $b_{j'_v}$ with b_{i_v} and $b_{j''_v}$ with b_{j_v}. Under this identification we have $\widehat{N}^{\widehat{P}} = \widetilde{N^P}$, where $\widehat{P}$ is the path from b_{i_v} to b_j in $\widehat{N}$ induced by P. Observe that $\widehat{N}$ has less inner vertices than N, and that the index set of the sources in $\widehat{N}$ is $I(i \to i_v)$. Therefore, by the induction hypothesis,

$$S_{t^P}(\widehat{x}_{\widehat{K}})\widehat{x}^{\widehat{P}}_{I(i\to i_v)} = \widehat{x}^{\widehat{P}}_{\widehat{K}} \tag{9.22}$$

for any $\widehat{K}$ of size k. Taking into account that $\widehat{x}_{I(i\to j_v)} = \widehat{M}(i_v, j_v)$,

$$\widehat{x}^{\widehat{P}}_{I(i\to i_v)} = \widehat{M}^{\widehat{P}}(j, j_v) = \widetilde{M^P}(j, j''_v), \qquad \widehat{x}^{\widehat{P}}_{I(i\to j_v)} = \widehat{M}^{\widehat{P}}(j, i_v) = \widetilde{M^P}(j, j'_v),$$

and using (9.22) for $\widehat{K} = K$, $\widehat{K} = K(i \to i_v) = K(i \to j'_v)$, $\widehat{K} = K(i \to j_v) = K(i \to j''_v)$ and $\widehat{K} = I(i \to j_v)$ we get both equalities above. □

9.3.2. Induced Poisson structures on $LG_k(n)$. Let us consider a subspace $LG^I_k(n) \subset LG_k(n)$ consisting of all $X \in LG_k(n)$ such that the Plücker coordinate x_I does not vanish identically; clearly, $X_N \in LG^I_k(n)$. Therefore, we can identify $LG^I_k(n)$ with the space $\mathrm{Rat}_{k,m}$ equipped with the 2-parametric family of Poisson brackets $\{\cdot,\cdot\}_{I_1,J_1,I_2,J_2}$.

The following result says that for any fixed $n_1 = |I_1| + |J_1|$, the families $\{\cdot,\cdot\}_{I_1,J_1,I_2,J_2}$ on different subspaces $LG^I_k(n)$ can be glued together to form the unique 2-parametric family of Poisson brackets on $LG_k(n)$ that makes all maps X_N Poisson.

THEOREM 9.22. (i) *For any fixed n_1, $0 \le n_1 \le n$, and any choice of parameters α and β there exists a unique Poisson bracket $\mathcal{P}^{n_1}_{\alpha,\beta}$ on $LG_k(n)$ such that for any network N with n_1 boundary vertices on the outer circle, $n - n_1$ boundary vertices on the inner circle, k sources, $n - k$ sinks and weights defined by* (8.4)*, the map $X_N\colon (\mathbb{R}\setminus 0)^{|E|} \to LG_k(n)$ is Poisson provided the parameters α_{ij} and β_{ij} defining the bracket $\{\cdot,\cdot\}_N$ on $(\mathbb{R}\setminus 0)^{|E|}$ satisfy relations* (8.9) *and* (8.11).

(ii) *For any $I \subset [1,n]$, $|I| = k$, and any n_1, $0 \le n_1 \le n$, the restriction of $\mathcal{P}^{n_1}_{\alpha,\beta}$ to the subspace $LG^I_k(n)$ coincides with the bracket $\{\cdot,\cdot\}_{I_1,J_1,I_2,J_2}$ with $I_1 = I\cap[1,n_1]$, $J_1 = [1,n_1]\setminus I_1$, $I_2 = I \setminus I_1$, $J_2 = [n_1+1,n]\setminus I_2$.*

PROOF. This result is an analog of Theorem 8.12, and one may attempt to prove it in a similar way. The main challenge in implementing such an approach is to check that the Poisson structures defined for two distinct subspaces $LG^I_k(n)$ and $LG^{I'}_k(n)$ coincide on the intersection $LG^I_k(n)\cap LG^{I'}_k(n)$ (cp. with Proposition 8.13). For the case of networks in an annulus, the direct check becomes too cumbersome. We suggest to bypass this difficulty in the following way.

Assume first that $|I \cap I'| = k - 1$ and take $i \in I \setminus I'$, $j \in I' \setminus I$. Denote by $\mathrm{Net}^{ij}_{I_1,J_1,I_2,J_2}$ the set of networks in $\mathrm{Net}_{I_1,J_1,I_2,J_2}$ satisfying the following two conditions: $M(i,j)$ does not vanish identically and there exists a path from b_i to b_j that does not intersect the cut. The set $\mathrm{Net}^{ji}_{I'_1,J'_1,I'_2,J'_2}$ is defined similarly, with the roles of i and j interchanged. Clearly, the path reversal introduced in Section 9.3.1 establishes a bijection between $\mathrm{Net}^{ij}_{I_1,J_1,I_2,J_2}$ and $\mathrm{Net}^{ji}_{I'_1,J'_1,I'_2,J'_2}$. Moreover, a suitable modification of Theorem 9.10 remains true for networks in $\mathrm{Net}^{ij}_{I_1,J_1,I_2,J_2}$: these networks represent all rational matrix function such that corresponding component of the matrix does not vanish identically. To see that we use the following construction. Let v be the neighbor of b_i in N and u be the neighbor of b_j in N. Add two new white vertices v' and v'' and two new black vertices u' and u''. Replace edge (b_i, v) by the edges (b_i, v') and (v', v) so that the weight of the obtained path is equal to the weight of the replaced edge. In a similar way, replace (u, b_j) by (u, u') and (u', b_j). Besides, add edges (v', v'') and (u'', u') of weight 1 and two parallel edges (v'', u''), one of weight 1, and the other of weight -1. Finally, resolve all the arising intersections with the help of the network N_{id}. Since the set of functions representable via networks in $\text{Net}^{ij}_{I_1,J_1,I_2,J_2}$ is dense in the space of all rational matrix functions, the 2-parametric family $\{\cdot,\cdot\}_{I_1,J_1,I_2,J_2}$ is defined uniquely already by the fact that M_N is Poisson for any $N \in \text{Net}^{ij}_{I_1,J_1,I_2,J_2}$. Recall that the boundary measurement map M_N factors through $\mathcal{F}_N$; clearly, the same holds for the Grassmannian boundary measurement map X_N. Besides, the path reversal map R^P commutes with the projection y: $\mathcal{E}_N \to \mathcal{F}_N$ and commutes with X_N up to S_{t^P}. Finally, Poisson brackets satisfying relations (9.5)-(9.8) and (9.9)-(9.11) commute with S_{t^P}. Therefore, Poisson structures $\{\cdot,\cdot\}_{I_1,J_1,I_2,J_2}$ and $\{\cdot,\cdot\}_{I_1',J_1',I_2',J_2'}$ coincide on $LG_k^I(n) \cap LG_k^{I'}(n)$. If $|I \cap I'| = r < k-1$, we consider a sequence $(I = I^{(0)}, I^{(1)}, \dots, I^{(k-r)} = I')$ such that $|I^{(t)} \cap I^{(t+1)}| = k-1$ for all $t = 0,\dots,k-r-1$ and apply to each pair $(I^{(t)}, I^{(t+1)})$ the same reasoning as above. □

9.4. Summary

- We extend the notion of a perfect planar network in a disk to the case of an annulus. An adequate description of boundary measurements requires introducing a cut with the endpoints on distinct connected components of the boundary of the annulus. Boundary measurements in this extended setting are rational functions in edge weights and a parameter associated to the cut, see Corollary 9.3.
- We introduce the space of face and trail weights and provide its cohomological interpretation, see Section 9.1.3.
- The family of universal Poisson brackets on the space of edge weights of a given network constructed in Chapter 8 induces a two-parameter family of Poisson brackets on boundary measurement matrices, see Theorem 9.4. This family depends on a mutual location of sources and sinks, but not on the network itself.
- We provide an explicit description of this family in Propositions 9.6 and 9.8. In particular, if all sources lie on the outer circle and all sinks lie on the inner circle, we recover a 2-parametric family of trigonometric R-matrices and the corresponding R-matrix brackets, see Theorem 9.15. An important tool in the proof of Theorem 9.4 is the realization theorem (see Theorem 9.10); it claims that any rational matrix function can be realized as the boundary measurement matrix of a perfect planar network.
- The boundary measurement map is extended to a map into the space of Grassmannian loops $LG_k(n)$ defined by a network with k sources, $n-k$ sinks, and exactly $n_1 \le n$ boundary vertices on the outer circle. The Poisson family on boundary measurement matrices allows us to equip the Grassmannian with a two-parameter family of Poisson brackets $\mathcal{P}^{n_1}_{\alpha,\beta}$ in such a way, that for any choice of a universal Poisson bracket on the edge weights there is a unique member of $\mathcal{P}^{n_1}_{\alpha,\beta}$ that makes the extended boundary measurement map Poisson, see Theorem 9.22. This family depends only on the number of sources and sinks and on the number of boundary

vertices on the outer circle, and otherwise does not depend on its mutual location.

- The main ingredient in the proof of Theorem 9.22 is the path reversal operation on networks. We prove that under certain conditions this operation preserves the Grassmannian boundary measurement map up to the action $\lambda \mapsto -\lambda$, see Theorem 9.17.

Bibliographical notes

Our exposition in this Chapter is based on [**GSV6**].

9.2. For a discussion of R-matrix Poisson brackets on $\mathrm{Rat}_{k,k}$ see [**FT**]. The trigonometric R-matrix is defined in [**BeD**].

9.3. Theorem 9.17 for perfect networks in a disk is proved (in a different way) in [**Po**]. Observe that for the networks in a disk, there is no cut and $t^P = 1$ identically, hence in this case X_N and R^P always commute.

CHAPTER 10

Generalized Bäcklund–Darboux transformations for Coxeter–Toda flows from a cluster algebra perspective

In this chapter we consider an example that ties together objects and concepts from the theory of cluster algebras, Poisson geometry of directed networks discussed in Chapter 9 and the theory of integrable systems.

10.1. Introduction

We will start by reviewing basic facts about *Toda flows* on GL_n. These are commuting Hamiltonian flows generated by conjugation-invariant functions on GL_n with respect to the standard Poisson–Lie structure. Toda flows (also known as *characteristic Hamiltonian systems*) are defined for an arbitrary standard semi-simple Poisson–Lie group, but we will concentrate on the GL_n case, where as a maximal algebraically independent family of conjugation-invariant functions one can choose $F_k : GL_n \ni X \mapsto \frac{1}{k} \operatorname{tr} X^k$, $k = 1, \dots, n-1$. The equation of motion generated by F_k has a *Lax form*:

$$dX/dt = \left[X, \ -\frac{1}{2}\left(\pi_+(X^k) - \pi_-(X^k)\right)\right], \tag{10.1}$$

where $\pi_+(A)$ and $\pi_-(A)$ denote strictly upper and lower parts of a matrix A.

Any double Bruhat cell $G^{u,v}$, $u, v \in S_n$, is a regular Poisson submanifold in GL_n invariant under the right and left multiplication by elements of the maximal torus (the subgroup of diagonal matrices) $\mathcal{H} \subset GL_n$. In particular, $G^{u,v}$ is invariant under the conjugation by elements of $\mathcal{H}$. The standard Poisson–Lie structure is also invariant under the conjugation action of $\mathcal{H}$ on GL_n. This means that Toda flows defined by (10.1) induce commuting Hamiltonian flows on $G^{u,v}/\mathcal{H}$ where $\mathcal{H}$ acts on $G^{u,v}$ by conjugation. In the case when $v = u^{-1} = (n\ 1\ 2 \dots n-1)$, $G^{u,v}$ consists of tridiagonal matrices with nonzero off-diagonal entries, $G^{u,v}/\mathcal{H}$ can be conveniently described as the set Jac of *Jacobi matrices* of the form

$$L = \begin{pmatrix} b_1 & 1 & 0 & \cdots & 0 \\ a_1 & b_2 & 1 & \cdots & 0 \\ & \ddots & \ddots & \ddots & \\ & & & b_{n-1} & 1 \\ 0 & & & a_{n-1} & b_n \end{pmatrix}, \quad a_1 \cdots a_{n-1} \neq 0, \ \det L \neq 0.$$

Lax equations (10.1) then become the equations of the *finite nonperiodic Toda hierarchy*:

$$dL/dt = [L, \pi_-(L^k)],$$

the first of which, corresponding to $k = 1$, is the celebrated *Toda lattice*

$$\begin{aligned} da_j/dt &= a_j(b_{j+1} - b_j), \quad j = 1, \dots, n-1, \\ db_j/dt &= (a_j - a_{j-1}), \quad j = 1, \dots, n, \end{aligned}$$

with the boundary conditions $a_0 = a_n = 0$. Recall that $\det L$ is a Casimir function for the standard Poisson–Lie bracket. The level sets of the function $\det L$ foliate Jac into $2(n-1)$-dimensional symplectic manifolds, and the Toda hierarchy defines a completely integrable system on every symplectic leaf. Note that although Toda flows on an arbitrary double Bruhat cell $G^{u,v}$ can be exactly solved via the so-called *factorization method*, in most cases the dimension of symplectic leaves in $G^{u,v}/\mathcal{H}$ exceeds $2(n-1)$, which means that conjugation-invariant functions do not form a Poisson commuting family rich enough to ensure Liouville complete integrability.

An important role in the study of Toda flows is played by the *Weyl function*

$$m(\lambda) = m(\lambda; X) = ((\lambda\mathbf{1} - X)^{-1}e_1, e_1) = \frac{q(\lambda)}{p(\lambda)}, \tag{10.2}$$

where $p(\lambda)$ is the characteristic polynomial of x and $q(\lambda)$ is the characteristic polynomial of the $(n-1) \times (n-1)$ submatrix of x formed by deleting the first row and column. Differential equations that describe the evolution of $m(\lambda; X)$ induced by Toda flows do not depend on the initial value $x(0)$ and are easy to solve: though nonlinear, they are also induced by *linear differential equations with constant coefficients* on the space

$$\left\{ M(\lambda) = \frac{Q(\lambda)}{P(\lambda)} : \deg P = \deg Q + 1 = n, \ P, Q \text{ coprime}, \ P(0) \neq 0 \right\} \tag{10.3}$$

by the map $M(\lambda) \mapsto m(\lambda) = -\frac{1}{H_0} M(-\lambda)$, where $H_0 = \lim_{\lambda\to\infty} \lambda M(\lambda) \neq 0$.

It is easy to see that $m(\lambda; X)$ is invariant under the action of $\mathcal{H}$ on $G^{u,v}$ by conjugation. Thus we have a map from $G^{u,v}/\mathcal{H}$ into the space

$$\mathcal{W}_n = \left\{ m(\lambda) = \frac{q(\lambda)}{p(\lambda)} : \deg p = \deg q + 1 = n, \ p, q \text{ monic and coprime}, \ p(0) \neq 0 \right\}.$$

In the tridiagonal case, this map, sometimes called the *Moser map*, is invertible: it is a classical result in the theory of moment problems that matrix entries of an element in Jac can be restored from its Weyl function $m(\lambda; X)$ via determinantal formulas for matrix entries of x in terms of Hankel determinants built from the coefficients of the Laurent expansion of $m(\lambda; X)$.

In this Chapter, we study double Bruhat cells $G^{u,v}$ that share common features with the tridiagonal case:

(i) the Toda hierarchy defines a completely integrable system on level sets of the determinant in $G^{u,v}/\mathcal{H}$, and

(ii) the Moser map $m_{u,v} : G^{u,v}/\mathcal{H} \to \mathcal{W}_n$ defined in the same way as in the tridiagonal case is invertible.

We will see that double Bruhat cells $G^{u,v}$ associated with any pair of Coxeter elements $u, v \in S_n$ enjoy these properties. (Recall that a *Coxeter element* in S_n is a product of $n-1$ distinct elementary transpositions.) We will call any such double Bruhat cell a *Coxeter double Bruhat cell*. Integrable equation induced on $G^{u,v}/\mathcal{H}$ by Toda flows will be called *Coxeter–Toda lattices*.

Since Coxeter–Toda flows associated with different choices of (u, v) lead to the same evolution of the Weyl function, and the corresponding Moser maps are invertible, one can construct transformations between different $G^{u,v}/\mathcal{H}$ that preserve the corresponding Coxeter–Toda flows and thus serve as *generalized Bäcklund–Darboux transformations* between them. Our goal is to describe these transformations from the cluster algebra point of view. To this end, we construct a cluster algebra of rank $2n - 2$ associated with an extension of the space (10.3)

$$\mathcal{R}_n = \left\{ \frac{Q(\lambda)}{P(\lambda)} : \deg P = n,\ \deg Q < n,\ P, Q \text{ are coprime},\ P(0) \neq 0 \right\}.$$

(Note that $\mathcal{W}_n$ is embedded into $\mathcal{R}_n$ as a codimension 1 subspace.) Distinguished clusters $\mathbf{x}_{u,v}$ in this algebra correspond to Coxeter double Bruhat cells, and are formed by certain collections of Hankel determinants built out of coefficients of the Laurent expansion of an element in $\mathcal{R}_n$. Sequences of cluster transformations connecting these distinguished clusters are then used as the main ingredient in the construction of generalized Bäcklund–Darboux transformations.

The insight necessary to implement this construction is drawn from two sources:

(i) the procedure for the inversion of the Moser map, that can be viewed as a generalization of the inverse moment problem, and

(ii) interpretation of functions in $\mathcal{R}_n$ as boundary measurement functions associated with a particular kind of networks in an annulus.

The chapter is organized as follows.

First, we describe a parametrization of a Coxeter double Bruhat cell. This is a particular case of the Berenstein-Fomin-Zelevinsky parametrization that we discussed in Section 1.2.4: for a generic element X in $G^{u,v}$, we consider a factorization of X into elementary bidiagonal factors consistent with the Gauss factorization of X, that is $X = X_- X_0 X_+$, where X_0 is the diagonal matrix $\operatorname{diag}(d_1, \ldots, d_n)$, X_+ is the product of $n-1$ elementary upper bidiagonal factors $E_i^+(c_i^+)$, $i = 1, \ldots, n-1$, with the order of factors in the product prescribed by v, and X_- is the product of $n-1$ elementary lower bidiagonal factors $E_i^-(c_i^-)$, $i = 1, \ldots, n-1$, with the order of factors in the product prescribed by u.

Elements in $G^{u,v}/\mathcal{H}$ are parametrized by d_i and $c_i = c_i^+ c_i^-$, $i = 1, \ldots, n-1$. We show that these parameters can be restored as monomial expressions in terms of an appropriately chosen collection of Hankel determinants built from the coefficients of the Laurent expansion of the Weyl function $m(\lambda)$. Both the choice of Hankel determinants and exponents entering monomial expressions for d_i, c_i are uniquely determined by the pair (u, v).

Next, the map $X \mapsto m(\lambda; X)$ is given a combinatorial interpretation in terms of weighted directed planar networks. To an elementary bidiagonal factorization of $X \in G^{u,v}$ there corresponds a network $N_{u,v}$ in a square (disk) with n sources located on one side of the square and n sinks located at the opposite side, both numbered bottom to top (cp. to Section 8.2.1). By gluing opposite sides of the square containing sinks and sources in such a way that each sink is glued to the corresponding source and adding two additional edges, one incoming and one outgoing, one obtains a weighted directed network in an annulus (outer and inner boundary circles of the annulus are formed by the remaining two sides of the square). The network we just described, $N_{u,v}^{\circ}$, has one sink and one source on the outer boundary of an annulus and, according to Sect. 9.1.2, the boundary measurement that corresponds to this network is a rational function $M(\lambda)$ in an auxiliary parameter

λ. We show that $-M(-\lambda)$ is equal to $m(\lambda; X)$ times the product of weights of the incoming and outgoing edges in $N^\circ_{u,v}$.

The determinantal formulae for the inverse of the Moser map are homogeneous of degree zero with respect to coefficients of the Laurent expansion, therefore the same formulae applied to $M(\lambda)$ also recover c_i, d_i. Thus, we can define a map $\rho_{u,v} : (\mathbb{C}^*)^{2n} \to G^{u,v}/\mathcal{H}$ in such a way that the through map

$$G^{u,v}/\mathcal{H} \overset{m_{u,v}}{\longrightarrow} \mathcal{W}_n \hookrightarrow \mathcal{R}_n \overset{\mathbf{x}_{u,v}}{\longrightarrow} (\mathbb{C}^*)^{2n} \overset{\rho_{u,v}}{\longrightarrow} G^{u,v}/\mathcal{H}$$

is the identity map.

We use the combinatorial data determined by the pair (u, v) (or, in a more transparent way, by the corresponding network $N^\circ_{u,v}$) to construct a cluster algebra $\mathcal{A} = \mathcal{A}_{u,v}$, with the (slightly modified) collection $\mathbf{x}_{u,v}$ serving as the initial cluster. The matrix $B_{u,v}$ that determines cluster transformations for the initial cluster is closely related to the incidence matrix of the graph dual to $N^\circ_{u,v}$. To construct $\mathcal{A}_{u,v}$, we start with a particular member of the family of Poisson structures described in Theorem 9.4. Initial cluster variables, viewed as functions on $\mathcal{R}_n$ form a coordinate system in which this Poisson structure takes a particular simple form: the Poisson bracket of logarithms of any two functions in the family is constant. This allows us to follow the strategy from Chapter 4 to construct $\mathcal{A}_{u,v}$ as a cluster algebra compatible with this Poisson bracket. We then show that $\mathcal{A}_{u,v}$ does not depend on the choice of Coxeter elements u, v, that is, that for any (u', v'), the initial seed of $\mathcal{A}_{u',v'}$ is a seed in the cluster algebra $\mathcal{A}_{u,v}$. Therefore, the change of coordinates $T^{u',v'}_{u,v} : \mathbf{x}_{u,v} \mapsto \mathbf{x}_{u',v'}$ is accomplished by a sequence of cluster transformations. Moreover, the ring of regular functions on $\mathcal{R}_n$ is isomorphic to the localization of the complex form of $\mathcal{A}$ with respect to the stable variables.

In the final section, we interpret generalized Bäcklund–Darboux transformations between Coxeter–Toda lattices corresponding to different pairs of Coxeter elements in terms of the cluster algebra $\mathcal{A}$ by observing that the map

$$\sigma^{u',v'}_{u,v} = \rho_{u',v'} \circ T^{u',v'}_{u,v} \circ \tau_{u,v} \; : G^{u,v}/\mathcal{H} \to G^{u',v'}/\mathcal{H}, \tag{10.4}$$

with $\tau_{u,v}$ being the right inverse of $\rho_{u,v}$, preserves flows generated by conjugation-invariant functions and makes the diagram

$$\begin{array}{ccc} G^{u,v}/\mathcal{H} & \xrightarrow{\quad \sigma^{u',v'}_{u,v} \quad} & G^{u',v'}/\mathcal{H} \\ & {\scriptstyle m_{u,v}} \searrow \qquad \swarrow {\scriptstyle m_{u',v'}} & \\ & \mathcal{W}_n & \end{array}$$

commutative. Besides, we explain how one represents generalized Bäcklund–Darboux transformations as equivalent transformations of the network $N^\circ_{u,v}$. Finally we show that classical Darboux transformations are also related to cluster algebra transformations via a formula similar to (10.4).

10.2. Coxeter double Bruhat cells

We start this section with describing a particular instance of the Berenstein-Fomin-Zelevinsky parametrization in the case of Coxeter double Bruhat cells in GL_n.

Denote $s_{[p,q]} = s_p s_{p+1} \cdots s_{q-1}$ for $1 \le p < q \le n$ and recall that every Coxeter element $v \in S_n$ can be written in the form

$$v = s_{[i_{k-1},i_k]} \cdots s_{[i_1,i_2]} s_{[1,i_1]} \tag{10.5}$$

for some subset $I = \{1 = i_0 < i_1 < \dots < i_k = n\} \subseteq [1,n]$. Besides, define $L = \{1 = l_0 < l_1 < \dots < l_{n-k} = n\}$ by $\{l_1 < \dots < l_{n-k-1}\} = [1,n] \setminus I$.

LEMMA 10.1. *Let v be given by (10.5), then*

$$v^{-1} = s_{[l_{n-k-1},l_{n-k}]} \cdots s_{[l_1,l_2]} s_{[1,l_1]}.$$

PROOF. We use induction on n. Denote the right-hand side of the above relation by $\bar v$. The index $n-1$ belongs either to I or to L. In the latter case $l_{n-k-1} = n-1$, and we have $v = s_{[i_{k-1},n]} \cdots s_{[i_1,i_2]} s_{[1,i_1]} = s_{[i_{k-1},n-1]} \cdots s_{[i_1,i_2]} s_{[1,i_1]}\, s_{n-1}$ and $\bar v = s_{n-1} s_{[l_{n-k-2},l_{n-k-1}]} \cdots s_{[l_1,l_2]} s_{[1,l_1]}$. Then $v = v' s_{n-1}$ and $\bar v = s_{n-1} \bar v'$, where v', $\bar v'$ are Coxeter elements in S_{n-1} corresponding to index sets $I \setminus \{n\} \cup \{n-1\}$ and $L \setminus \{n\}$, and hence $v \bar v = v' \bar v' = 1$ by the induction hypothesis. Otherwise, if $n-1$ belongs to I, we interchange the roles of v and $\bar v$ and use the same argument. □

LEMMA 10.2. *The permutation matrix that corresponds to a Coxeter element v is*

$$\tilde v = \sum_{j=1}^{k} e_{i_{j-1} i_j} + \sum_{j=1}^{n-k} e_{l_j l_{j-1}}.$$

PROOF. We use the same inductive argument as in the proof of Lemma 10.1. Assuming that $n-1 \in L$, the relation $v = v' s_{n-1}$ and the induction hypothesis imply $\tilde v = (e_{1i_1} + \dots + e_{i_{k-1} n-1} + e_{l_1 1} + \dots + e_{n-1 l_{n-k-2}} + e_{nn})(e_{11} + \dots + e_{n-2 n-2} + e_{nn-1} + e_{n-1 n}) = e_{1i_1} + \dots + e_{i_{k-1} n} + e_{l_1 1} + \dots + e_{n-1 l_{n-k-2}} + e_{nn-1}$ as claimed. □

Let now (u,v) be a pair of Coxeter elements and

$$\begin{aligned} I^+ &= \{1 = i_0^+ < i_1^+ < \dots < i_{k^+}^+ = n\},\\ I^- &= \{1 = i_0^- < i_1^- < \dots < i_{k^-}^- = n\},\\ L^+ &= \{1 = l_0^+ < l_1^+ < \dots < l_{n-k^+-1}^+ < l_{n-k^+}^+ = n\},\\ L^- &= \{1 = l_0^- < l_1^- < \dots < l_{n-k^--1}^- < l_{n-k^-}^- = n\} \end{aligned} \tag{10.6}$$

be subsets of $[1,n]$ that correspond to v and u^{-1} in the way just described. For a set of complex parameters $c_1^-, \dots, c_{n-1}^-; c_1^+, \dots, c_{n-1}^+; d_1, \dots, d_n$, define matrices $D = \operatorname{diag}(d_1, \dots, d_n)$,

$$C_j^+ = \sum_{\alpha = i_{j-1}^+}^{i_j^+ - 1} c_\alpha^+ e_{\alpha,\alpha+1},\ j \in [1,k^+], \quad C_j^- = \sum_{\alpha = i_{j-1}^-}^{i_j^- - 1} c_\alpha^- e_{\alpha+1,\alpha},\ j \in [1,k^-], \tag{10.7}$$

and

$$\bar C_j^+ = \sum_{\alpha = l_{j-1}^+}^{l_j^+ - 1} c_\alpha^+ e_{\alpha,\alpha+1},\ j \in [1,n-k^+], \quad \bar C_j^- = \sum_{\alpha = l_{j-1}^-}^{l_j^- - 1} c_\alpha^+ e_{\alpha+1,\alpha},\ j \in 1,n-k^-].$$

LEMMA 10.3. *A generic element $X \in G^{u,v}$ can be written as*

$$X = (\mathbf{1} - C_1^-)^{-1} \cdots (\mathbf{1} - C_{k^-}^-)^{-1} D (\mathbf{1} - C_{k^+}^+)^{-1} \cdots (\mathbf{1} - C_1^+)^{-1}, \tag{10.8}$$

and its inverse can be factored as

$$X^{-1} = (\mathbf{1} + \bar{C}^+_{n-k^+})^{-1} \cdots (\mathbf{1} + \bar{C}^+_1)^{-1} D^{-1} (\mathbf{1} + \bar{C}^-_1)^{-1} \cdots (\mathbf{1} + \bar{C}^-_{k^-})^{-1}. \tag{10.9}$$

PROOF. It is easy to see that

$$\begin{aligned} (\mathbf{1} - C^+_j)^{-1} &= E^+_{i^+_{j-1}}(c^+_{i^+_{j-1}}) \cdots E^+_{i^+_j-1}(c^+_{i^+_j-1}),\\ (\mathbf{1} - C^-_j)^{-1} &= E^-_{i^-_j-1}(c^-_{i^-_j-1}) \cdots E^-_{i^-_{j-1}}(c^-_{i^-_{j-1}}),\\ (\mathbf{1} + \bar{C}^+_j)^{-1} &= E^+_{l^+_{j-1}}(-c^+_{l^+_{j-1}}) \cdots E^+_{l^+_j-1}(-c^+_{l^+_j-1}),\\ (\mathbf{1} + \bar{C}^-_j)^{-1} &= E^-_{l^-_j-1}(-c^-_{l^-_j-1}) \cdots E^-_{l^-_{j-1}}(-c^-_{l^-_{j-1}}) \end{aligned} \tag{10.10}$$

with $E^\pm_i$ defined by (8.3). Then, by (1.12) and (10.5), a generic $X \in G^{u,v}$ can be written as in (10.8). Next, the same reasoning as in the proof of Lemma 10.1 implies that

$$\begin{aligned} (\mathbf{1} - C^+_1) \cdots (\mathbf{1} - C^+_{k^+}) &= \left(\left(\prod_{s=i^+_{k-1}}^{i^+_k-1} E^+_s(c^+_s) \right) \cdots \left(\prod_{s=1}^{i^+_1-1} E^+_s(c^+_s) \right) \right)^{-1}\\ &= \left(\prod_{s=l^+_{n-k-1}}^{l^+_{n-k}-1} E^+_s(-c^+_s) \right) \cdots \left(\prod_{s=1}^{l^+_1-1} E^+_s(-c^+_s) \right)\\ &= (\mathbf{1} + \bar{C}^+_{n-k^+})^{-1} \cdots (\mathbf{1} + \bar{C}^+_1)^{-1} \end{aligned}$$

and, similarly,

$$(\mathbf{1} - C^-_{k^-}) \cdots (\mathbf{1} - C^-_1) = (\mathbf{1} + \bar{C}^-_1)^{-1} \cdots (\mathbf{1} + \bar{C}^-_{k^-})^{-1}.$$

Therefore,

$$\begin{aligned} X^{-1} &= (\mathbf{1} - C^+_1) \cdots (\mathbf{1} - C^+_{k^+}) D^{-1} (\mathbf{1} - C^-_{k^-}) \cdots (\mathbf{1} - C^-_1)\\ &= (\mathbf{1} + \bar{C}^+_{n-k^+})^{-1} \cdots (\mathbf{1} + \bar{C}^+_1)^{-1} D^{-1} (\mathbf{1} + \bar{C}^-_1)^{-1} \cdots (\mathbf{1} + \bar{C}^-_{k^-})^{-1}. \end{aligned}$$

□

The network $N_{u,v}$ that corresponds to factorization (10.8) is obtained by the concatenation (left to right) of $2n-1$ building blocks (as depicted in Fig. 8.5) that correspond to elementary matrices

$$\begin{gathered} E^-_{i^-_2-1}(c^-_{i^-_2-1}), \dots, E^-_1(c^-_1), E^-_{i^-_3-1}(c^-_{i^-_3-1}), \dots, E^-_{i^-_2}(c^-_{i^-_2}), \dots,\\ E^-_{n-1}(c^-_{n-1}), \dots, E^-_{i^-_{k^--1}}(c^-_{i^-_{k^--1}}), D, E^+_{i^+_{k^+-1}}(c^+_{i^+_{k^+-1}}), \dots, E^+_{n-1}(c^+_{n-1}),\\ \dots, E^+_{i^+_2}(c^+_{i^+_2}), \dots E^+_{i^+_3-1}(c^+_{i^+_3-1}), E^+_1(c^+_1) \cdots E^+_{i^+_2-1}(c^+_{i^+_2-1}). \end{gathered}$$

This network has $4(n-1)$ internal vertices and $5n-4$ horizontal edges. One can use the gauge group action described in Section 9.1.3 to change the weights of all horizontal edges except for those belonging to block D to 1. Note that the weights on the remaining edges become Laurent monomials in the initial weights of the network.

Let us now introduce some combinatorial data that will be useful in the following sections.

Let us fix a pair (u, v) of Coxeter elements, and hence, fix the sets $I^\pm$ given by (10.6). For any $i \in [1, n]$ define integers $\varepsilon_i^\pm$ and $\zeta_i^\pm$ by setting

$$\varepsilon_i^\pm = \begin{cases} 0 & \text{if } i = i_j^\pm \text{ for some } 0 < j \le k_\pm \\ 1 & \text{otherwise} \end{cases} \tag{10.11}$$

and

$$\zeta_i^\pm = i(1 - \varepsilon_i^\pm) - \sum_{\beta=1}^{i-1} \varepsilon_\beta^\pm; \tag{10.12}$$

note that by definition, $\varepsilon_1^\pm = 1$, $\zeta_1^\pm = 0$. (Here and in what follows a relation involving variables with superscripts $\pm$ is a shorthand for two similar relations: the one obtained by simultaneously replacing each $\pm$ by $+$, and the other, by $-$.) Further, put $M_i^\pm = \{\zeta_\alpha^\pm \ : \ \alpha = 1, \ldots, i\}$ and

$$k_i^\pm = \max\{j : i_j^\pm \le i\}. \tag{10.13}$$

Finally, put

$$\varepsilon_i = \varepsilon_i^+ + \varepsilon_i^- \tag{10.14}$$

and

$$\varkappa_i = i + 1 - \sum_{\beta=1}^{i} \varepsilon_\beta. \tag{10.15}$$

REMARK 10.4. It is easy to see that there exist distinct pairs (u, v) and (u', v') that produce the same n-tuple ε. The ambiguity occurs when $\varepsilon_i = 1$ for some $i \in [2, n-1]$. By (10.14), this situation corresponds either to $\varepsilon_i^+ = 1$, $\varepsilon_i^- = 0$, or to $\varepsilon_i^+ = 0$, $\varepsilon_i^- = 1$. Consequently, the number of pairs (u, v) with the identical n-tuple ε equals 2 power the number of times ε_i takes value 1.

LEMMA 10.5. (i) *The n-tuples $\varepsilon^\pm = (\varepsilon_i^\pm)$ and $\zeta^\pm = (\zeta_i^\pm)$ uniquely determine each other.*

(ii) *For any $i \in [1, n]$,*

$$\zeta_i^\pm = \begin{cases} j & \text{if } i = i_j^\pm \text{ for some } 0 < j \le k_\pm \\ -\sum_{\beta=1}^{i-1} \varepsilon_\beta^\pm & \text{otherwise} \end{cases}.$$

(iii) *For any $i \in [1, n]$,*

$$k_i^\pm = i - \sum_{\beta=1}^{i} \varepsilon_\beta^\pm.$$

(iv) *For any $i \in [1, n]$,*

$$M_i^\pm = [k_i^\pm - i + 1, k_i^\pm] = [1 - \sum_{\beta=1}^{i} \varepsilon_\beta^\pm, i - \sum_{\beta=1}^{i} \varepsilon_\beta^\pm].$$

PROOF. (i) Follows form the fact that the transformation $\varepsilon^\pm \mapsto \zeta^\pm$ defined by (10.12) is given by a lower–triangular matrix with a non-zero diagonal.

(ii) By (10.12), the first equality is equivalent to

$$i_j^\pm = j + \sum_{\beta=1}^{i_j^\pm - 1} \varepsilon_\beta^\pm. \tag{10.16}$$

By (10.11), the latter can be interpreted as counting the first $i_j^\pm$ elements of $(\varepsilon_i^\pm)$: exactly j of them are equal to 0, and all the other are equal to 1.

The second equality follows trivially from (10.11) and (10.12).

(iii) For $i = i_j^\pm$, follows immediately from (10.13) and (10.16). For $i \neq i_j^\pm$, the same counting argument used in (10.16) gives

$$i = k_i^\pm + \sum_{\beta=1}^{i} \varepsilon_\beta^\pm .$$

(iv) Follows from parts (ii) and (iii). □

REMARK 10.6. (i) If $v = s_{n-1} \cdots s_1$, then X is a lower Hessenberg matrix, and if $u = s_1 \cdots s_{n-1}$, then X is an upper Hessenberg matrix.

(ii) If $v = s_{n-1} \cdots s_1$ *and* $u = s_1 \cdots s_{n-1}$, then $G^{u,v}$ consists of tri-diagonal matrices with non-zero off-diagonal entries (*Jacobi matrices*). In this case $I^+ = I^- = [1, n]$, $\varepsilon_1^\pm = 1$ and $\varepsilon_i^\pm = 0$ for $i = 2, \ldots, n$.

(iii) If $u = v = s_{n-1} \cdots s_1$ (which leads to $I^+ = [1, n], I^- = \{1, n\}$), then elements of $G^{u,v}$ have a structure of recursion operators arising in the theory of orthogonal polynomials on the unit circle.

(iv) The choice $u = v = (s_1 s_3 \cdots)(s_2 s_4 \cdots)$ (the so-called *bipartite Coxeter element*) gives rise to a special kind of pentadiagonal matrices X (called *CMV matrices*), which serve as an alternative version of recursion operators for orthogonal polynomials on the unit circle.

EXAMPLE 10.7. Let $n = 5$, $v = s_4 s_3 s_1 s_2$ and $u = s_3 s_2 s_1 s_4$. The network $N_{u,v}$ that corresponds to factorization (10.8) is shown in Figure 10.1.

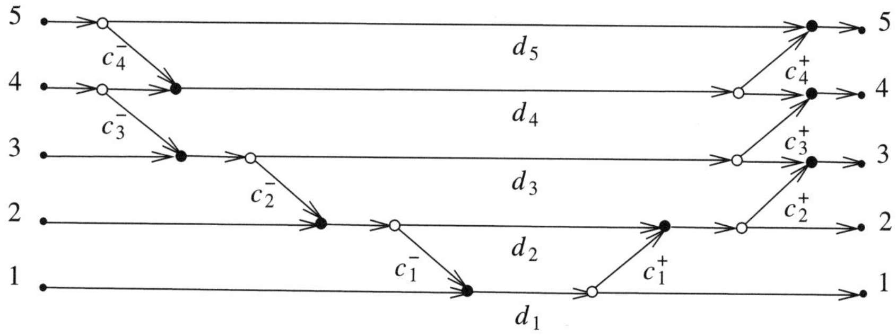

FIGURE 10.1. Network representation for elements in $G^{s_3 s_2 s_1 s_4, s_4 s_3 s_1 s_2}$

A generic element $X \in G^{u,v}$ has a form

$$X = (x_{ij})_{i,j=1}^5 = \begin{pmatrix} d_1 & x_{11}c_1^+ & x_{12}c_2^+ & 0 & 0 \\ c_1^- x_{11} & d_2 + c_1^- x_{12} & x_{22}c_2^+ & 0 & 0 \\ c_2^- x_{21} & c_2^- x_{22} & d_3 + c_2^- x_{23} & d_3 c_3^+ & 0 \\ c_3^- x_{31} & c_3^- x_{32} & c_3^- x_{33} & d_4 + c_3^- x_{34} & d_4 c_4^+ \\ 0 & 0 & 0 & c_4^- d_4 & d_5 + c_4^- x_{45} \end{pmatrix}.$$

One finds by a direct observation that $k^+ = 3$ and $I^+ = \{i_0^+, i_1^+, i_2^+, i_3^+\} = \{1, 3, 4, 5\}$, and hence $L^+ = \{l_0^+, l_1^+, l_2^+\} = \{1, 2, 5\}$. Next, $u^{-1} = s_4 s_1 s_2 s_3$, therefore, $k^- = 2$ and $I^- = \{i_0^-, i_1^-, i_2^-\} = \{1, 4, 5\}$, and hence $L^- = \{l_0^-, l_1^-, l_2^-, l_3^-\} =$

$\{1, 2, 3, 5\}$. Further,

$$\varepsilon^+ = (1, 1, 0, 0, 0), \qquad \varepsilon^- = (1, 1, 1, 0, 0),$$

and hence

$$\zeta^+ = (0, -1, 1, 2, 3), \qquad \zeta^- = (0, -1, -2, 1, 2).$$

Therefore,

$$(k_i^+)_{i=1}^5 = (0, 0, 1, 2, 3), \qquad (k_i^-)_{i=1}^5 = (0, 0, 0, 1, 2),$$

and hence

$$(M_i^+)_{i=1}^5 = ([0, 0], [-1, 0], [-1, 1], [-1, 2], [-1, 3]),$$
$$(M_i^-)_{i=1}^5 = ([0, 0], [-1, 0], [-2, 0], [-2, 1], [-2, 2]).$$

Finally, $\varepsilon = (2, 2, 1, 0, 0)$ and $\varkappa = (0, -1, -1, 0, 1)$. By Remark 10.4, there is one more pair of Coxeter elements such that produces the same 5-tuples: $v' = s_4 s_1 s_2 s_3$ and $u' = s_2 s_1 s_3 s_4$. We will use this example as our running example in the next section.

10.3. Inverse problem

10.3.1. In this section we show how an element X of a Coxeter double Bruhat cell $G^{u,v}$ that admits factorization (10.8) can be restored from its Weyl function (10.2) up to a conjugation by a diagonal matrix.

Recall various useful representations for the Weyl function $m(\lambda; X)$:

$$m(\lambda; X) = ((\lambda \mathbf{1} - X)^{-1} e_1, e_1) = \frac{q(\lambda)}{p(\lambda)} = \sum_{j=0}^{\infty} \frac{h_j(X)}{\lambda^{j+1}}. \tag{10.17}$$

Here e_i denotes the vector $(\delta_{i\alpha})_{\alpha=1}^n$ of the standard basis in $\mathbb{C}^n$, $(\,\cdot, \cdot)$ is the standard inner product, $p(\lambda)$ is the characteristic polynomial of X, $q(\lambda)$ is the characteristic polynomial of the $(n-1) \times (n-1)$ submatrix of X formed by deleting the first row and column, and

$$h_j(X) = (X^j)_{11} = (X^j e_1, e_1), \quad j \in \mathbb{Z},$$

is the *jth moment* of X. (Only moments with nonnegative indices are present in (10.17), however, $h_j(X)$ for $j < 0$, that we will need below, are also well-defined, since X is invertible.) In what follows, when it does not lead to a confusion, we occasionally omit the argument and write h_j instead of $h_j(X)$.

To solve the inverse problem, we generalize the approach used previously in the cases of symmetric or Hessenberg X. The main idea stems from the classical moments problem: one considers the space $\mathbb{C}[\lambda, \lambda^{-1}]/\det(\lambda - X)$ equipped with the so-called *moment functional* — a bi-linear functional $\langle\ ,\ \rangle$ on Laurent polynomials in one variable, uniquely defined by the property

$$\langle \lambda^i, \lambda^j \rangle = h_{i+j}. \tag{10.18}$$

X is then realized as a matrix of the operator of multiplication by λ relative to appropriately selected bases $\{p_i^+(\lambda)\}_{i=0}^{n-1}$, $\{p_i^-(\lambda)\}_{i=0}^{n-1}$ bi-orthogonal with respect to the moment functional:

$$\langle p_i^-(\lambda), p_j^+(\lambda) \rangle = \delta_{ij}.$$

For example, the classical tridiagonal case corresponds to the orthogonalization of the sequence $1, \lambda, \dots, \lambda^{n-1}$. Elements of $G^{s_{n-1}\cdots s_1, s_{n-1}\cdots s_1}$ (cf. Remark 10.6(iii)) result from the bi-orthogonalization of sequences $1, \lambda, \dots, \lambda^{n-1}$ and $\lambda^{-1}, \dots, \lambda^{1-n}$, while CMV matrices (Remark 10.6(iv)) correspond to the bi-orthogonalization of sequences $1, \lambda, \lambda^{-1}, \lambda^2, \dots$ and $1, \lambda^{-1}, \lambda, \lambda^{-2}, \dots$.

For any $l \in \mathbb{Z}$, $i \in \mathbb{N}$ define Hankel matrices

$$\mathcal{H}_i^{(l)} = (h_{\alpha+\beta+l-i-1})_{\alpha,\beta=1}^{i},$$

and Hankel determinants

$$\Delta_i^{(l)} = \det \mathcal{H}_i^{(l)}; \tag{10.19}$$

we assume that $\Delta_0^l = 1$ for any $l \in \mathbb{Z}$.

REMARK 10.8. (i) Let X be an $n \times n$ matrix, then it follows from the Cayley-Hamilton theorem that, for $i > n$, the columns of $\mathcal{H}_i^{(l)}$ are linearly dependent and so $\Delta_i^{(l)} = 0$.

(ii) In what follows we will frequently use the identity

$$\Delta_{i+1}^{(l)} \Delta_{i-1}^{(l)} = \Delta_i^{(l-1)} \Delta_i^{(l+1)} - \left(\Delta_i^{(l)}\right)^2, \tag{10.20}$$

which is a particular case of Jacobi's determinantal identity (4.27). In particular, for $i = n$, (10.20) and the first part of the Remark imply $\Delta_n^{(l-1)} \Delta_n^{(l+1)} = \left(\Delta_n^{(l)}\right)^2$ for any l.

The main result of this Section is

THEOREM 10.9. *If $X \in G^{u,v}$ admits factorization* (10.8), *then*

$$\begin{aligned} d_i &= \frac{\Delta_i^{(\varkappa_i+1)} \Delta_{i-1}^{(\varkappa_{i-1})}}{\Delta_i^{(\varkappa_i)} \Delta_{i-1}^{(\varkappa_{i-1}+1)}}, \\ c_i^+ c_i^- &= \frac{\Delta_{i-1}^{(\varkappa_{i-1})} \Delta_{i+1}^{(\varkappa_{i+1})}}{\left(\Delta_i^{(\varkappa_i+1)}\right)^2} \left(\frac{\Delta_{i+1}^{(\varkappa_{i+1}+1)}}{\Delta_{i+1}^{(\varkappa_{i+1})}}\right)^{\varepsilon_{i+1}} \left(\frac{\Delta_{i-1}^{(\varkappa_{i-1}+1)}}{\Delta_{i-1}^{(\varkappa_{i-1})}}\right)^{2-\varepsilon_i} \end{aligned} \tag{10.21}$$

for any $i \in [1, n]$.

REMARK 10.10. Formulae (10.21) allow us to restore an element $X \in G^{u,v}$ from its Weyl function $m(\lambda; X)$ only modulo the diagonal conjugation. Indeed, it is clear from (10.2) that $m(\lambda; X) = m(\lambda; TXT^{-1})$ for any invertible diagonal matrix $T = \operatorname{diag}(t_1, \dots, t_n)$. On the other hand, under the action $X \mapsto TXT^{-1}$, factorization parameters $d_i, c_i^\pm$ in (10.8) are transformed as follows: $d_i \mapsto d_i$, $c_i^\pm \mapsto (t_i/t_{i+1})^{\pm 1} c_i^\pm$, thus leaving the left-hand sides in (10.21) unchanged.

10.3.2. The rest of the section is devoted to the proof of Theorem 10.9. The proof relies on properties of polynomials of the form

$$\mathcal{P}_i^{(l)}(\lambda) = \det \begin{bmatrix} h_{l-i+1} & h_{l-i+2} & \cdots & h_{l+1} \\ \cdots & \cdots & \cdots & \cdots \\ h_l & h_{l+1} & \cdots & h_{l+i} \\ 1 & \lambda & \cdots & \lambda^i \end{bmatrix}. \tag{10.22}$$

To prove the first equality in (10.21) we need two auxiliary lemmas.

LEMMA 10.11. *Let $m \in [1, n-1]$ and X_m be the $m \times m$ submatrix of $X \in G^{u,v}$ obtained by deleting $n - m$ last rows and columns. Then*

$$h_\alpha(X_m) = h_\alpha(X) \tag{10.23}$$

for $\alpha \in [\varkappa_m - m + 1, \varkappa_m + m]$.

PROOF. It is enough to prove the claim for $X \in G^{u,v}$ that admits factorization (10.8). It is clear that X_m does not depend on parameters $c^{\pm}_m, \dots, c^{\pm}_{n-1}$, $d_{m+1}, \dots, d_n$. Moreover, $X_m \in G^{u_m,v_m}$, where u_m and v_m are obtained from u and v, respectively, by deleting all transpositions s_i with $i \geq m$. Consequently, the network N_{u_m,v_m} can be obtained from the network $N_{u,v}$ by deleting all the edges above the horizontal line joining the mth source with the mth sink. Note also that if $\alpha > 0$ then $h_\alpha(X)$ is the sum of path weights over all paths from the first source to the first sink in the network obtained by the concatenation of α copies $N_{u,v}$. Thus, $h_\alpha(X_m) = h_\alpha(X)$ as long as none of the paths involved reaches above the mth horizontal level. The smallest positive power of X such that in the corresponding network there is a path joining the first source to the first sink and reaching above the mth horizontal level is $r = r^+ + r^-$, where $r^{\pm} = \min\{j : i^{\pm}_j \geq m+1\}$. By (10.13), $r^{\pm} = k^{\pm}_m + 1$. Therefore, (10.23) holds for $\alpha \in [0, k^+_m + k^-_m + 1]$. By Lemma 10.5(iii), (10.14) and (10.15), the latter interval coincides with $[0, \varkappa_m + m]$.

Next, consider the network $\bar{N}_{u^{-1},v^{-1}}$ that represents X^{-1} corresponding to factorization (10.9). Note that this network differs from $N_{u^{-1},v^{-1}}$. In particular, in $\bar{N}$ all "north-east" edges are to the left of any "south-east" edge. Once again, the network $\bar{N}_{u_m^{-1},v_m^{-1}}$ is obtained from the network $\bar{N}_{u^{-1},v^{-1}}$ by deleting all the edges above the horizontal line joining the mth sink with the mth source. The smallest positive power of X^{-1} such that in the corresponding network obtained by concatenation of copies of $\bar{N}_{u^{-1},v^{-1}}$ there is a path joining the first source to the first sink and reaching above the mth horizontal level is $\bar{r} = \bar{r}^+ + \bar{r}^- - 1$, where $\bar{r}^+ = \min\{j : l^+_j \geq m+1\}$ and $\bar{r}^- = \min\{j : l^-_j \geq m+1\}$. The difference in the formulas for r and $\bar{r}$ stems from the difference in the structure of the networks N and $\bar{N}$: the latter already contains paths from the first source to the first sink that reach above the first horizontal level. Consequently, it is possible that $\bar{r} = 1$ for some $m > 1$, whereas $r > 1$ for any $m > 1$.

One can define combinatorial parameters $\bar{\varepsilon}^{\pm}_i$ and $\bar{k}^{\pm}_i$ similarly to (10.11) and (10.13) based on the sets $L^{\pm}$ rather than on $I^{\pm}$ (cp. (10.6)). It follows immediately from definitions that $\bar{\varepsilon}^{\pm}_i = 1 - \varepsilon^{\pm}_i$ for $i \in [2, n-1]$ and $\bar{\varepsilon}^{\pm}_1 = \varepsilon^{\pm}_1 = 1$. One can prove, similarly to Lemma 10.5(iii), that $\bar{k}^{\pm}_i = i - \sum_{\beta=1}^{i} \bar{\varepsilon}^{\pm}_\beta$, which translates to $\bar{k}^{\pm}_i = \sum_{\beta=1}^{i} \varepsilon^{\pm}_\beta - 1$. Since $\bar{r}^{\pm} = \bar{k}^{\pm}_m + 1$, we get $\bar{r} = \sum_{\beta=1}^{m} \varepsilon_\beta - 1$, and hence, by (10.15), $\bar{r} = m - \varkappa_m$. If $\bar{r} = 1$, then $\varkappa_m - m + 1 = 0$, and the interval $[0, \varkappa_m + m]$ coincides with $[\varkappa_m - m + 1, \varkappa_m + m]$. Otherwise we can concatenate up to $\bar{r} - 1 = m - 1 - \varkappa_m$ networks $\bar{N}_{u^{-1},v^{-1}}$, and hence (10.23) holds additionally for $\alpha \in [\varkappa_m - m + 1, -1]$. □

LEMMA 10.12. *Let* $m \in [1, n-1]$, *then*

$$\det(\lambda - X_m) = \frac{1}{\Delta_m^{(\varkappa_m)}} \mathcal{P}_m^{(\varkappa_m)}(\lambda). \tag{10.24}$$

In particular,

$$d_1 \cdots d_m = \frac{\Delta_m^{(\varkappa_m + 1)}}{\Delta_m^{(\varkappa_m)}}. \tag{10.25}$$

PROOF. Let $\det(\lambda - X_m) = \lambda^m + \sum_{i=0}^{m-1} a_{mi}\lambda^i$. Then the Hamilton-Cayley theorem implies

$$h_{\alpha+m}(X_m) + \sum_{i=0}^{m-1} a_{mi}h_{\alpha+i}(X_m) = 0$$

for any $\alpha \in \mathbb{Z}$. By Lemma 10.11, this relation remains valid if we replace $h_{\alpha+i}(X_m)$ with $h_{\alpha+i} = h_{\alpha+i}(X)$ for $i = 0, \dots, m$, as long as $\varkappa_m - m+1 \le \alpha \le \varkappa_m$. This means that, after the right multiplication of the matrix used in the definition (10.22) of $\mathcal{P}_m^{(\varkappa_m)}$ by the unipotent matrix $\mathbf{1} + \sum_{\beta=0}^{m-1} a_{m\beta}e_{\beta+1,m+1}$, one gets a matrix of the form

$$\begin{pmatrix} \mathcal{H}_m^{(\varkappa_m)} & 0 \\ 1\ \lambda\ \cdots\ \lambda^{m-1} & \det(\lambda - X_m) \end{pmatrix},$$

and (10.24) follows. Since $\det X_m = d_1 \cdots d_m$, (10.25) drops out immediately from (10.24) and (10.22) after substitution $\lambda = 0$. □

REMARK 10.13. Combining Remark 10.8(ii) with (10.25) for $m = n$ and taking into account that $\det X = d_1 \cdots d_n$, we see that for any l

$$\frac{\Delta_n^{(l+1)}}{\Delta_n^{(l)}} = \det X,$$

which implies that for any l

$$\Delta_n^{(l)} = \Delta_n^{(n-1)} \det X^{l+1-n}. \tag{10.26}$$

Now, the first formula in (10.21) is an easy consequence of (10.25). To be in a position to prove the second formula in (10.21), we first need the following statement. For any $i \in [1, n]$ define subspaces

$$\mathcal{L}_i^+ = \operatorname{span}\{e_1^T, \dots, e_i^T\}, \qquad \mathcal{L}_i^- = \operatorname{span}\{e_1, \dots, e_i\}.$$

Besides, put

$$\gamma_i^\pm = (-1)^{(i-1)\varepsilon_i^\pm} d_i^{-\varepsilon_i^\pm} \prod_{j=1}^{i-1} c_j^\pm d_j^{\bar\varepsilon_j^\pm - \varepsilon_i^\pm}, \quad i \in [2, n], \tag{10.27}$$

where $\bar\varepsilon_j^\pm$ are defined in the proof of Lemma 10.11, and $\gamma_1^\pm = 1$.

LEMMA 10.14. *For any $i \in [1, n]$ one has*

$$\gamma_i^+ e_i^T = e_1^T X^{\zeta_i^+} \mod \mathcal{L}_{i-1}^+ \tag{10.28}$$

and

$$\gamma_i^- e_i = X^{\zeta_i^-} e_1 \mod \mathcal{L}_{i-1}^-. \tag{10.29}$$

In particular,

$$\mathcal{L}_i^+ = \operatorname{span}\{e_1^T X^{\zeta_1^+}, \dots, e_1^T X^{\zeta_i^+}\}, \qquad \mathcal{L}_i^- = \operatorname{span}\{X^{\zeta_1^-} e_1, \dots, X^{\zeta_i^-} e_1\}.$$

PROOF. We prove here only (10.28). The case of (10.29) can be treated similarly.

For any X given by (10.8), consider an upper triangular matrix

$$V = D(\mathbf{1} - C_{k^+}^+)^{-1}(\mathbf{1} - C_{k^+-1}^+)^{-1} \cdots (\mathbf{1} - C_1^+)^{-1}.$$

Note that V is the upper triangular factor in the Gauss factorization of X.

By (10.7), $e_r^T C_j^+ = 0$ for $r < i_{j-1}^+$ and $r \geq i_j^+$, and hence

$$e_r^T(1 - C_j^+)^{-1} = \begin{cases} e_r^T, & r < i_{j-1}^+, \\ e_r^T \mod \mathcal{L}_{r-1}^+, & r \geq i_j^+. \end{cases}$$

Thus, for $j \in [1, k^+]$,

$$\begin{aligned} e_{i_{j-1}^+}^T V &= d_{i_{j-1}^+} e_{i_{j-1}^+}^T (\mathbf{1} - C_{k^+}^+)^{-1} \cdots (\mathbf{1} - C_1^+)^{-1} \\ &= d_{i_{j-1}^+} e_{i_{j-1}^+}^T (\mathbf{1} - C_j^+)^{-1} \mod \mathcal{L}_{i_j^+ - 1}^+ \\ &= d_{i_{j-1}^+} c_{i_{j-1}^+}^+ \cdots c_{i_j^+ - 1}^+ e_{i_j^+}^T \mod \mathcal{L}_{i_j^+ - 1}^+. \end{aligned}$$

A similar argument shows that $e_r^T V \in \mathcal{L}_{i_j^+ - 1}^+$ for $r < i_{j-1}^+$. This implies

$$\begin{aligned} e_1^T V^j &= \left(\prod_{\beta=0}^{j-1} d_{i_\beta^+} c_{i_\beta^+}^+ \cdots c_{i_{\beta+1}^+ - 1}^+ \right) e_{i_j^+}^T \mod \mathcal{L}_{i_j^+ - 1}^+ \\ &= \left(\prod_{r=1}^{i_j^+ - 1} c_r^+ d_r^{\bar{\varepsilon}_r^+} \right) e_{i_j^+}^T \mod \mathcal{L}_{i_j^+ - 1}^+. \end{aligned}$$

Besides, $e_{i_{j-1}^+}^T X V^{-1} = e_{i_{j-1}^+}^T \mod \mathcal{L}_{i_{j-1}^+}^+$, hence the above relation can be re-written as

$$e_1^T X^j = \left(\prod_{r=1}^{i_j^+ - 1} c_r^+ d_r^{\bar{\varepsilon}_r^+} \right) e_{i_j^+}^T \mod \mathcal{L}_{i_j^+ - 1}^+. \tag{10.30}$$

On the other hand, for $l \in [i_{j-1}^+, i_j^+ - 1]$, define $m \geq 0$ so that $l + m + 1$ is the smallest index greater than l that belongs to the index set L^+. Then

$$\begin{aligned} e_l^T V^{-1} &= e_l^T (\mathbf{1} - C_j^+) \cdots (\mathbf{1} - C_{k^+}^+) D^{-1} \mod \mathcal{L}_{l+m}^+ \\ &= \left((-1)^{m+1} c_l^+ \cdots c_{l+m}^+ d_{l+m+1}^{-1} \right) e_{l+m+1}^T \mod \mathcal{L}_{l+m}^+. \end{aligned}$$

The latter equality implies

$$e_1^T V^{-\alpha} = \left((-1)^{l_\alpha^+ - 1} c_1^+ \cdots c_{l_\alpha^+ - 1}^+ d_{l_1^+}^{-1} \cdots d_{l_\alpha^+}^{-1} \right) e_{l_\alpha^+}^T \mod \mathcal{L}_{l_\alpha^+ - 1}^+ \tag{10.31}$$

for any $l_\alpha^+ \in L^+$ distinct from 1 and n. Note now that $l_\alpha = i$ if and only if $\alpha = \sum_{\beta=1}^{i-1} \varepsilon_i^+$ (this can be considered as an analog or (10.16)). Furthermore, $d_{l_1^+}^{-1} \cdots d_{l_\alpha^+}^{-1} = \prod_{j=2}^{i} d_j^{-\varepsilon_j^+}$. Thus, one can re-write (10.31) as

$$e_1^T V^{-\sum_{j=1}^{i-1} \varepsilon_i^+} = (-1)^{i-1} \prod_{j=1}^{i-1} c_j^+ d_{j+1}^{-\varepsilon_{j+1}^+} e_i^T \mod \mathcal{L}_{i-1}^+.$$

Together with $e_l^T V X^{-1} = e_l^T \mod \mathcal{L}_{l+m}^+$ this leads to

$$e_1^T X^{-\sum_{j=1}^{i-1} \varepsilon_i^+} = (-1)^{i-1} \prod_{j=1}^{i-1} c_j^+ d_{j+1}^{-\varepsilon_{j+1}^+} e_i^T \mod \mathcal{L}_{i-1}^+.$$

Combining this relation with (10.30) and Lemma 10.5(ii), one gets (10.28). □

EXAMPLE 10.15. We illustrate (10.28) using Example 10.7 and Fig. 10.1. If $j > 0$ then to find i such that $e_1^T X^j = \gamma_i^+ e_i^T \mod \mathcal{L}_{i-1}^+$ it is enough to find the highest sink that can be reached by a path starting from the source 1 in the network obtained by concatenation of j copies of $N_{u,v}$. Thus, we conclude from Fig. 10.1, that

$$e_1^T X = d_1 c_1^+ c_2^+ e_3^T \mod \mathcal{L}_2^+, \quad e_1^T X^2 = d_1 c_1^+ c_2^+ d_3 c_3^+ e_4^T \mod \mathcal{L}_3^+,$$
$$e_1^T X^3 = d_1 c_1^+ c_2^+ d_3 c_3^+ d_4 c_4^+ e_5^T \mod \mathcal{L}_4^+.$$

Similarly, using the network $\bar{N}_{u^{-1},v^{-1}}$ shown in Fig. 10.2, one observes that $e_1^T X^{-1} = -c_1^+ d_2^{-1} e_2^T \mod \mathcal{L}_1^+$. These relations are in agreement with (10.28).

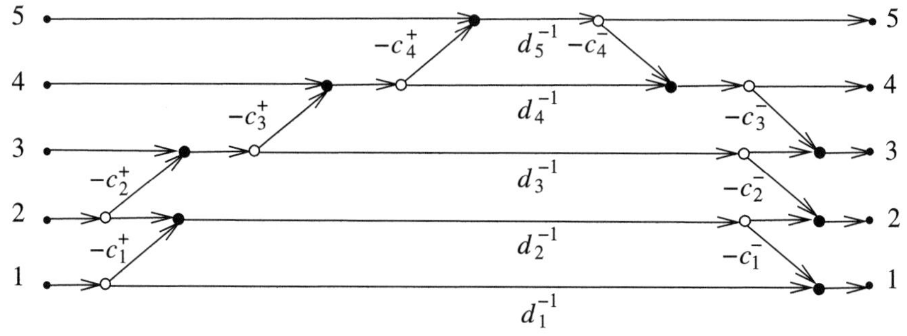

FIGURE 10.2. Network $\bar{N}_{u^{-1},v^{-1}}$ for the double Bruhat cell $G^{u,v}$ from Example 10.7

Define Laurent polynomials

$$p_i^{\pm}(\lambda) = \frac{(-1)^{(i-1)\varepsilon_i^{\pm}}}{\gamma_i^{\pm}\Delta_{i-1}^{(\varkappa_{i-1})}} \lambda^{k_i^{\pm}-i+1} \mathcal{P}_{i-1}^{(\varkappa_{i-1}-\varepsilon_i^{\pm})}(\lambda), \quad i \in [1, n].$$

COROLLARY 10.16. (i) *One has*

$$e_1^T p_i^+(X) = e_i^T, \quad p_i^-(X) e_1 = e_i, \quad i \in [1, n].$$

(ii) *For any eigenvalue λ of X, the column-vector $(p_i^+(\lambda))_{i=1}^n$ and the row-vector $(p_i^-(\lambda))_{i=1}^n$ are, respectively, right and left eigenvectors of X corresponding to λ.*

PROOF. (i) We will only give a proof for $p_i^+(\lambda)$. By Lemma 10.14, $e_1^T X^{\zeta_\alpha^+}$, $\alpha = 1, \ldots, i-1$, form a basis of $\mathcal{L}_{i-1}^+$, hence, taking into account (10.28), we get

$$\gamma_i^+ e_i^T = e_1^T \left(X^{\zeta_i^+} + \sum_{\alpha=1}^{i-1} \pi_\alpha X^{\zeta_\alpha^+} \right)$$

for some coefficients π_α. By Lemma 10.5(iv), this can be re-written as

$$\gamma_i^+ e_i^T = e_1^T X^{k_i^+ - i + 1} \sum_{\alpha=1}^{i} \tilde{\pi}_\alpha X^{\alpha-1},$$

where either $\tilde{\pi}_i = 1$ (if $\varepsilon_i^+ = 0$), or $\tilde{\pi}_1 = 1$ (if $\varepsilon_i^+ = 1$). Define a polynomial $p(\lambda) = \sum_{\alpha=1}^{i} \tilde{\pi}_\alpha \lambda^{\alpha-1}$. By Lemma 10.14, vectors $X^\alpha e_1$, $\alpha \in M_{i-1}^-$, span the subspace $\mathcal{L}_{i-1}^-$.

Therefore, by Lemma 10.5(iv), $\left(X^{k_i^+-i+1}p(X)X^\alpha e_1, e_1\right) = 0$ for $\alpha \in [k_{i-1}^- - i + 2, k_{i-1}^-]$. This system of linear equations determines $p(\lambda)$ uniquely as

$$p(\lambda) = \frac{(-1)^{(i-1)\varepsilon_i^+}}{\Delta_{i-1}^{(k_i^+ + k_{i-1}^- - i + 1 + \varepsilon_i^+)}} \mathcal{P}_{i-1}^{(k_i^+ + k_{i-1}^- - i+1)}(\lambda),$$

which by (10.15) and Lemma 10.5(iii) gives

$$p(\lambda) = \frac{(-1)^{(i-1)\varepsilon_i^+}}{\Delta_{i-1}^{(\varkappa_{i-1})}} \mathcal{P}_{i-1}^{(\varkappa_{i-1} - \varepsilon_i^+)}(\lambda) = \gamma_i^+ \lambda^{-k_i^+ + i - 1} p_i^+(\lambda).$$

It remains to notice that $\gamma_i^+ e_i^T = e_1^T X^{k_i^+ - i + 1} p(X)$, and the result follows.

(ii) Let $z^\lambda = (z_i^\lambda)_{i=1}^n$ be a right eigenvector of X corresponding to an eigenvalue λ. Then $e_1^T p_i^+(X) z^\lambda = e_i^T z^\lambda$, which means that $z_i^\lambda = p_i^+(\lambda) e_1^T z^\lambda = p_i^+(\lambda) z_1^\lambda$. Therefore, $z_1^\lambda \neq 0$, $p_i^+(\lambda) = \frac{z_i^\lambda}{z_1^\lambda}$ and $(p_i^+(\lambda))_{i=1}^n$ is a right eigenvector of X. The case of $(p_i^-(\lambda))_{i=1}^n$ can be treated in the same way.

□

REMARK 10.17. The statement of Corollary 10.16(i) is equivalent to saying that Laurent polynomials $p_i^\pm(\lambda)$ form a bi-orthonormal family with respect to the moment functional (10.18) obtained by the Gram process applied to the sequences $1, \lambda^{\zeta_1^+}, \lambda^{\zeta_2^+}, \dots$ and $1, \lambda^{\zeta_1^-}, \lambda^{\zeta_2^-}, \dots$. Indeed,

$$\langle p_i^+(\lambda), p_j^-(\lambda)\rangle = e_1^T p_i^+(X) p_j^-(X) e_1 = (e_i, e_j) = \delta_{ij}.$$

Finally, we can complete the proof of Theorem 10.9. To prove the second relation in (10.21), observe that by Lemma 10.14 and Corollary 10.16(i),

$$\gamma_i^+ = \left(X^{\zeta_i^+} p_i^-(X) e_1, e_1\right) = \frac{(-1)^{(i-1)\varepsilon_i^-}}{\gamma_i^- \Delta_{i-1}^{(\varkappa_{i-1})}} \left(X^{\zeta_i^+ + k_i^- - i + 1} \mathcal{P}_{i-1}^{(\varkappa_{i-1} - \varepsilon_i^-)}(X) e_1, e_1\right).$$

Since by (10.12), (10.14), (10.15) and Lemma 10.5(iii), $\zeta_i^+ + k_i^- - i + 1 = \varkappa_i - (i-1)\varepsilon_i^+$, the above equality gives

$$\gamma_i^+ = \frac{(-1)^{(i-1)\varepsilon_i^-}}{\gamma_i^- \Delta_{i-1}^{(\varkappa_{i-1})}} (-1)^{(i-1)\varepsilon_i^+} \Delta_i^{(\varkappa_i)},$$

and so

$$\gamma_i^+ \gamma_i^- = (-1)^{(i-1)\varepsilon_i} \frac{\Delta_i^{(\varkappa_i)}}{\Delta_{i-1}^{(\varkappa_{i-1})}}.$$

Consider the ratio

$$\frac{\gamma_{i+1}^+ \gamma_{i+1}^-}{\gamma_i^+ \gamma_i^-} = (-1)^{i\varepsilon_{i+1} - (i-1)\varepsilon_i} \frac{\Delta_{i-1}^{(\varkappa_{i-1})} \Delta_{i+1}^{(\varkappa_{i+1})}}{\left(\Delta_i^{(\varkappa_i)}\right)^2}.$$

Taking into account (10.27), we obtain

$$c_i^+ c_i^- = \frac{\Delta_{i-1}^{(\varkappa_{i-1})} \Delta_{i+1}^{(\varkappa_{i+1})}}{\left(\Delta_i^{(\varkappa_i)}\right)^2} \frac{d_{i+1}^{\varepsilon_{i+1}}}{d_i^{2-\varepsilon_i}} (d_1 d_2 \cdots d_{i-1})^{\varepsilon_i - \varepsilon_{i+1}},$$

which together with the first relation in (10.21) gives the second one.

10.4. Cluster algebra

10.4.1. Let $N_{u,v}$ be the network associated with $X \in GL_n$ and the factorization scheme (1.12). We will now construct a network $N^\circ_{u,v}$ in an annulus as follows:

(i) For each $i \in [1, n]$, add an edge that is directed from the ith sink on the right to the ith source on the left in such a way that moving from the ith source to the ith sink in $N_{u,v}$ and then returning to the ith source in along the new edge, one traverses a closed contour in the counter-clockwise direction. These n new edges do not intersect and to each of them we assign weight 1.

(ii) Place the resulting network in the interior of an annulus in such a way that the cut intersects n new edges, and the inner boundary of the annulus is inside the domain bounded by the top horizontal path in $N_{u,v}$ and the nth new edge.

(iii) Place one source and one sink on the outer boundary of the annulus, the former slightly to the right and the latter slightly to the left of the cut. Split the first (the outermost) new edge into three similarly directed edges by adding two vertices, a black one slightly to the right and a white one slightly to the left of the cut. Add an edge with weight w_{in} directed from the source to the new black vertex and another edge with weight w_{out} directed from the new white vertex to the sink.

It is important to note that the gauge group is rich enough to assure the possibility of assigning weights as described above, with unit weights at prescribed edges.

EXAMPLE 10.18. The network $N^\circ_{u,v}$ that corresponds to $N_{u,v}$ discussed in Example 10.7 is shown in Figure 10.3.

We equip the space $\mathcal{F}_{N_{u,v}}$ (respectively, $\mathcal{F}_{N^\circ_{u,v}}$) with the Poisson bracket from the 2-parametric family described in Proposition 8.19 (respectively, Proposition 9.5) with $\alpha = 1/2$ and $\beta = -1/2$. In what follows we call this Poisson bracket *standard*, since by Theorem 8.10, the corresponding Sklyanin bracket on square matrices is associated with the standard R-matrix.

Now let X be an element in a Coxeter double Bruhat cell $G^{u,v}$, $N_{u,v}$ be the network that corresponds to the factorization (10.8) and $N^\circ_{u,v}$ be the corresponding network in an annulus. Then $N^\circ_{u,v}$ has $2(n-1)$ bounded faces f_{0i}, f_{1i}, $i \in [1, n-1]$, which we enumerate as follows: each face f_{0i} contains a piece of the cut and each face f_{1i} does not, and the value of i is assigned according to the natural bottom to top order inherited from $N_{u,v}$. There are also three unbounded faces: two of them, adjacent to the outer boundary of the annulus, will be denoted f_{00}, f_{10}, where the first index is determined using the same convention as for bounded faces. The third unbounded face is adjacent to the inner boundary. It will be denoted by f_{0n}.

Recall that faces of $N^\circ_{u,v}$ correspond to the vertices of the directed dual network $(N^\circ_{u,v})^*$ (in Section 8.4.1 we have defined the directed dual network for networks in a disk; the definition for networks in an annulus is completely similar). To describe adjacency properties of $(N^\circ_{u,v})^*$, let us first consider inner vertices of $N^\circ_{u,v}$. There are altogether $4n-2$ inner vertices. For every $i \in [1, n-1]$, the ith level contains two black and two white vertices. One of the black vertices is an endpoint of an edge directed from the $(i+1)$th level; it is denoted $v_b^-(i)$. The other one is an endpoint of an edge directed from the $(i-1)$th level (or from the source, for $i = 1$); it is denoted $v_b^+(i)$. Similarly, white vertices are startpoints of the edges directed towards the $(i+1)$th and the $(i-1)$th levels (or towards the sink, for $i = 1$); they

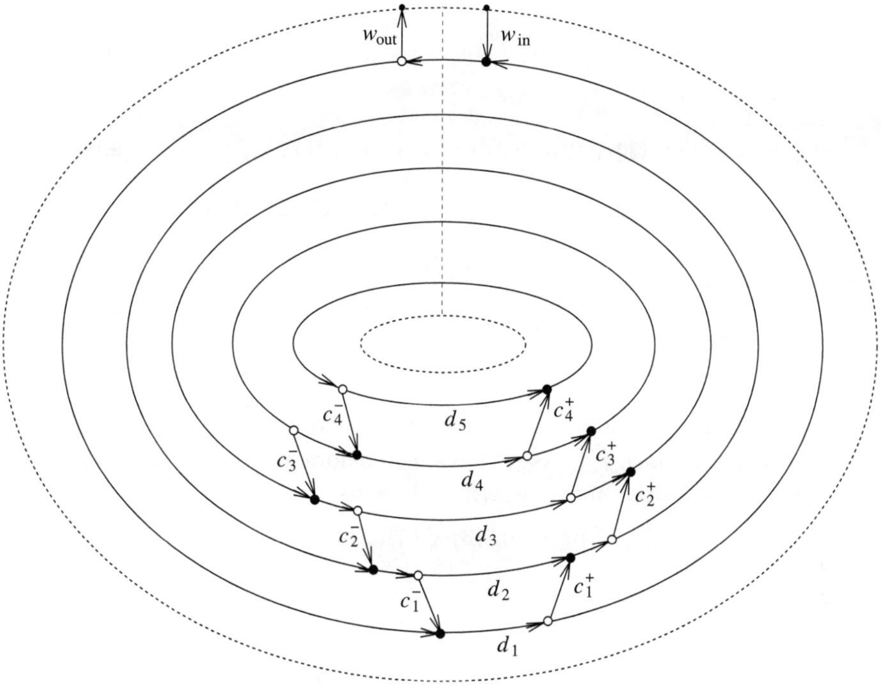

FIGURE 10.3. Network $N_{u,v}^{\circ}$ for $N_{u,v}$ from Example 10.7

are denoted $v_w^+(i)$ and $v_w^-(i)$, respectively. The nth level contains only $v_b^+(n)$ and $v_w^-(n)$.

Now we can describe faces f_{0i}, $i \in [0,n]$, and f_{1i}, $i \in [0, n-1]$, by listing their vertices in the counterclockwise order. Below we use the following convention: if a vertex appears in the description of a face with the exponent 0, this means that this vertex does not belong to the boundary of the face. With this in mind, we obtain

$$f_{1i} = \left(v_b^-(i)v_w^-(i)^{\varepsilon_i^-} v_b^+(i)^{\varepsilon_i^+} v_w^+(i)v_b^+(i+1)v_w^+(i+1)^{\bar\varepsilon_{i+1}^+} v_b^-(i+1)^{\bar\varepsilon_{i+1}^-} v_w^-(i+1)\right),$$
$$f_{0i} = \left(v_w^+(i)v_b^+(i)^{\bar\varepsilon_i^+} v_w^-(i)^{\bar\varepsilon_i^-} v_b^-(i)v_w^-(i+1)v_b^-(i+1)^{\varepsilon_{i+1}^-} v_w^+(i+1)^{\varepsilon_{i+1}^+} v_b^+(i+1)\right)$$

for $i \in [2, n-2]$ and

$$f_{10} = \left(\text{source } v_b^+(1)v_w^+(1)v_b^-(1)v_w^-(1) \text{ sink}\right),$$
$$f_{00} = \left(\text{sink } v_w^-(1))v_b^+(1) \text{ source}\right),$$
$$f_{11} = \left(v_b^-(1)v_w^+(1)v_b^+(2)v_w^+(2)^{\bar\varepsilon_2^+} v_b^-(2)^{\bar\varepsilon_2^-} v_w^-(2)\right),$$
$$f_{01} = \left(v_w^+(1)v_b^+(1)v_w^-(1)v_b^-(1)v_w^-(2)v_b^-(2)^{\varepsilon_2^-} v_w^+(2)^{\varepsilon_2^+} v_b^+(2)\right),$$
$$f_{1n-1} = \left(v_b^-(n-1)v_w^-(n-1)^{\varepsilon_{n-1}^-} v_b^+(n-1)^{\varepsilon_{n-1}^+} v_w^+(n-1)v_b^+(n)v_w^-(n)\right),$$
$$f_{0n-1} = \left(v_w^+(n-1)v_b^+(n-1)^{\bar\varepsilon_{n-1}^+} v_w^-(n-1)^{\bar\varepsilon_{n-1}^-} v_b^-(n-1)v_w^-(n)v_b^+(n)\right),$$
$$f_{0n} = \left(v_b^+(n)v_w^-(n)\right).$$

EXAMPLE 10.19. Vertices and faces of $N_{u,v}^\circ$ from Example 10.18 are shown in Figure 10.4. Consider face f_{13}. As we have seen before in Example 10.7, $\varepsilon_3^+ = \varepsilon_4^+ = \varepsilon_4^- = 0$ and $\varepsilon_3^- = 1$, so the above description yields

$$f_{13} = \left(v_b^-(3)v_w^-(3)v_w^+(3)v_b^+(4)v_w^+(4)v_b^-(4)v_w^-(4)\right).$$

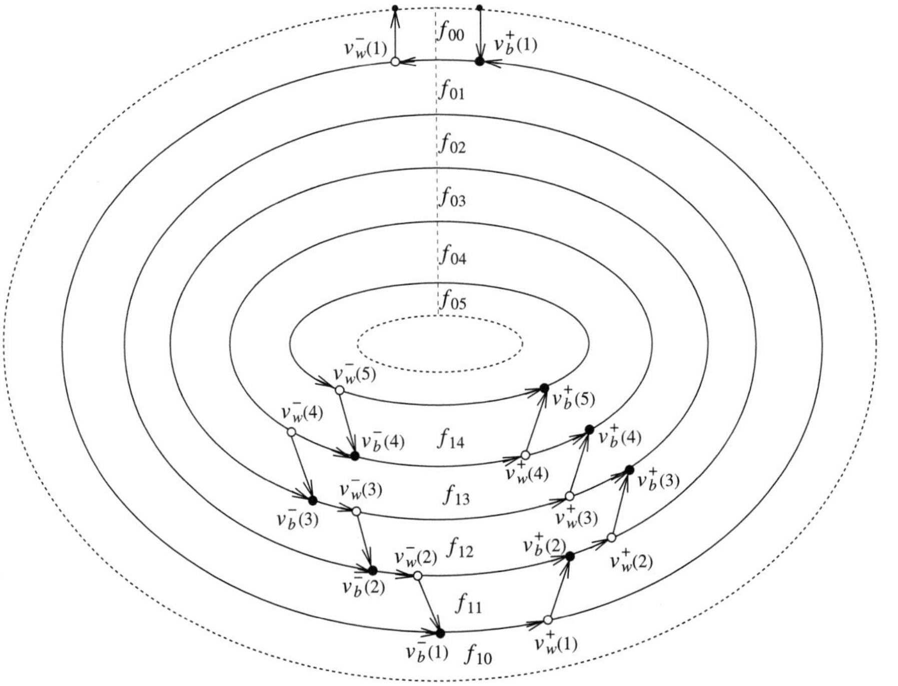

FIGURE 10.4. Vertices and faces of $N_{u,v}^\circ$ for Example 10.18

In our description of $(N_{u,v}^\circ)^*$ below we use the following convention: whenever we say that there are $\alpha < 0$ edges directed from vertex f to vertex f', it means that there are $|\alpha|$ edges directed from f' to f. It is easy to see that for any $i \in [0, n-1]$, faces f_{0i} and f_{1i} have two common edges: $v_w^-(i+1) \to v_b^-(i)$ and $v_w^+(i) \to v_b^+(i+1)$.

The startpoints of each one of the edges are white, the endpoints are black, and in both cases face f_{1i} lies to the right of the edge. This means that in $(N^\circ_{u,v})^*$ there are two edges directed from f_{0i} to f_{1i}. Similarly, the description above shows that $(N^\circ_{u,v})^*$ has

(i) $1-\varepsilon_{i+1}$ edges directed from f_{1i+1} to f_{1i},
(ii) $2-\varepsilon_{i+1}$ edges directed from f_{1i} to f_{0i+1},
(iii) $1-\varepsilon_{i+1}$ edges directed from f_{0i+1} to f_{0i},
(iv) ε_{i+1} edges directed from f_{1i+1} to f_{0i}

for $i \in [1, n-2]$, one edge directed from f_{1n-1} to f_{0n} and one edge directed from f_{0n} to f_{0n-1}. Finally, for $|i-j|>1$ and $a, b \in \{0,1\}$, vertices f_{ai}, f_{bj} in $(N^\circ_{u,v})^*$ are not connected by edges.

Next, we associate with every face f_{st} in $N^\circ_{u,v}$ a face weight y_{st}. We will see below that y_{st} are related to parameters $c_i^\pm$, d_i via a monomial transformation. At this point, however, let us examine the standard Poisson bracket on $\mathcal{F}_{N^\circ_{u,v}}$. Since $N^\circ_{u,v}$ has one source and one sink, both lying on the same connected component of the boundary, and the other connected component of the boundary does not carry boundary vertices, we have $\mathcal{F}_{N^\circ_{u,v}} = \mathcal{F}^f_{N^\circ_{u,v}}$ in the notation of Section 9.1.3. Therefore, the standard Poisson bracket on $\mathcal{F}_{N^\circ_{u,v}}$ is completely described by Proposition 8.19. Together with the above description of $(N^\circ_{u,v})^*$ this implies the following Poisson relations for face weights:

$$
\begin{aligned}
&\{y_{0i}, y_{1i}\} = 2y_{0i}y_{1i}, \quad \{y_{1i}, y_{1i+1}\} = -(1-\varepsilon_{i+1})y_{1i}y_{1i+1},\\
&\{y_{1i}, y_{0i+1}\} = (2-\varepsilon_{i+1})y_{1i}y_{0i+1}, \quad \{y_{0i}, y_{0i+1}\} = -(1-\varepsilon_{i+1})y_{0i}y_{0i+1},\\
&\{y_{0i}, y_{1i+1}\} = -\varepsilon_{i+1}y_{0i}y_{1i+1}
\end{aligned}
\tag{10.32}
$$

for $i \in [2, n-2]$ and

$$
\begin{aligned}
&\{y_{00}, y_{10}\} = y_{00}y_{10}, \quad \{y_{10}, y_{01}\} = 2y_{10}y_{01},\\
&\{y_{10}, y_{11}\} = -y_{10}y_{11}, \quad \{y_{00}, y_{01}\} = -y_{00}y_{01},\\
&\{y_{0n}, y_{0n-1}\} = y_{0n}y_{0n-1}, \quad \{y_{0n}, y_{1n-1}\} = -y_{0n}y_{1n-1},
\end{aligned}
\tag{10.33}
$$

and the rest of the brackets are zero.

Denote by $M(\lambda)$ the *boundary measurement* for the network $N^\circ_{u,v}$, and put $H_0 = w_{in}w_{out}$. We have the following

PROPOSITION 10.20. *Let $X \in G^{u,v}$ be given by* (10.8), *then*

$$M(\lambda) = H_0\left((\lambda\mathbf{1}+X)^{-1}e_1, e_1\right) = -H_0 m(-\lambda; X).$$

PROOF. Clearly, $M(\lambda)$ is a power series in λ^{-1} with the coefficient of λ^{-k} equal to $(-1)^{k-1}$ times the sum of weights of all paths from the source to the sink that cross the cut exactly k times. Moreover, the leading term of $M(\lambda)$ is $H_0\lambda^{-1}$. Denote $M(\lambda) = \sum_{k=0}^\infty (-1)^k H_k \lambda^{-k-1}$. Since the weight of every path has a factor H_0, computing H_k/H_0 is equivalent to computing the boundary measurement between the first source and the first sink in the planar network obtained by concatenation of k copies of $N_{u,v}$. Therefore, $H_k/H_0 = (X^k e_1, e_1) = h_k$, and

$$M(\lambda) = H_0 \sum_{k=0}^{\infty} (-1)^k h_k \lambda^{-k-1} = H_0 \left(\lambda^{-1} \sum_{k=0}^{\infty} (-\lambda X)^k e_1, e_1 \right)$$
$$= H_0 \left((\lambda \mathbf{1} + X)^{-1} e_1, e_1 \right).$$

□

Let $\mathcal{R}_n$ denote the space of rational functions of the form Q/P, where P is a monic polynomial of degree n, Q is a polynomial of degree at most $n-1$, P and Q are co-prime and $P(0) \neq 0$.

PROPOSITION 10.21. *The space of boundary measurements associated with the network $N^\circ_{u,v}$ is dense in $\mathcal{R}_n$.*

PROOF. Let us first prove that any boundary measurement indeed belongs to $\mathcal{R}_n$. By Proposition 10.20, the roots of P are exactly the eigenvalues of $-X$, hence the degree of P equals to n. Next, since $M(\lambda) = \sum_{k=0}^{\infty} (-1)^k H_k \lambda^{-k-1}$, the value of $M(\lambda)$ at infinity equals zero, and hence the degree of Q is at most $n-1$.

By Proposition 10.20 and (10.17), the coprimality statement is equivalent to saying that X and its submatrix obtained by deleting the first row and column have no common eigenvalues. Suppose this is not true, and $\tilde\lambda$ is a common eigenvalue. Denote $\tilde X = X - \tilde\lambda \mathbf{1}$. Then $\det \tilde X = \tilde X^{[2,n]}_{[2,n]} = 0$, and, by the Jacobi determinantal identity,

$$\tilde X^{[2,n]}_{[1,n-1]} \tilde X^{[1,n-1]}_{[2,n]} = \tilde X^{[2,n-1]}_{[2,n-1]} \det \tilde X + \tilde X^{[2,n]}_{[2,n]} \tilde X^{[1,n-1]}_{[1,n-1]} = 0.$$

Thus either $\tilde X^{[1,n-1]}_{[2,n]}$ or $\tilde X^{[2,n]}_{[1,n-1]}$ is zero. Assume the latter is true (the other case can be treated similarly). Consider the classical adjoint $\widehat X$ of $\tilde X$. Since $\tilde X$ is degenerate, $\widehat X$ has rank one. Since $\widehat X_{11} = \widehat X_{1n} = 0$, either the first row or the first column of $\widehat X$ has all zero entries. Assume the latter is true. This means that every $(n-1) \times (n-1)$ minor based on the last $n-1$ rows of $\tilde X$ equals zero, and so the $n \times (n-1)$ submatrix of $\tilde X$ obtained by deleting the first column does not have the full rank. Then there is a non-zero vector w with $w_1 = 0$ such that $\tilde X w = 0$. Therefore, w is linearly independent with the eigenvector $(p_i^+(\lambda))_{i=1}^n$ of X constructed in Corollary 10.16(ii), whose first component is equal to 1. We conclude that the dimension of the eigenspace of X corresponding to $\tilde\lambda$ is greater than one. However, due to Lemma 10.14 and invertibility of X, e_1 is a cyclic vector for X, which implies that all eigenspaces of X are one-dimensional. This completes the proof of coprimality by contradiction. The case when the first row of $\widehat X$ is zero can be treated similarly.

To prove that $P(0) \neq 0$, we denote $P(\lambda) = \lambda^n + p_{n-1}\lambda^{n-1} + \cdots + p_0$. Then relation $Q(\lambda) = M(\lambda)P(\lambda)$ yields

$$\sum_{i=0}^{n} (-1)^i p_i H_{k+i} = 0, \quad k \geq 0, \tag{10.34}$$

with $p_n = 1$. Relations (10.34) for $k \in [0, n-1]$ provide a system of linear equations for $p_0, \ldots, p_{n-1}$. The determinant of this system equals $H_0^n \Delta_n^{(n-1)}$. It is well-known that the coprimality of P and Q is equivalent to the non-vanishing of $\Delta_n^{(n-1)}$. So,

$p_0 = P(0)$ can be restored uniquely as $(-1)^n\Delta_n^{(n)}/\Delta_n^{(n-1)}$, which by Remark 10.13 is equal to $(-1)^n \det X$. It remains to recall that $\det X = d_1 \cdots d_n \neq 0$.

The density statement follows easily from Theorem 10.9: given $M(\lambda)$, one builds Hankel determinants (10.19) and makes use of formulas (10.21) to restore X, provided H_0 and all determinants in the denominator do not vanish. □

REMARK 10.22. (i) Since $p_0 \neq 0$, equations (10.34) extended to $k = -1, -2, \dots$ can be used as a recursive definition of $H_{-1}, H_{-2}, \dots$.

(ii) Since $m(-\lambda; X) = \frac{q(\lambda)}{p(\lambda)}$, where $p(\lambda) = \sum_{i=0}^n (-1)^i p_i \lambda^i$ is the characteristic polynomial of $-X$, the Cayley-Hamilton theorem implies that for any $k \in \mathbb{Z}$,

$$\sum_{i=0}^{n}(-1)^i p_i h_{k+i} = (\sum_{i=0}^{n}(-1)^i p_i X^{k+i}e_1, e_1) = (X^k p(-X)e_1, e_1) = 0.$$

Therefore,

$$H_k = h_k H_0 \tag{10.35}$$

for any $k \in \mathbb{Z}$.

(iii) Denote $Q(\lambda) = q_{n-1}\lambda^n + \cdots + q_0$. Similarly to (10.34) one gets

$$(-1)^{j+1}q_j = \sum_{i=j+1}^{n}(-1)^i p_i H_{i-j-1}, \quad j \in [0, n-1]. \tag{10.36}$$

The following proposition is a particular case of Proposition 9.6.

PROPOSITION 10.23. *The standard Poisson bracket on $\mathcal{F}_{N^\circ_{u,v}}$ induces a Poisson bracket on $\mathcal{R}_n$. This bracket is given by*

$$\{M(\lambda), M(\mu)\} = -(\lambda\, M(\lambda) - \mu\, M(\mu))\, \frac{M(\lambda) - M(\mu)}{\lambda - \mu}. \tag{10.37}$$

REMARK 10.24. Using Proposition 10.20, one can deduce from (10.37) the Poisson brackets for the Weyl function $m(\lambda) = m(\lambda; X)$:

$$\{m(\lambda), m(\mu)\} = -(\lambda m(\lambda) - \mu m(\mu)) \left(\frac{m(\lambda) - m(\mu)}{\lambda - \mu} + m(\lambda)m(\mu)\right). \tag{10.38}$$

Thus a combination of Theorem 2.1 and Propositions 10.20 and 10.23 provides a network-based proof of the fact that the standard Poisson–Lie structure on GL_n induces the Poisson bracket (10.38) on Weyl functions. This fact plays a useful role in the study of a multi-Hamiltonian structure of Toda flows.

10.4.2. To compute face weights in terms of factorization parameters $c_i^\pm, d_i$, we introduce new notation that makes formulas (10.21) more convenient. First of all, for any $l \in \mathbb{Z}$, $i \in \mathbb{N}$ define, similarly to (10.19), Hankel determinants

$$\mathit{\Delta}_i^{(l)} = \det(H_{\alpha+\beta+l-i-1})_{\alpha,\beta=1}^i; \tag{10.39}$$

we assume that $\mathit{\Delta}_0^l = 1$ for any $l \in \mathbb{Z}$. It follows from (10.35) that $\mathit{\Delta}_i^{(l)} = H_0^i \Delta_i^{(l)}$.

Let us fix a pair of Coxeter elements (u, v) and denote $\varepsilon = (\varepsilon_i)_{i=1}^n$ and

$$x_{0i} = x_{0i}(\varepsilon) = \mathit{\Delta}_i^{(\varkappa_i)}, \quad x_{1i} = x_{1i}(\varepsilon) = \mathit{\Delta}_i^{(\varkappa_i+1)}, \quad c_i = c_i^+ c_i^-. \tag{10.40}$$

Then formulae (10.21) become

$$d_i = \frac{x_{1i}x_{0i-1}}{x_{0i}x_{1i-1}}, \quad c_i = \frac{x_{0i-1}x_{0i+1}}{x_{1i}^2}\left(\frac{x_{1i+1}}{x_{0i+1}}\right)^{\varepsilon_{i+1}}\left(\frac{x_{1i-1}}{x_{0i-1}}\right)^{2-\varepsilon_i}. \tag{10.41}$$

We now compute the face weights for $N^\circ_{u,v}$:

$$(10.42)\qquad \begin{aligned} y_{0i} &= c_i^{-1} = x_{0i-1}^{1-\varepsilon_i} x_{0i+1}^{\varepsilon_{i+1}-1} x_{1i}^2 x_{1i-1}^{\varepsilon_i-2} x_{1i+1}^{-\varepsilon_{i+1}},\\ y_{1i} &= \frac{c_i d_i}{d_{i+1}} = x_{1i-1}^{1-\varepsilon_i} x_{1i+1}^{\varepsilon_{i+1}-1} x_{0i}^{-2} x_{0i-1}^{\varepsilon_i} x_{0i+1}^{2-\varepsilon_{i+1}}, \end{aligned}$$

for $i \in [1, n-1]$ and
(10.43)

$$y_{00} = \frac{1}{H_0} = x_{01}^{-1}, \quad y_{10} = \frac{H_0^2}{H_1} = x_{01}^2 x_{11}^{-1}, \quad y_{0n} = d_n = x_{1n} x_{0n-1} x_{0n}^{-1} x_{1n-1}^{-1}.$$

We will re-write (10.42) for $i = n-1$ in a slightly different way. Recall that $\varepsilon_n = 0$. Due to (10.26),

$$(10.44)\qquad \begin{aligned} x_{0n}^2 x_{1n}^{-1} &= \Delta_n^{(\varkappa_n-1)} = \Delta_n^{(n-1)} (\det X)^{\varkappa_n - n},\\ x_{0n} &= \Delta_n^{(\varkappa_n)} = \Delta_n^{(n-1)} (\det X)^{\varkappa_n - n+1}. \end{aligned}$$

Thus, (10.42) yields

$$\begin{aligned} y_{0n-1} &= x_{0n-2}^{1-\varepsilon_{n-1}} x_{1n-1}^2 x_{1n-2}^{\varepsilon_{n-1}-2} \left(\Delta_n^{(n-1)}\right)^{-1} \left(\frac{1}{\det X}\right)^{\varkappa_n - n+1},\\ y_{1n-1} &= x_{1n-2}^{1-\varepsilon_{n-1}} x_{0n-1}^{-2} x_{0n-2}^{\varepsilon_{n-1}} \Delta_n^{(n-1)} \left(\frac{1}{\det X}\right)^{n-\varkappa_n}. \end{aligned}$$

Define
(10.45)

$$\mathbf{x} = \mathbf{x}(\varepsilon) = (x_i)_{i=1}^{2n} = \left(x_{01}, x_{11}, \dots, x_{0n-1}, x_{1n-1}, \Delta_n^{(n-1)}, \frac{\Delta_n^{(n-2)}}{\Delta_n^{(n-1)}} = \frac{1}{\det X} \right)$$

and $\mathbf{y} = \mathbf{y}(\varepsilon) = (y_i)_{i=1}^{2n} = (y_{00}, y_{10}, \dots, y_{0n-1}, y_{1n-1})$. Then $y_i = \prod_{i=1}^{2n} x_j^{a_{ij}}$, where $A = (a_{ij})_{i=1}^{2n}$ is an $n \times n$ block lower–triangular matrix with 2×2 blocks:

$$A = \begin{pmatrix} V_1 & 0 & 0 & 0 & 0\\ U & V_2 & 0 & 0 & 0\\ -V_2^T & U & V_3 & 0 & 0\\ 0 & \ddots & \ddots & \ddots & 0\\ 0 & 0 & -V_{n-1}^T & U & V_n \end{pmatrix}$$

with

$$(10.46)\qquad \begin{aligned} U &= \begin{pmatrix} 0 & 2\\ -2 & 0 \end{pmatrix}, \quad V_1 = \begin{pmatrix} -1 & 0\\ 2 & -1 \end{pmatrix}, \quad V_n = \begin{pmatrix} -1 & \varkappa_n - n + 1\\ 1 & n - \varkappa_n \end{pmatrix},\\ V_i &= \begin{pmatrix} \varepsilon_i - 1 & -\varepsilon_i\\ 2 - \varepsilon_i & \varepsilon_i - 1 \end{pmatrix}, \quad i \in [2, n-1]. \end{aligned}$$

The matrix A is invertible, since $\det V_i = 1$, $i \in [1, n-1]$ and $\det V_n = -1$.

REMARK 10.25. Note that the expression for x_{2n} in terms of face weights is independent of ε:

$$x_{2n} = (y_{00} y_{10})^n (y_{01} y_{11})^{n-1} \cdots (y_{0n-1} y_{1n-1}).$$

10.4.3. Given a pair of Coxeter elements (u, v), we want to define a cluster algebra with the compatible Poisson bracket given by (10.37). To this end, we use the strategy developed in Chapter 4. The first step consists in finding a coordinate system on $\mathcal{R}_n$ such that written in terms of their logarithms, the Poisson bracket (10.37) becomes constant. Having in mind Proposition 3.37, we require this coordinate system to be given by a collection of regular functions on $\mathcal{R}_n$. Clearly, H_i, $i \geq 0$, are regular on $\mathcal{R}_n$, and hence so are $\Delta_n^{(n-1)}$ and $\Delta_n^{(n)}$. Besides, it was explained in the proof of Lemma 10.21 that $\Delta_n^{(n-1)}$ and $\Delta_n^{(n)}$ do not vanish on $\mathcal{R}_n$, hence $(\Delta_n^{(n-1)})^{-1}$ and $(\Delta_n^{(n)})^{-1}$ are regular as well. Consequently, by Remark 10.22(i), H_i are regular functions on $\mathcal{R}_n$ for $i < 0$, and hence so are Hankel determinants (10.39) for any $l \in \mathbb{Z}$, $i \in \mathbb{N}$. Therefore, components of $\mathbf{x}(\varepsilon)$ are regular functions on $\mathcal{R}_n$ and they are connected by an invertible monomial transformation to face weights $\mathbf{y}(\varepsilon)$ that satisfy Poisson relations (10.32), (10.33) of the required kind. Therefore we can use $\mathbf{x}(\varepsilon)$ as an initial extended cluster (this notation deviates slightly from that used in the previous chapters, where $\mathbf{x}$ was reserved for the cluster, and the corresponding extended cluster was denoted $\widetilde{\mathbf{x}}$; however, this should not cause any confusion). Now, following Theorem 4.5, we have to compute the matrix that defines cluster transformations, based on the coefficient matrix of the bracket (10.37).

Define a $2n \times 2n$ matrix

$$B(\varepsilon) = -\begin{pmatrix} U & V_2 & 0 & 0 \\ -V_2^T & U & V_3 & 0 \\ 0 & \ddots & \ddots & \ddots \\ 0 & 0 & -V_n^T & -\frac{1}{2}U \end{pmatrix} \tag{10.47}$$

with 2×2 block coefficients given by (10.46). Denote by $\tilde{B}(\varepsilon)$ the $(2n-2) \times 2n$ submatrix of $B(\varepsilon)$ formed by the first $2n-2$ rows and consider the cluster algebra $\mathcal{A}_\varepsilon$ of rank $2n-2$ with the initial seed $\Sigma(\varepsilon) = (\mathbf{x}(\varepsilon), \tilde{B}(\varepsilon))$, so that x_i, $i \in [1, 2n-2]$, are cluster variables and x_{2n-1}, x_{2n} are stable variables.

LEMMA 10.26. *Poisson structure* (10.37) *is compatible with the cluster algebra* $\mathcal{A}_\varepsilon$.

PROOF. Let us first revisit standard Poisson structure on $\mathcal{F}_{N_{u,v}^\circ}$ described by (10.32), (10.33). It is easy to see that in terms of the components of the vector $\mathbf{y} = \mathbf{y}(\varepsilon)$, this bracket can be written as $\{y_i, y_j\} = \omega_{ij} y_i y_j$, where the matrix $\Omega = \Omega(\varepsilon) = (\omega_{ij})_{i,j=1}^{2n}$ is given by

$$\Omega = \begin{pmatrix} \frac{1}{2}U & V_1 & 0 & 0 \\ -V_1^T & U & V_2 & 0 \\ 0 & \ddots & \ddots & \ddots \\ 0 & 0 & -V_{n-1}^T & U \end{pmatrix}. \tag{10.48}$$

Therefore, the matrix of coefficients of the Poisson bracket (10.37) written in coordinates $\mathbf{x}(\varepsilon)$ is $\Omega^{\mathbf{x}} = A^{-1}\Omega(A^T)^{-1}$.

Note that Ω defined by (10.48) is invertible. To see that, observe that the block-entries of Ω satisfy relations $V_i^T U V_i = U$, $i \in [1, n-1]$, and $U^2 = 4\mathbf{1}_2$, which implies that Ω can be factored as

$$(10.49)\qquad \Omega = \begin{pmatrix} \mathbf{1}_2 & 0 & 0 & 0 \\ \frac12 V_1^T U & \mathbf{1}_2 & 0 & 0 \\ 0 & \ddots & \ddots & \ddots \\ 0 & 0 & \frac12 V_{n-1}^T U & \mathbf{1}_2 \end{pmatrix} \begin{pmatrix} \frac12 U & V_1 & 0 & 0 \\ 0 & \frac12 U & V_2 & 0 \\ 0 & \ddots & \ddots & \ddots \\ 0 & 0 & 0 & \frac12 U \end{pmatrix}.$$

Therefore, $\det \Omega = 1$, and hence $\Omega^\times$ is invertible.

To find its inverse $A^T\Omega^{-1}A$, observe that if we define

$$J = \begin{pmatrix} 0 & \mathbf{1}_2 & 0 & 0 \\ 0 & 0 & \mathbf{1}_2 & 0 \\ 0 & \ddots & \ddots & \ddots \\ 0 & 0 & 0 & 0 \end{pmatrix},$$

then $A = \Omega J^T + V_n \otimes E_{nn}$ with $E_{nn} = (\delta_{in}\delta_{jn})_{i,j=1}^n$. We then have

$$\begin{aligned} A^T\Omega^{-1}A = J\Omega^T J^T - J\left(V_n \otimes E_{nn}\right) + \left(V_n^T \otimes E_{nn}\right) J^T \\ + \left(V_n^T \otimes E_{nn}\right)\Omega^{-1}\left(V_n \otimes E_{nn}\right) = B(\varepsilon), \end{aligned}$$

since by (10.49), the lower-right 2×2 block of Ω^{-1} equals $-\frac12 U$ and $V_n^T U V_n = -U$.

So, $B(\varepsilon)$ is non-degenerate and skew-symmetric. Thus we can invoke Theorem 4.5. According to (4.4), compatibility will follow from the condition $\tilde B(\varepsilon)\Omega^\times = (D\ \ 0)$, where D is a $(2n-2)\times(2n-2)$ diagonal matrix. Since $B^{-1}(\varepsilon) = \Omega^\times$, this condition is obviously satisfied with $D = \mathbf{1}_{2n-2}$. □

Our goal is to prove

THEOREM 10.27. (i) *The cluster algebra $\mathcal{A}_\varepsilon$ does not depend on ε.*

(ii) *The localization of $\mathcal{A}_\varepsilon$ with respect to the stable variables x_{2n-1}, x_{2n} coincides with the ring of regular functions on $\mathcal{R}_n$.*

PROOF. First, we will compute cluster transformations of the initial extended cluster $\mathbf{x}(\varepsilon)$ in directions $(0i)$ and $(1i)$. The transformed variables are denoted $\bar x_{0i}$ and $\bar x_{1i}$, respectively. By (10.47), for $i \in [1, n-2]$ the transformations in question are determined by the matrix

$$\begin{pmatrix} \varepsilon_i - 1 & 2-\varepsilon_i & 0 & -2 & 1-\varepsilon_{i+1} & \varepsilon_{i+1} \\ -\varepsilon_i & \varepsilon_i - 1 & 2 & 0 & \varepsilon_{i+1}-2 & 1-\varepsilon_{i+1} \end{pmatrix}.$$

Therefore, we have to consider the following cases.

Case 1: $\varepsilon_i = \varepsilon_{i+1} = 0$. Then by (10.15), $\varkappa_{i+1} = \varkappa_i + 1$ and $\varkappa_{i-1} = \varkappa_i - 1$, so $x_{0i} = \Delta_i^{(\varkappa_i)}$ is transformed into

$$\bar x_{0i} = \frac{\Delta_{i-1}^{(\varkappa_i-1)}\left(\Delta_i^{(\varkappa_i+1)}\right)^2 + \Delta_{i+1}^{(\varkappa_i+1)}\left(\Delta_{i-1}^{(\varkappa_i)}\right)^2}{\Delta_i^{(\varkappa_i)}}.$$

Using (10.20) and (10.35), we re-write the numerator as

$$\begin{aligned} &\Delta_{i-1}^{(\varkappa_i-1)}\left(\Delta_i^{(\varkappa_i)}\Delta_i^{(\varkappa_i+2)} - \Delta_{i-1}^{(\varkappa_i+1)}\Delta_{i+1}^{(\varkappa_i+1)}\right) + \Delta_{i+1}^{(\varkappa_i+1)}\left(\Delta_{i-1}^{(\varkappa_i)}\right)^2 \\ &= \Delta_i^{(\varkappa_i)}\Delta_{i-1}^{(\varkappa_i-1)}\Delta_i^{(\varkappa_i+2)} + \Delta_{i+1}^{(\varkappa_i+1)}\left(\left(\Delta_{i-1}^{(\varkappa_i)}\right)^2 - \Delta_{i-1}^{(\varkappa_i-1)}\Delta_{i-1}^{(\varkappa_i+1)}\right) \\ &= \Delta_i^{(\varkappa_i)}\left(\Delta_{i-1}^{(\varkappa_i-1)}\Delta_i^{(\varkappa_i+2)} - \Delta_{i-2}^{(\varkappa_i)}\Delta_{i+1}^{(\varkappa_i+1)}\right), \end{aligned}$$

and so

$$\bar{x}_{0i} = \Delta_{i-1}^{(\varkappa_i-1)} \Delta_i^{(\varkappa_i+2)} - \Delta_{i-2}^{(\varkappa_i)} \Delta_{i+1}^{(\varkappa_i+1)}. \tag{10.50}$$

Similarly, $x_{1i} = \Delta_i^{(\varkappa_i+1)}$ is transformed into

$$\bar{x}_{1i} = \frac{\Delta_{i-1}^{(\varkappa_i)} \left(\Delta_{i+1}^{(\varkappa_i+1)}\right)^2 + \Delta_{i+1}^{(\varkappa_i+2)} \left(\Delta_i^{(\varkappa_i)}\right)^2}{\Delta_i^{(\varkappa_i+1)}},$$

which can be re-written as

$$\bar{x}_{1i} = \Delta_i^{(\varkappa_i-1)} \Delta_{i+1}^{(\varkappa_i+2)} - \Delta_{i-1}^{(\varkappa_i)} \Delta_{i+2}^{(\varkappa_i+1)}. \tag{10.51}$$

Case 2: $\varepsilon_i = \varepsilon_{i+1} = 2$. This case is similar to Case 1. We have $\varkappa_{i+1} = \varkappa_i - 1$ and $\varkappa_{i-1} = \varkappa_i + 1$, hence

$$\bar{x}_{0i} = \frac{\Delta_{i-1}^{(\varkappa_i+1)} \left(\Delta_{i+1}^{(\varkappa_i)}\right)^2 + \Delta_{i+1}^{(\varkappa_i-1)} \left(\Delta_i^{(\varkappa_i+1)}\right)^2}{\Delta_i^{(\varkappa_i)}} = \Delta_i^{(\varkappa_i+2)} \Delta_{i+1}^{(\varkappa_i-1)} - \Delta_{i-1}^{(\varkappa_i+1)} \Delta_{i+2}^{(\varkappa_i)} \tag{10.52}$$

and

$$\bar{x}_{1i} = \frac{\Delta_{i+1}^{(\varkappa_i)} \left(\Delta_{i-1}^{(\varkappa_i+1)}\right)^2 + \Delta_{i-1}^{(\varkappa_i+2)} \left(\Delta_i^{(\varkappa_i)}\right)^2}{\Delta_i^{(\varkappa_i+1)}} = \Delta_{i-1}^{(\varkappa_i+2)} \Delta_i^{(\varkappa_i-1)} - \Delta_{i-2}^{(\varkappa_i+1)} \Delta_{i+1}^{(\varkappa_i)}. \tag{10.53}$$

Case 3: $\varepsilon_i = 0$, $\varepsilon_{i+1} = 2$. We have $\varkappa_{i+1} = \varkappa_{i-1} = \varkappa_i - 1$, and so x_{0i} is transformed into

$$\bar{x}_{0i} = \frac{\left(\Delta_{i-1}^{(\varkappa_i)} \Delta_{i+1}^{(\varkappa_i)}\right)^2 + \Delta_{i-1}^{(\varkappa_i-1)} \Delta_{i+1}^{(\varkappa_i-1)} \left(\Delta_i^{(\varkappa_i+1)}\right)^2}{\Delta_i^{(\varkappa_i)}}.$$

The numerator of the above expression can be re-written as

$$\begin{aligned}
&\left(\Delta_{i-1}^{(\varkappa_i)} \Delta_{i+1}^{(\varkappa_i)}\right)^2 + \left(\Delta_i^{(\varkappa_i)} \Delta_i^{(\varkappa_i-2)} - \left(\Delta_i^{(\varkappa_i-1)}\right)^2\right) \left(\Delta_i^{(\varkappa_i+1)}\right)^2 \\
&= \left(\left(\Delta_{i-1}^{(\varkappa_i)} \Delta_{i+1}^{(\varkappa_i)}\right)^2 - \left(\Delta_i^{(\varkappa_i+1)} \Delta_i^{(\varkappa_i-1)}\right)^2\right) + \Delta_i^{(\varkappa_i)} \Delta_i^{(\varkappa_i-2)} \left(\Delta_i^{(\varkappa_i+1)}\right)^2 \\
&= \Delta_i^{(\varkappa_i)} \left(\Delta_i^{(\varkappa_i-2)} \left(\Delta_i^{(\varkappa_i+1)}\right)^2 - \Delta_i^{(\varkappa_i)} \left(\Delta_{i-1}^{(\varkappa_i)} \Delta_{i+1}^{(\varkappa_i)} + \Delta_i^{(\varkappa_i+1)} \Delta_i^{(\varkappa_i-1)}\right)\right),
\end{aligned}$$

and so

$$\bar{x}_{0i} = \Delta_i^{(\varkappa_i-2)} \left(\Delta_i^{(\varkappa_i+1)}\right)^2 - \Delta_i^{(\varkappa_i)} \left(\Delta_{i-1}^{(\varkappa_i)} \Delta_{i+1}^{(\varkappa_i)} + \Delta_i^{(\varkappa_i+1)} \Delta_i^{(\varkappa_i-1)}\right). \tag{10.54}$$

On the other hand,

$$\bar{x}_{1i} = \frac{\Delta_{i-1}^{(\varkappa_i)} \Delta_{i+1}^{(\varkappa_i)} + \left(\Delta_i^{(\varkappa_i)}\right)^2}{\Delta_i^{(\varkappa_i+1)}} = \Delta_i^{(\varkappa_i-1)}. \tag{10.55}$$

Case 4: $\varepsilon_i = 2$, $\varepsilon_{i+1} = 0$. This case is similar to Case 3. We have $\varkappa_{i+1} = \varkappa_{i-1} = \varkappa_i + 1$, hence

$$\bar{x}_{0i} = \frac{\Delta_{i-1}^{(\varkappa_i+1)} \Delta_{i+1}^{(\varkappa_i+1)} + \left(\Delta_i^{(\varkappa_i+1)}\right)^2}{\Delta_i^{(\varkappa_i)}} = \Delta_i^{(\varkappa_i+2)} \tag{10.56}$$

and

$$\bar{x}_{1i} = \frac{\left(\Delta_{i-1}^{(\varkappa_i+1)}\Delta_{i+1}^{(\varkappa_i+1)}\right)^2 + \Delta_{i-1}^{(\varkappa_i+2)}\Delta_{i+1}^{(\varkappa_i+2)}\left(\Delta_i^{(\varkappa_i)}\right)^2}{\Delta_i^{(\varkappa_i+1)}} \tag{10.57}$$
$$= \Delta_i^{(\varkappa_i+3)}\left(\Delta_i^{(\varkappa_i)}\right)^2 - \Delta_i^{(\varkappa_i+1)}\left(\Delta_{i-1}^{(\varkappa_i+1)}\Delta_{i+1}^{(\varkappa_i+1)} + \Delta_i^{(\varkappa_i)}\Delta_i^{(\varkappa_i+2)}\right).$$

Case 5: $\varepsilon_i = 1$, $\varepsilon_{i+1} = 2$. We have $\varkappa_{i+1} = \varkappa_i - 1$, $\varkappa_{i-1} = \varkappa_i$, so x_{0i} is transformed via (10.52) and x_{1i} via (10.55).

Case 6: $\varepsilon_i = 2$, $\varepsilon_{i+1} = 1$. We have $\varkappa_{i+1} = \varkappa_i$, $\varkappa_{i-1} = \varkappa_i + 1$, so x_{0i} is transformed via (10.56) and x_{1i} via (10.53).

Case 7: $\varepsilon_i = 0$, $\varepsilon_{i+1} = 1$. We have $\varkappa_{i+1} = \varkappa_i$, $\varkappa_{i-1} = \varkappa_i - 1$, so x_{0i} is transformed via (10.50) and x_{1i} via (10.55).

Case 8: $\varepsilon_i = 1$, $\varepsilon_{i+1} = 0$. We have $\varkappa_{i+1} = \varkappa_i + 1$, $\varkappa_{i-1} = \varkappa_i$, so x_{0i} is transformed via (10.56) and x_{1i} via (10.51).

Case 9: $\varepsilon_i = \varepsilon_{i+1} = 1$. We have $\varkappa_{i+1} = \varkappa_{i-1} = \varkappa_i$, so x_{0i} is transformed via (10.56) and x_{1i} via (10.55).

Now, let $i = n-1$. In this situation transformations of the initial cluster are determined by the matrix

$$\begin{pmatrix} \varepsilon_{n-1}-1 & 2-\varepsilon_{n-1} & 0 & -2 & 1 & n-\varkappa_n-1 \\ -\varepsilon_{n-1} & \varepsilon_{n-1}-1 & 2 & 0 & -1 & \varkappa_n - n \end{pmatrix}.$$

Note that $\varkappa_n$ does not exceed $n-1$, so the last two elements in the first row are always nonnegative, and the last two elements in the second row are always negative. By (10.44), they contribute to the corresponding relations $\Delta_n^{(\varkappa_n)}$ and $\Delta_n^{(\varkappa_n-1)}$, respectively. Therefore, we have to consider the following cases.

Case 10: $\varepsilon_{n-1} = 0$. Then $\varkappa_{n-1} = \varkappa_n - 1 = \varkappa_{n-2}+1$, and x_{0n-1} is transformed into

$$\bar{x}_{0n-1} = \frac{\left(\Delta_{n-1}^{(\varkappa_n)}\right)^2\Delta_{n-2}^{(\varkappa_n-2)} + \Delta_n^{(\varkappa_n)}\left(\Delta_{n-2}^{(\varkappa_n-1)}\right)^2}{\Delta_{n-1}^{(\varkappa_n-1)}};$$

Similarly to Case 1, this gives (10.50) for $i = n-1$.

On the other hand,

$$\bar{x}_{1n-1} = \frac{\left(\Delta_{n-1}^{(\varkappa_n-1)}\right)^2 + \Delta_n^{(\varkappa_n-1)}\Delta_{n-2}^{(\varkappa_n-1)}}{\Delta_{n-1}^{(\varkappa_n)}},$$

which gives (10.55) for $i = n-1$. We thus see that the transformations in this case are exactly the same as in Case 7.

Case 11: $\varepsilon_{n-1} = 1$. Then $\varkappa_{n-1} = \varkappa_n - 1 = \varkappa_{n-2}$, and hence

$$\bar{x}_{0n-1} = \frac{\left(\Delta_{n-1}^{(\varkappa_n)}\right)^2 + \Delta_n^{(\varkappa_n)}\Delta_{n-2}^{(\varkappa_n)}}{\Delta_{n-1}^{(\varkappa_n-1)}},$$

which gives (10.56) for $i = n-1$.

Similarly,

$$\bar{x}_{1n-1} = \frac{\left(\Delta_{n-1}^{(\varkappa_n-1)}\right)^2 + \Delta_n^{(\varkappa_n-1)}\Delta_{n-2}^{(\varkappa_n-1)}}{\Delta_{n-1}^{(\varkappa_n)}},$$

which gives (10.55) for $i = n-1$. We thus see that the transformations in this case are exactly the same as in Case 9.

Case 12: $\varepsilon_{n-1} = 2$. Then $\varkappa_{n-1} = \varkappa_n - 1 = \varkappa_{n-2} - 1$, and x_{0n-1} transforms exactly as in the previous case.

On the other hand,

$$\bar{x}_{0n-1} = \frac{\left(\Delta_{n-1}^{(\varkappa_n-1)}\right)^2 \Delta_{n-2}^{(\varkappa_n+1)} + \Delta_n^{(\varkappa_n-1)}\left(\Delta_{n-2}^{(\varkappa_n)}\right)^2}{\Delta_{n-1}^{(\varkappa_n)}}.$$

Similarly to Case 2, this gives (10.53) for $i = n-1$. We thus see that the transformations in this case are exactly the same as in Case 6.

Let (u', v') be an arbitrary pair of Coxeter elements, ε' be the corresponding n-tuple built by (10.11) and (10.14).

LEMMA 10.28. *For any Coxeter elements u', v', the seed $\Sigma(\varepsilon') = (\mathbf{x}(\varepsilon'), \tilde{B}(\varepsilon'))$ belongs to $\mathcal{A}_\varepsilon$.*

PROOF. First we will show that, in certain cases, mutations of the seed $\Sigma(\varepsilon)$ transform it into a seed equivalent to $\Sigma(\varepsilon')$ for an appropriately chosen ε'. These situations are listed in the table below. In this table, only the entries at which ε and ε' differ are specified. In the first four rows i is assumed to be less than $n-1$. The second column describes the direction of the seed mutation: under the mutation in direction (s, i), the cluster variable x_{si} is being transformed. It should also be understood that each mutation is followed by the permutation of variables with indices $(0, i)$ and $(1, i)$ in the new cluster, which results in permuting columns and rows $2i-1$ and $2i$ in the matrix obtained via the corresponding matrix mutation. In particular, if $\mathbf{x}(\varepsilon')$ is obtained from $\mathbf{x}(\varepsilon)$ via the cluster transformation in direction (s, i), then $\mathbf{x}(\varepsilon)$ is obtained from $\mathbf{x}(\varepsilon')$ via the cluster transformation in direction $(1-s, i)$.

ε	Direction	ε'
$\varepsilon_i = 0, \varepsilon_{i+1} = 2$	$(1, i)$	$\varepsilon'_i = 1, \varepsilon'_{i+1} = 1$
$\varepsilon_i = 2, \varepsilon_{i+1} = 0$	$(0, i)$	$\varepsilon'_i = 1, \varepsilon'_{i+1} = 1$
$\varepsilon_i = 1, \varepsilon_{i+1} = 0$	$(0, i)$	$\varepsilon'_i = 0, \varepsilon'_{i+1} = 1$
$\varepsilon_i = 2, \varepsilon_{i+1} = 1$	$(0, i)$	$\varepsilon'_i = 1, \varepsilon'_{i+1} = 2$
$\varepsilon_{n-1} = 0$	$(0, n-1)$	$\varepsilon'_{n-1} = 1$
$\varepsilon_{n-1} = 1$	$(1, n-1)$	$\varepsilon'_{n-1} = 2$

We will only provide justification for rows one and five of the table. The remaining cases can be treated similarly. If $\varepsilon_i = 0, \varepsilon_{i+1} = 2$, let ε' be defined by $\varepsilon'_i = 1, \varepsilon'_{i+1} = 1$ and $\varepsilon'_j = \varepsilon_j$ for $j \neq i, i+1$ Then it is easy to check that the matrix mutation in the direction $(1, i)$ followed by the permutation of rows and columns $2i-1$ and $2i$ transforms $B(\varepsilon)$ into $B(\varepsilon')$. Note also that when ε is replaced by ε', the corresponding sequence $\varkappa = (\varkappa_i)_{i=1}^n$ transforms into a sequence $\varkappa'$ that differs from $\varkappa$ only in the component $\varkappa'_i = \varkappa_i - 1$. This means, that $\mathbf{x}' = \mathbf{x}(\varepsilon')$ differs from $\mathbf{x}(\varepsilon)$ only in components $x_{0i}(\varepsilon') = \Delta_i^{(\varkappa_i-1)} = \bar{x}_{1i}(\varepsilon)$ (cf. (10.54) in Case 3) and $x_{1i}(\varepsilon') = \Delta_i^{(\varkappa_i)} = x_{0i}(\varepsilon)$. Thus, we see that the seed mutation in direction $(1, i)$ of the initial seed of $\mathcal{A}_\varepsilon$ transforms it into a seed equivalent to the initial seed of $\mathcal{A}_{\varepsilon'}$.

Now consider the case $\varepsilon_{n-1} = 0$, $\varepsilon'_{n-1} = 1$. Then $\varkappa_{n-1} = \varkappa_n - 1$, $\varkappa'_{n-1} = \varkappa_{n-1} - 1$ and $\varkappa'_n = \varkappa_n - 1$. The fact that the matrix mutation in direction $(1, n-1)$

followed by the permutation of rows and columns $2n-1$ and $2n$ transforms $B(\varepsilon)$ into $B(\varepsilon')$ becomes easy to check once we recall that, by (10.15), $n-1-\varkappa_n$ is always nonnegative. As was shown in Case 11 above, $\bar{x}_{1n-1}(\varepsilon) = \Delta_{n-1}^{(\varkappa_n-2)} = \Delta_{n-1}^{(\varkappa'_{n-1})} = x_{0n-1}(\varepsilon')$. Also $\bar{x}_{0n-1}(\varepsilon) = \Delta_{n-1}^{(\varkappa_{n-1})} = \Delta_{n-1}^{(\varkappa'_{n-1}+1)} = x_{1n-1}(\varepsilon')$, which completes the check.

To complete the proof of the lemma, it suffices to show that, for any ε', the seed $\Sigma(\varepsilon')$ is a seed in $\mathcal{A}_{\varepsilon^{(0)}}$ for $\varepsilon^{(0)} = (2,0,\dots,0)$. This can be done by induction on $\sum_{i=2}^{n-1}\varepsilon'_i$. Indeed, if $\varepsilon'_{n-1} \neq 0$, then $\Sigma(\varepsilon')$ can be obtained via a single mutation from $\Sigma(\varepsilon)$, where ε differs from ε' only in the $(n-1)$st component: $\varepsilon_{n-1} = \varepsilon'_{n-1} - 1$ (see the last two rows of the above table). Otherwise, if $i \in [2, n-2]$ is the largest index such that $\varepsilon'_i \neq 0$, then, using the table again, we see that $\Sigma(\varepsilon')$ can be obtained via a sequence of mutations from $\Sigma(\varepsilon)$, where $\varepsilon = (\varepsilon'_1,\dots,\varepsilon'_{i-1},\varepsilon'_i-1,0,\dots,0)$. The intermediate transformations of the n-tuple ε in this case are $(\varepsilon'_1,\dots,\varepsilon'_{i-1},\varepsilon'_i-1,0,\dots,0,1,0)$, $(\varepsilon'_1,\dots,\varepsilon'_{i-1},\varepsilon'_i-1,0,\dots,1,0,0)$, $\dots$, $(\varepsilon'_1,\dots,\varepsilon'_{i-1},\varepsilon'_i-1,1,0,\dots,0,0)$. □

The first statement of Theorem 10.27 follows immediately. We can now drop the dependence on ε in the cluster algebra $\mathcal{A}_\varepsilon$ and denote it simply by $\mathcal{A}$.

LEMMA 10.29. *For any $k \leq 2n-2$, H_k is a cluster variable in $\mathcal{A}$.*

PROOF. Consider the extended cluster $\mathbf{x}(\varepsilon)$ that corresponds to $\varepsilon = (2,0,\dots,0)$. In this case $x_{2j-1} = \Delta_j^{(j-1)}$, $j \in [1,n]$, and $x_{2j} = \Delta_j^{(j)}$, $j \in [1, n-1]$. The matrix $B(\varepsilon)$ can be conveniently represented by a planar graph Γ, whose vertices are represented by nodes of a $2 \times n$ rectangular grid. Vertices in the top row (listed left to right) correspond to cluster variables $x_1, x_3, \dots, x_{2n-1}$, and vertices in the bottom row correspond to cluster variables $x_2, x_4, \dots, x_{2n}$. We will label the jth vertex in the sth row by (s,j), $s = 0,1$, $j \in [2,n]$ ($s=0$ corresponds to the top row, and $s=1$ to the bottom row). In accordance with (10.47), Γ has edges $(i,j) \to (s,j-1)$ for $s=0,1$ and any $j \in [2, n-1]$, edges $(0,n) \to (0,n-1)$, $(1,n-1) \to (1,n)$, $(1,n-1) \to (0,n)$, $(1,n) \to (0,n)$ and double edges $(0,j) \to (1,j)$ for $j \in [1, n-1]$ and $(1,j) \to (0,j+1)$ for $j \in [1, n-2]$:

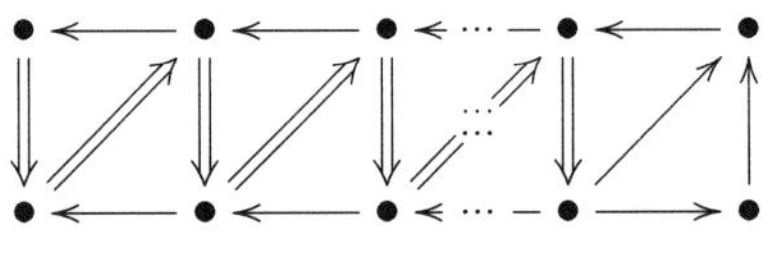

Denote by T_p the cluster transformation in direction p. Let us consider the result of the composition

$$T^{(1)} = T_{2n-4} \circ \cdots \circ T_4 \circ T_2 \circ T_{2n-3} \circ \cdots \circ T_3 \circ T_1.$$

An application of T_1 transforms $x_1 = H_0$ into $\tilde{x}_1 = \frac{1}{H_0}(H_1^2 + \Delta_2^{(1)}) = H_2$ and the graph Γ into

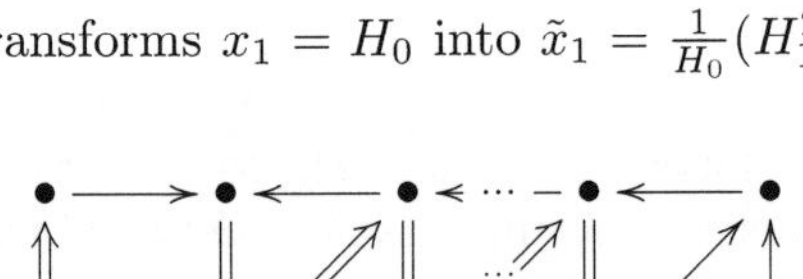

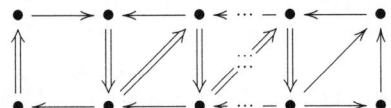

Next, an application of T_3 transforms x_3 into $\tilde{x}_3 = \Delta_2^{(3)}$ (here we use (10.20) with $i = l = 2$) and the graph Γ into

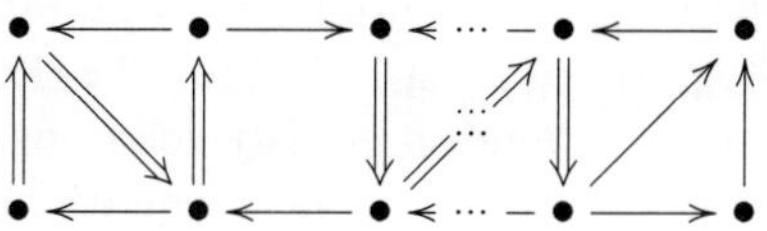

Continuing in the same fashion and using on the jth step relation (10.20) with $i = l = j$, we conclude that an application of $T_{2n-3} \circ \cdots \circ T_3 \circ T_1$ to the initial cluster transforms Γ into

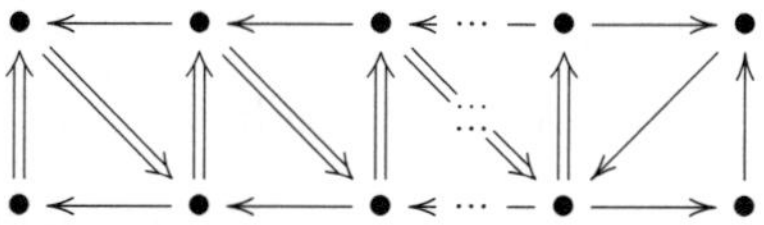

with the variable x_{2i-1} replaced with $\tilde{x}_{2i-1} = \Delta_i^{(i+1)}$ for all $i \in [1, n-1]$. Similarly, the subsequent application of $T_{2n-4} \circ \cdots \circ T_4 \circ T_2$ transforms Γ into

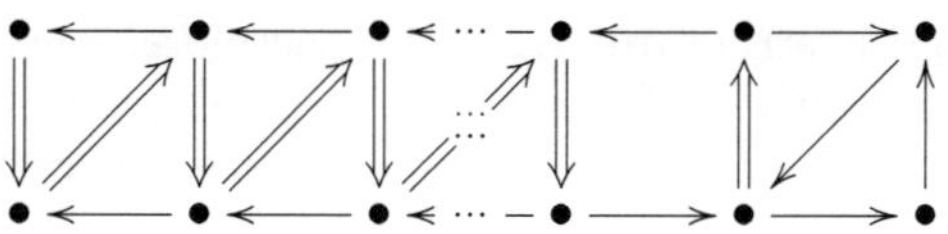

and replaces x_{2i} with $\tilde{x}_{2i} = \Delta_i^{(i+2)}$ for all $i \in [1, n-2]$.

Note that the transformation $x_i \mapsto \tilde{x}_i$ for $i \in [1, 2n-3]$ is described by the shift $H_l \mapsto H_{l+2}$ in the corresponding formulas. In particular, $\tilde{x}_1 = H_2$ and $\tilde{x}_2 = H_3$. Moreover, the subgraph of the resulting graph spanned by vertices $(0,1), \ldots, (0, n-1), (1,1), \ldots, (1, n-2)$ is obtained from the corresponding subgraph in Γ by deleting the double edge between $(1, n-2)$ and $(0, n-1)$. This means that applying the composition

$$T^{(2)} = T_{2n-6} \circ \cdots \circ T_4 \circ T_2 \circ T_{2n-5} \circ \cdots \circ T_3 \circ T_1.$$

to the cluster we just obtained will result in another shift $H_l \to H_{l+2}$ in the formulas defining x_i for $i \in [1, 2n-5]$. In particular, we will recover H_4 and H_5. Similarly, continuing with transformations

$$T^{(s)} = T_{2(n-s)-2} \circ \cdots \circ T_4 \circ T_2 \circ T_{2(n-s)-1} \circ \cdots \circ T_3 \circ T_1, \quad s = 3, \ldots n-1,$$

we will recover H_l for $l \le 2n-2$.

To recover H_l for $l < 0$, we act in an analogous fashion, starting with the cluster corresponding to $\varepsilon = (2, 2, \ldots, 2, 0)$ and repeatedly applying a composition of cluster transformations

$$T_{2n-3} \circ \cdots \circ T_3 \circ T_1 \circ T_{2n-2} \circ \cdots \circ T_4 \circ T_2.$$

□

REMARK 10.30. The strategy of the proof of Lemma 10.29 can be also used to show that $H_k x_{2n}^{k+2-2n}$ is a cluster variable in $\mathcal{A}$ for any $k > 2n-2$.

To prove the second statement of Theorem 10.27, we would like to apply Proposition 3.37 in the situation when $V = \mathcal{R}_n$, $\mathcal{A}$ is the cluster algebra discussed above and $\mathbf{x}$ is given by (10.45). Clearly $\mathcal{R}_n$ is Zariski open in $\mathbb{C}^{2n}$, as the complement to the union of hypersurfaces $\Delta_n^{(n-1)} = 0$ and $p_0 = 0$. Condition (ii) is satisfied by construction and by the invertibility of Ω, and condition (iii) follows from Cases 1–12 discussed above. It remains to check condition (i).

Observe that the ring of regular functions on $\mathcal{R}_n$ is generated by $2(n+1)$ functions $p_0, \dots, p_{n-1}$, $q_0, \dots, q_{n-1}$, p_0^{-1}, $(\Delta_n^{(n-1)})^{-1}$. Clearly, the last two generators belong to $\mathcal{A}^*$. Recall that coefficients p_i satisfy relations (10.34). These relations for $k \in [-2, n-3]$ provide a system of linear equations, and by Lemma 10.29, the coefficients of this system belong to $\mathcal{A}^*$. The determinant of the system is $\Delta_n^{(n-3)} = \Delta_n^{(n-1)}/p_0^2$, so it does not vanish on $\mathcal{R}_n$. Therefore, coefficients $p_0, \dots, p_{n-1}$ belong to $\mathcal{A}^*$. Finally, coefficients $q_0, \dots, q_{n-1}$ belong to $\mathcal{A}^*$ due to relations (10.36) that involve $p_0, \dots, p_{n-1}$ and H_k for $k < n$. □

10.5. Coxeter–Toda lattices

10.5.1. The goal of this section is to establish a connection between the cluster algebra $\mathcal{A}$ defined above and transformations of Coxeter–Toda flows. First, we review the basic facts about the Toda flows on GL_n.

Double Bruhat cells are regular Poisson submanifolds of GL_n equipped with the standard Poisson–Lie structure (1.22). Furthermore,

(i) any symplectic leaf of GL_n is of the form $S^{u,v}a$, where $S^{u,v} \subset G^{u,v}$ is a certain distinguished symplectic leaf and a is an element of the Cartan subgroup, and

(ii) the dimension of symplectic leaves in $G^{u,v}$ equals $l(u) + l(v) + \operatorname{corank}(uv^{-1} - \mathrm{Id})$.

Conjugation-invariant functions on GL_n form a Poisson-commuting family. Any such function F generates a Hamiltonian flow described by *the Lax equation*

$$dX/dt = \left[X \, , \, -\frac{1}{2} R\left(X \nabla F(X)\right) \right]. \tag{10.58}$$

The resulting family of equations is called *the hierarchy of Toda flows* If one chooses $F(X) = F_k(X) = \frac{1}{k}\operatorname{tr}(X^k)$, then equation (10.58) becomes (10.1). Functions $F_1, \dots, F_{n-1}$ form a maximal family of algebraically independent conjugation-invariant functions on GL_n.

For an element $h \in GL_n$, denote by C_h the action of h on GL_n by conjugation: $C_h(X) = hXh^{-1}$. For any smooth function f on GL_n we have

$$\nabla(f \circ C_h) = Ad_{h^{-1}}(\nabla f).$$

Furthermore, if h belongs to $\mathcal{H}$, then it is easy to see that

$$R(Ad_{h^{-1}}(\xi)) = Ad_{h^{-1}}(R(\xi))$$

for any $\xi \in gl_n$. Together, these observations imply that for any $h \in \mathcal{H}$ and any pair of smooth functions f_1, f_2 on GL_n,

$$\{f_1 \circ C_h, f_2 \circ C_h\} = \{f_1, f_2\} \circ C_h.$$

In other words, the action of $\mathcal{H}$ on GL_n by conjugation is Poisson with respect to the standard Poisson–Lie structure. Since the action preserves double Bruhat cells, the standard Poisson–Lie structure induces a Poisson structure on $G^{u,v}/\mathcal{H}$, and the Toda hierarchy induces the family of commuting Hamiltonian flows on $G^{u,v}/\mathcal{H}$.

REMARK 10.31. (i) The Lax equation (10.58) can be solved explicitly via *the factorization method*, which we will not review here.

(ii) Written in terms of matrix entries, equations (10.58) have exactly the same form as equations of the Toda hierarchy on gl_n, where the relevant Poisson structure is the Lie–Poisson structure associated with the R-matrix Lie bracket $[\xi, \eta]_R = \frac{1}{2}([R(\xi), \eta] + [\xi, R(\eta)])$. In fact, viewed as equations on the algebra of $n \times n$ matrices, the Toda hierarchy becomes a family of bi-Hamiltonian flows with compatible linear and quadratic Poisson brackets given by, respectively, Lie–Poisson and the extension of the Poisson–Lie brackets. However, we will not need the linear Poisson structure in the current paper.

Now, consider the Toda hierarchy defined by (10.1). Equations on x induce an evolution of the corresponding Weyl function $m(\lambda; X)$, which can be most conveniently described in terms of its Laurent coefficients h_i. The following proposition is well known in the case of the usual (tridiagonal) Toda flows.

PROPOSITION 10.32. *If $x = X(t)$ satisfies the Lax equation* (10.1), *then coefficients $h_i(X) = (X^i e_1, e_1)$ of the Laurent expansion of the Weyl function $m(\lambda; X)$ evolve according to equations*

$$dh_i(X)/dt = h_{i+k}(X) - h_k(X)h_i(X).$$

PROOF. If x satisfies (10.1), then so does x^i. By rewriting x^k as $\pi_+(X^k) + \pi_-(X^k) + \pi_0(X^k)$, we get

$$\begin{aligned}\frac{dh_i(X)}{dt} &= \left(\left[X^i, -\frac{1}{2}\left(\pi_+(X^k) - \pi_-(X^k)\right)\right] e_1, e_1\right) \\ &= \frac{1}{2}\left(X^i\left(X^k - 2\pi_+(X^k) - \pi_0(X^k)\right) e_1, e_1\right) \\ &\quad - \frac{1}{2}\left(\left(-X^k + 2\pi_-(X^k) + \pi_0(X^k)\right) X^i e_1, e_1\right) \\ &= \left(X^{i+k} e_1, e_1\right) - \left(\pi_0(X^k) e_1, e_1\right)\left(X^i e_1, e_1\right) = h_{i+k}(X) - h_k(X)h_i(X).\end{aligned}$$

□

Now, let (u, v) be a pair of Coxeter elements. Coxeter–Toda flows on $G^{u,v}/\mathcal{H}$ are induced by the restriction of the Toda hierarchy to $G^{u,v}$. To get a more detailed description of Coxeter–Toda flows, we choose parameters $c_i = c_i^+ c_i^-$, d_i that correspond to the factorization (10.8) of a generic element in $G^{u,v}$ as coordinates on the open dense set in $G^{u,v}/\mathcal{H}$. Indeed, c_i, d_i are invariant under conjugation by diagonal matrices (cf. Remark 10.10) and are clearly independent as functions on $G^{u,v}/\mathcal{H}$.

LEMMA 10.33. *The standard Poisson–Lie structure on GL_n induces the following Poisson brackets for variables c_i, d_i:*

(10.59)

$$\{c_i, c_{i+1}\} = (\varepsilon_{i+1} - 1)c_i c_{i+1}, \quad \{d_i, d_j\} = 0, \quad \{c_i, d_i\} = -c_i d_i, \quad \{c_i, d_{i+1}\} = c_i d_{i+1},$$

and the rest of the brackets are zero.

PROOF. In view of Theorem 8.10, it is sufficient to compute Poisson brackets for c_i, d_i induced by Poisson brackets (10.32), (10.33) for face weights of the network $N_{u,v}^\circ$. The first equation is an easy consequence of the equality $y_{0i} = c_i^{-1}$, $i \in [1, n-1]$, (cf. (10.42)) and Poisson relations for y_{0i} described in (10.32), (10.33).

By (10.42), (10.43), $y_{0i} y_{1i} = d_i/d_{i+1}$ for $i \in [0, n-1]$ (here $d_0 = 1$). Therefore,

$$\{\log d_i/d_{i+1}, \log d_j/d_{j+1}\} = \{\log y_{0i} y_{1i}, \log y_{0j} y_{1j}\}, \quad i, j \in [0, n-1],$$

which equals the sum of the entries of the 2×2 block of Ω in rows $2i+1, 2i+2$ and columns $2j+1, 2j+2$. By (10.48), each such block is proportional either to U, or to V_k, or to V_k^T, $k\in[1,n-1]$, given by (10.46). It is easy to see that the sum of the entries for each of these matrices equals zero, and hence $\{d_i/d_{i+1}, d_j/d_{j+1}\} = 0$ for all $i,j\in[0,n-1]$. In particular, this holds for $i=0$, which can be re-written as $\{d_1, d_j/d_{j+1}\}=0$ for all $j\in[0,n-1]$. Taking into account that $d_j = d_1(d_2/d_1)\cdots(d_i/d_{i-1})$, we get the second formula in (10.59).

Similarly,

$$\{\log c_i, \log d_{j+1}/d_j\} = \{\log 1/c_i, \log d_j/d_{j+1}\} = \{\log y_{0i}, \log y_{0j}y_{1j}\},$$

for $i\in[1,n-1]$, $j\in[0,n-1]$, which equals the sum of the two upper entries of the 2×2 block of Ω in rows $2i+1, 2i+2$ and columns $2j+1, 2j+2$. By (10.48), if such a block is nontrivial, it is equal either to U, or to V_k, or to $-V_k^T$, $k\in[1,n-1]$, given by (10.46). Since the sum of the two upper entries equals 2 for U and -1 in the other two cases, we conclude that $\{\log c_i, \log d_{j+1}/d_j\} = 2\delta_{i,j} - \delta_{i,j+1} - \delta_{i,j-1}$ for $i\in[1,n-1]$, $j\in[0,n-1]$. In particular, for $j=0$ one gets $\{\log c_i, \log d_1\} = -\delta_{i1}$ for $i\in[1,n-1]$. Re-writing d_j via d_1 and d_{i+1}/d_i as before, one gets $\{\log c_i, \log d_j\} = -\delta_{i,j}+\delta_{i,j-1}$, $i\in[1,n-1]$, $j\in[1,n]$, which is equivalent to the last two equations in (10.59). □

Due to their invariance under conjugation by elements of $\mathcal{H}$, Hamiltonians $F_k(X) = \frac{1}{k}\operatorname{tr} X^k$ of the Toda flows, when restricted to a Coxeter double Bruhat cell $G^{u,v}$, can be expressed as functions of c_i, d_i, which, in turn, serve as Hamiltonians for Coxeter–Toda flows on $G^{u,v}/\mathcal{H}$. The easiest way to write down F_k as a function of c_i, d_i explicitly is to observe that $\operatorname{tr} X^k$ is equal to the sum of weights of all paths that start and end at the same level in the planar network obtained by concatenation of k copies of $N_{u,v}$. In the case $k=1$, we only need to use $N_{u,v}$ itself, which leads to the following formula for F_1: define I^- and I^+ by (10.6) and denote $I^-\cup I^+ = \{1=i_1,\dots,i_m=n\}$, then

$$F_1 = F_1(c,d) = d_1 + \sum_{l=1}^{k-1}\sum_{j=i_l+1}^{i_{l+1}} \left(d_j + c_{j-1}d_{j-1} + \dots c_{j-1}\cdots c_{i_l}d_{i_l}\right). \tag{10.60}$$

One can use (10.60), (10.59) to write equations of the first Coxeter–Toda flow generated by F_1 on $G^{u,v}/\mathcal{H}$ as a system of evolution equations for c_i, d_i.

EXAMPLE 10.34. (i) For our running Example 10.7, $I^-\cup I^+ = \{1,3,4,5\}$, so (10.60) becomes

$$F_1 = d_1 + d_2 + c_1d_1 + d_3 + c_2d_2 + c_2c_1d_1 + d_4 + c_3d_3 + d_5 + c_4d_4.$$

(ii) Let $v = s_{n-1}\cdots s_1$, then $I^-\cup I^+ = [1,n]$ and formula (10.60) reads $F_1(c,d) = d_1 + d_2 + c_1d_1 + \dots + d_n + c_{n-1}d_{n-1}$. If, in addition, $u=v^{-1}$, then $\varepsilon = (2,0,\dots,0)$ and F_1 and (10.59) generate Hamiltonian equations

$$\begin{aligned} dd_i/dt &= \{d_i, F_1\} = \{d_i, c_id_i + c_{i-1}d_{i-1}\} = d_i(c_id_i - c_{i-1}d_{i-1}),\\ dc_i/dt &= \{c_i, F_1\} = \{c_i, d_i + d_{i+1} + c_{i-1}d_{i-1} + c_id_i + c_{i+1}d_{i+1}\}\\ &= c_i(d_{i+1} - d_i + c_{i-1}d_{i-1} - c_id_i). \end{aligned}$$

Then a change of variables $r_{2i-1} = d_i$, $i\in[1,n]$, and $r_{2i} = c_id_i$, $i\in[1,n-1]$, results in the equations of the *open Volterra lattice*:

$$dr_i/dt = r_i(r_{i+1} - r_{i-1}),\quad i\in[1,2n-1];\ r_0 = r_{2n} = 0.$$

Another change of variables, $a_i = c_i d_i^2$, $b_i = d_i + c_{i-1}d_{i-1}$, leads to equations of motion of the Toda lattice that were presented in the introduction. Note that a_i, b_i are, resp., subdiagonal and diagonal matrix entries in a lower Hessenberg representative of an element in $G^{u,v}/\mathcal{H}$ defined by parameters c_i, d_i.

(iii) If $u = v = s_{n-1}\cdots s_1$, then $\varepsilon = \{2, 1, \ldots, 1, 0\}$, and Hamiltonian equations generated by F_1 and (10.59) produce the system

$$dd_i/dt = d_i(c_i d_i - c_{i-1}d_{i-1}), \quad dc_i/dt = c_i(d_{i+1} - d_i + c_{i+1}d_{i+1} - c_i d_i).$$

After the change of variables $\tilde{c}_i = c_i d_i$ this system turns into the *relativistic Toda lattice*

$$dd_i/dt = d_i(\tilde{c}_i - \tilde{c}_{i-1}), \quad d\tilde{c}_i/dt = \tilde{c}_i(d_{i+1} - d_i + \tilde{c}_{i+1} - \tilde{c}_{i-1}).$$

Proposition 10.32 combined with Theorem 10.9 suggests a method to solve Coxeter–Toda lattices explicitly, following the strategy that was originally applied to the usual Toda lattice. In order to find a solution with initial conditions $c_i(0), d_i(0)$ to the Coxeter–Toda equation on $G^{u,v}/\mathcal{H}$ generated by the Hamiltonian F_k, we first define

$$m^0(\lambda) = m(\lambda; X(0)) = \sum_{i=0}^{\infty} \frac{h_i^0}{\lambda^{i+1}}$$

to be the Weyl function of any representative $x(0) \in G^{u,v}$ of the element in $G^{u,v}/\mathcal{H}$ with coordinates $c_i(0), d_i(0)$. Let $M(\lambda; t) = \sum_{i=0}^{\infty} H_i(t)\lambda^{-i-1}$ be the solution to a linear system on $\mathcal{R}_n$ described in terms of Laurent coefficients $H_i(t)$ by

$$dH_i(t)/dt = H_{i+k}(t), \quad i = 0, 1, \ldots,$$

with initial conditions $H_i(0) = h_i^0$. For $i < 0$, define $H_i(t)$ via (10.34), where $(-1)^{n-i}p_i$ are coefficients of the characteristic polynomial of $x(0)$.

PROPOSITION 10.35. *The solution with initial conditions $c_i(0), d_i(0)$ to the kth Coxeter–Toda equation on $G^{u,v}/\mathcal{H}$ is given by formulas* (10.21) *with $h_i = h_i(t) = H_i(t)/H_0(t)$, $i \in \mathbb{Z}$.*

PROOF. An easy calculation shows that $h_i = h_i(t) = H_i(t)/H_0(t)$, $i \geq 0$, give the solution to the system presented in Proposition 10.32 with initial conditions $h_i(0) = h_i^0$. Thus the function $m(\lambda, t) = \sum_{i=0}^{\infty} h_i(t)\lambda^{-i-1}$ evolves in the way prescribed by the kth Toda flow and therefore coincides with $m(\lambda; X(t))$, where $x(t)$ is the solution of (10.1) with the initial condition $x(0)$. Since coefficients of the characteristic polynomial are preserved by Toda flows, Remark 10.22(ii) implies that for $i < 0$ we also have $h_i(t) = h_i(X(t))$. Finally, since the Moser map is invertible on $G^{u,v}/\mathcal{H}$, we see that the system in Proposition 10.32 is, in fact, equivalent to the kth Toda flow on $G^{u,v}/\mathcal{H}$ which completes the proof. □

We see that for any pair of Coxeter elements (u, v), the Coxeter–Toda flows are equivalent to the same evolution of Weyl functions. We want to exploit this fact to construct, for any two pairs (u, v) and (u', v') of Coxeter elements, a transformation between $G^{u,v}/\mathcal{H}$ and $G^{u',v'}/\mathcal{H}$ that is Poisson and maps the kth Coxeter–Toda flow into the kth Coxeter–Toda flow. We call such a transformation a *generalized Bäcklund–Darboux transformation.* The term "Bäcklund transformation" has been used broadly over the years for any transformation that maps solutions of one nonlinear equation into solutions of another. To justify the use of Darboux's name, we recall that traditionally a Bäcklund–Darboux transformation consists in interchanging factors in some natural factorization of the Lax operator associated with a given integrable system. In the case of Coxeter–Toda flows, the same number and type of elementary factors appears in the Lax matrix associated with any Coxeter double Bruhat cell. Hence we use the term "generalized Bäcklund–Darboux transformation" even though in our case, re-arrangement of factors is accompanied by a transformation of factorization parameters.

Let us fix two pairs, (u, v) and (u', v'), of Coxeter elements and let $\varepsilon = (\varepsilon_i)_{i=1}^n$, $\varepsilon' = (\varepsilon_i')_{i=1}^n$ be the corresponding n-tuples defined by (10.11), (10.14). We construct a map $\sigma_{u,v}^{u',v'} : G^{u,v}/\mathcal{H} \to G^{u',v'}/\mathcal{H}$ using the following procedure. Consider the cluster algebra $\mathcal{A}$ defined in Section 10.4. Fix a seed $\Sigma(\varepsilon) = (\mathbf{x}(\varepsilon), \tilde{B}(\varepsilon))$ in $\mathcal{A}$, where $\mathbf{x}(\varepsilon)$ is given by (10.45) and $B(\varepsilon)$ by (10.47). Let $T_\varepsilon^{\varepsilon'}$ be the sequence of cluster transformations defined in the proof of Lemma 10.28 that transforms $\Sigma(\varepsilon)$ into the seed $\Sigma(\varepsilon')$. Next, for an element in $G^{u,v}/\mathcal{H}$ with coordinates c_i, d_i, consider its representative $x \in G^{u,v}$, the corresponding Weyl function $m(\lambda; X)$ and the sequence of moments $h_i(X)$, $i \in \mathbb{Z}$. Apply transformation $\tau_{u,v}$ by assigning values to cluster variables in the cluster $\mathbf{x}(\varepsilon)$ according to formulas (10.39), (10.40), (10.45) with H_i replaced by $h_i(X)$. Then apply transformation $T_\varepsilon^{\varepsilon'}$ to $\mathbf{x}(\varepsilon)$ to obtain the cluster $\mathbf{x}(\varepsilon')$. Finally, apply transformation $\rho_{u',v'}$ by using equations (10.41) with ε replaced by ε' and components of $\mathbf{x}(\varepsilon)$ replaced by those of $\mathbf{x}(\varepsilon')$ to compute parameters c_i', d_i' that serve as coordinates of an element in $G^{u',v'}/\mathcal{H}$. This concludes the construction of $\sigma_{u,v}^{u',v'}$.

THEOREM 10.36. *The map $\sigma_{u,v}^{u',v'} : G^{u,v}/\mathcal{H} \to G^{u',v'}/\mathcal{H}$ is a birational transformation that preserves the Weyl function, maps Coxeter–Toda flows on $G^{u,v}/\mathcal{H}$ into matching Coxeter–Toda flows on $G^{u',v'}/\mathcal{H}$ and is Poisson with respect to Poisson structures on $G^{u,v}/\mathcal{H}$ and $G^{u',v'}/\mathcal{H}$ induced by the standard Poisson–Lie bracket on GL_n.*

PROOF. Moments $h_j(X)$ are polynomial functions of c_i, d_i for $i \geq 0$ and rational functions of c_i, d_i for $i < 0$. Values we assign to cluster variables in $\mathbf{x}(\varepsilon)$ are thus rational functions of c_i, d_i. This, combined with the rationality of $T_\varepsilon^{\varepsilon'}$ and equations (10.41), shows that the map $\sigma_{u,v}^{u',v'}$ is rational. It is easy to see that its inverse is $\sigma_{u',v'}^{u,v}$ which implies birationality. The claim that $\sigma_{u,v}^{u',v'}$ preserves the Weyl function is simply a re-statement of Lemma 10.28, which implies that if clusters $\mathbf{x}(\varepsilon)$ and $\mathbf{x}(\varepsilon')$ are obtained from a function $M(\lambda) \in \mathcal{R}_n$ according to (10.39), (10.40), (10.45), then $T_\varepsilon^{\varepsilon'}$ transforms $\mathbf{x}(\varepsilon)$ into $\mathbf{x}(\varepsilon')$. The rest of the statement of the theorem is a consequence of the invariance of the Weyl function, since Poisson structures on $G^{u,v}/\mathcal{H}$ and $G^{u',v'}/\mathcal{H}$ induce the same Poisson bracket on $\mathcal{R}_n$ compatible with $\mathcal{A}$ and, by Proposition 10.32, Coxeter–Toda flows generated by Hamiltonians F_k on $G^{u,v}/\mathcal{H}$ and $G^{u',v'}/\mathcal{H}$ induce the same evolution of the Weyl function.

□

To illustrate Theorem 10.36, in the table below we list elementary generalized Bäcklund–Darboux transformations that correspond to cluster transformations from a fixed cluster $\mathbf{x}(\varepsilon)$ into an adjacent cluster $\mathbf{x}(\varepsilon')$. The table can be viewed in parallel with the table in the proof of Lemma 10.28. Expressions for transformed variables c_j', d_j' are obtained by combining formulas for cluster transformations with equations (10.41). Variables that are not listed are left unchanged.

ε	ε'	Transformation	Inverse
$\varepsilon_i = 0$	$\varepsilon'_i = 1$	$c'_i = \frac{c_i d_i}{d_{i+1}}$	$c_i = \frac{c'_i d'_{i+1}}{d'_i (1+c'_i)^2}$
$\varepsilon_{i+1} = 2$	$\varepsilon'_{i+1} = 1$	$d'_i = \frac{d_i d_{i+1}}{d_{i+1}+c_i d_i}$, $d'_{i+1} = d_{i+1} + c_i d_i$	$d_i = d'_i(1+c'_i)$, $d_{i+1} = \frac{d'_{i+1}}{1+c'_i}$
$\varepsilon_i = 2$	$\varepsilon'_i = 1$	$c'_i = \frac{c_i d_{i+1}}{d_i(1+c_i)^2}$, $c'_{i+1} = c_{i+1}(1+c_i)$	$c_i = \frac{c'_i d'_i}{d'_{i+1}}$, $c_{i+1} = \frac{c'_{i+1} d'_{i+1}}{d'_{i+1}+c'_i d'_i}$
$\varepsilon_{i+1} = 0$	$\varepsilon'_{i+1} = 1$	$d'_{i+1} = \frac{d_{i+1}}{1+c_i}$, $d'_i = d_i(1+c_i)$	$d_{i+1} = d'_{i+1} + d'_i c'_i$, $d_i = \frac{d'_i d'_{i+1}}{d'_{i+1}+d'_i c'_i}$
$\varepsilon_i = 1$	$\varepsilon'_i = 0$	$c'_i = \frac{c_i d_{i+1}}{d_i(1+c_i)^2}$, $c'_{i+1} = c_{i+1}(1+c_i)$	$c_i = \frac{c'_i d'_i}{d'_{i+1}}$, $c_{i+1} = \frac{c'_{i+1} d'_{i+1}}{d'_{i+1}+c'_i d'_i}$
$\varepsilon_{i+1} = 0$	$\varepsilon'_{i+1} = 1$	$d'_i = d_i(1+c_i)$, $d'_{i+1} = \frac{d_{i+1}}{1+c_i}$	$d_{i+1} = d'_{i+1} + c'_i d'_i$, $d_i = \frac{d'_i d'_{i+1}}{d'_{i+1}+c'_i d'_i}$
$\varepsilon_i = 2$	$\varepsilon'_i = 1$	$c'_i = \frac{c_i d_{i+1}}{d_i(1+c_i)^2}$, $c'_{i-1} = c_{i-1}(1+c_i)$	$c_i = \frac{c'_i d'_i}{d'_{i+1}}$, $c_{i-1} = \frac{c'_{i-1} d'_{i+1}}{d'_{i+1}+c'_i d'_i}$
$\varepsilon_{i+1} = 1$	$\varepsilon'_{i+1} = 2$	$d'_i = d_i(1+c_i)$, $d'_{i+1} = \frac{d_{i+1}}{1+c_i}$	$d_{i+1} = d'_{i+1} + c'_i d'_i$, $d_i = \frac{d'_i d'_{i+1}}{d'_{i+1}+c'_i d'_i}$
$\varepsilon_{n-1} = 0$	$\varepsilon'_{n-1} = 1$	$c'_{n-1} = \frac{c_{n-1} d_{n-1}}{d_n}$	$c_{n-1} = \frac{c'_{n-1} d'_n}{d'_i (1+c'_{n-1})^2}$
		$d'_{n-1} = \frac{d_n d_{n-1}}{d_n + c_{n-1} d_{n-1}}$, $d'_n = d_n + c_{n-1} d_{n-1}$	$d_{n-1} = d'_{n-1}(1+c'_{n-1})$, $d_n = \frac{d'_n}{1+c'_{n-1}}$
$\varepsilon_{n-1} = 1$	$\varepsilon'_{n-1} = 2$	$c'_{n-2} = \frac{c_{n-2} d_n}{d_n + c_{n-1} d_{n-1}}$, $c'_{n-1} = \frac{c_{n-1} d_{n-1}}{d_n}$	$c_{n-2} = c'_{n-2}(1+c'_{n-1})$, $c_{n-1} = \frac{c'_{n-1} d'_n}{d'_{n-1}(1+c'_{n-1})^2}$
		$d'_{n-1} = \frac{d_n d_{n-1}}{d_n + c_{n-1} d_{n-1}}$, $d'_n = d_n + c_{n-1} d_{n-1}$	$d_{n-1} = d'_{n-1}(1+c'_{n-1})$, $d_n = \frac{d'_n}{1+c'_{n-1}}$

Elementary generalized Bäcklund–Darboux transformations can be conveniently interpreted in terms of equivalent transformations of perfect networks. The three types of equivalent transformations are shown in Figure 10.5. Instead of attempting to describe the general case, we will provide an example.

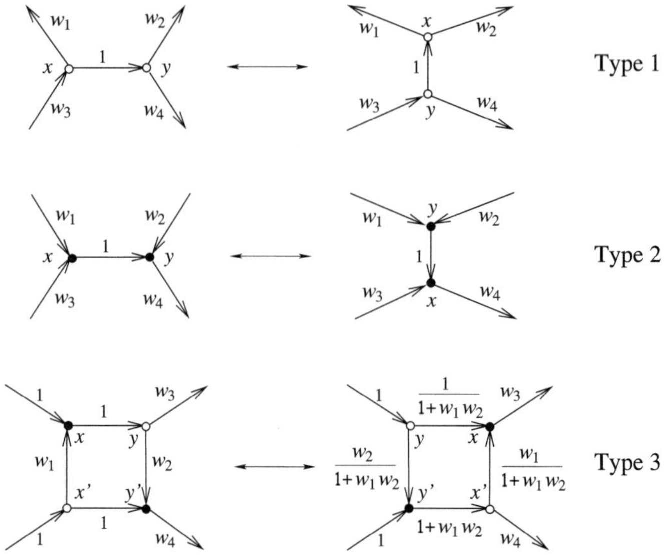

FIGURE 10.5. Equivalent transformations of perfect networks

EXAMPLE 10.37. Consider the network from Example 10.18. Recall that $\varepsilon = (2,2,1,0,0)$ and set $i = 2$. So, $\varepsilon_2 = 2$ and $\varepsilon_3 = 1$, which corresponds to the fourth row of the above table. The corresponding transformation consists of the following steps:

(i) Type 2 transformation with $x = v_b^+(3)$, $y = v_b^-(3)$ and $w_1 = w_4 = 1$, $w_2 = c_2^+$, $w_3 = c_3^-$.

(ii) Type 3 transformation with $x = v_b^+(3)$, $y = v_w^-(3)$, $x' = v_w^+(2)$, $y' = v_w^-(2)$ and $w_1 = c_2^+$, $w_2 = c_2^-$, $w_3 = d_3$, $w_4 = 1$.

(iii) Type 1 transformation with $x = v_w^+(2)$, $y = v_w^-(2)$ and $w_1 = c_2^+/(1+c_2)$, $w_2 = 1$, $w_3 = d_2$, $w_4 = c_1^-$.

(iv) The gauge group action at $v_b^+(3)$ that takes the triple of weights $(d_3, c_2^+/(1+c_2), 1/(1+c_2))$ to $(1, d_3c_2^+/(1+c_2), d_3/(1+c_2))$.

(v) The gauge group action at $v_w^-(2)$ that takes the triple of weights $(1+c_2, 1, c_1^-)$ to $(1, 1+c_2, c_1^-(1+c_2))$.

(vi) The gauge group action at $v_w^+(2)$ that takes the triple of weights $(1+c_2, d_3c_2^+/(1+c_2), d_2)$ to $(d_2(1+c_2), d_3c_2^+/[d_2(1+c_2)], 1)$.

Thus, at the end we have $(c_2^-)' = c_2^-/(1+c_2)$, $(c_2^+)' = d_3c_2^+/[d_2(1+c_2)]$, and hence $c_2' = d_3c_2/[d_2(1+c_2)]$. Besides, $(c_1^-)' = c_1^-(1+c_2)$, $c_1^+)' = c_1^+$, and hence $c_1' = c_1(1+c_2)$. Finally, $d_2' = d_2(1+c_2)$ and $d_3' = d_3/(1+c_2)$. All these expressions coincide with those given in the fourth row of the table.

Transformations of the relevant part of the network during the first two steps are shown in Figure 10.6.

Transformations of the relevant part of the network during the remaining four steps are shown in Figure 10.7.

10.5.2. It is natural to ask whether the classical *Darboux transformation* $X = X_-X_0X_+ \mapsto D(X) = X_0X_+X_-$ can also be interpreted in terms of the cluster

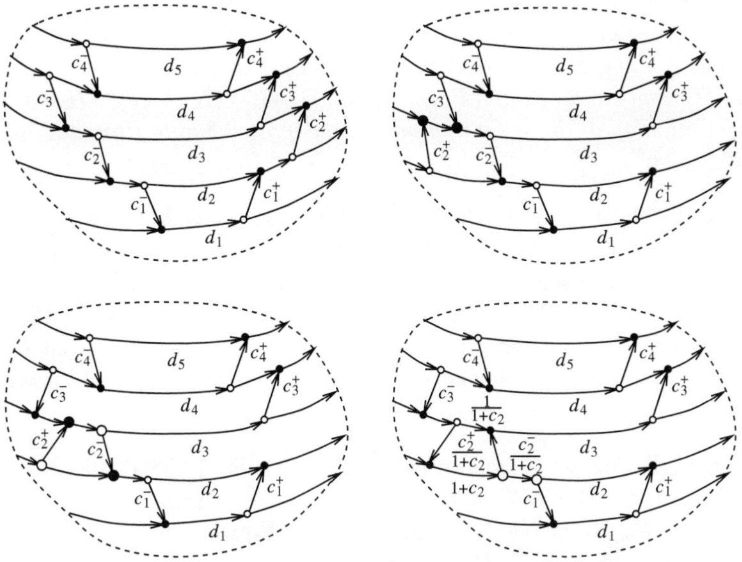

FIGURE 10.6. Elementary generalized Bäcklund–Darboux transformation: steps (i) and (ii)

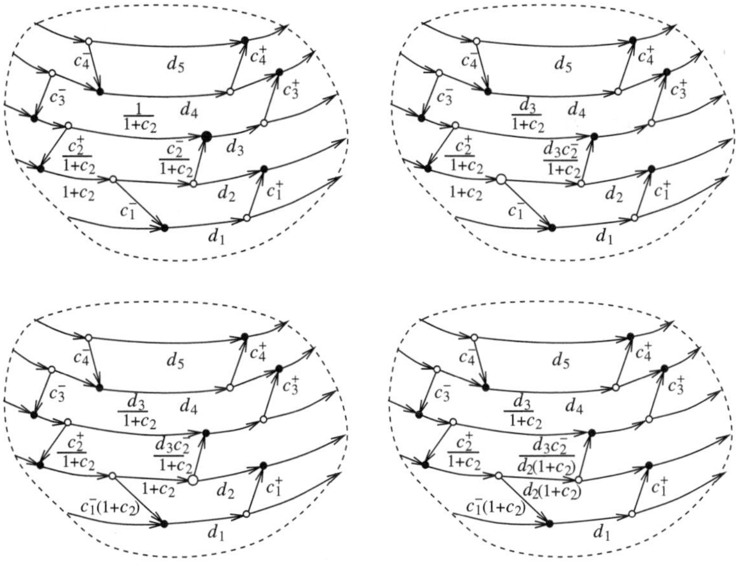

FIGURE 10.7. Elementary generalized Bäcklund–Darboux transformation: steps (iii)-(vi)

algebra $\mathcal{A}$. The transformation D constitutes a step in the *LU-algorithm* for computing eigenvalues of a matrix x. We collect some relevant simple facts about the transformation $\mathcal{D}$ in the proposition below.

PROPOSITION 10.38. *Let $x \in \mathcal{N}_-\mathcal{B}_+$. Then*
(i) for any $i \in \mathbb{Z}$, $h_i(D(X)) = h_{i+1}(X)/h_1(X)$;
(ii) for any $u, v \in S_n$, if $x \in G^{u,v}$ then $D(X) \in G^{u,v}$;
(iii) D descends to a rational Poisson map $\mathcal{D} : G^{u,v}/\mathcal{H} \to G^{u,v}/\mathcal{H}$ that coincides with a time-one map of the Hamiltonian flow generated by the Hamiltonian $F(X) = \frac{1}{2}\operatorname{tr}\log^2 X$.

PROOF. (i) For $i \geq 0$, we have $h_{i+1}(X) = \left(X_-(X_0X_+X_-)^iX_0X_+e_1, e_1\right) = d_1\left(D(X)^ie_1, e_1\right) = h_1(X)h_i(D(X))$. The case $i < 0$ can be treated similarly.

(ii) It suffices to observe that if $Y_1 \in \mathcal{N}_-$ and $Y_2 \in \mathcal{B}_+$ than both statements $Y_1Y_2 \in G^{u,v}$ and $Y_2Y_1 \in G^{u,v}$ are equivalent to $Y_1 \in \mathcal{B}_+v\mathcal{B}_+, Y_2 \in \mathcal{B}_-u\mathcal{B}_-$.

(iii) Claim (ii) implies that D descends to a rational map from $G^{u,v}/\mathcal{H}$ to $G^{u,v}/\mathcal{H}$. The rest of the claim is an immediate corollary of well-known general results. □

For a pair of Coxeter elements (u, v), Proposition 10.38(i) allows us to completely describe the action of $\mathcal{D}$ on $G^{u,v}/\mathcal{H}$ in terms of a simple map on $\mathcal{R}_n$. Namely, define $\eta : \mathcal{R}_n \to \mathcal{R}_n$ by $\eta(M(\lambda)) = \lambda M(\lambda) - H_0$. Equivalently, η can be described by $\eta\left(\sum_{i=0}^{\infty} H_i\lambda^{-i-1}\right) = \sum_{i=0}^{\infty} H_{i+1}\lambda^{-i-1}$. Then Proposition 10.38(i) implies that on $G^{u,v}/\mathcal{H}$

$$\mathcal{D} = \rho_{u,v} \circ \mathbf{x}_{u,v} \circ \eta \circ m_{u,v},$$

where maps $\rho_{u,v}$, $\mathbf{x}_{u,v}$, $m_{u,v}$ were defined in the Introduction.

To tie together the cluster algebra $\mathcal{A}$ and the Darboux transformation $\mathcal{D}$, we will only need to fix the stable variable x_{1n} to be equal to 1. In view of (10.45), this means that we are dealing with double Bruhat cells in SL_n rather than in GL_n.

To emphasize a similarity between the classical Darboux transformation $\mathcal{D}$ and the generalized Bäcklund–Darboux transformation $\sigma_{u,v}^{u',v'}$, we express the former similarly to (10.4).

PROPOSITION 10.39. *$\mathcal{D} = \rho_{u,v} \circ T_{\mathcal{D}} \circ \tau_{u,v}$, where $T_{\mathcal{D}}$ is a sequence of cluster transformations in $\mathcal{A}$.*

PROOF. Note that in the graphical representation of the matrix $B(\varepsilon)$ that we employed in the proof of Lemma 10.29, the condition $x_{1n} = 1$ amounts to erasing the rightmost lower vertex and all corresponding edges in the graph Γ. Consider the cluster corresponding to $\varepsilon = (2, 0, \dots, 0)$. Then the proof of Lemma 10.29 shows that an application of $T_{2n-3} \circ \dots \circ T_1$ leads from the cluster with the graph Γ with cluster variables $\Delta_{i-1}^{(i)}$, $i \in [1, n]$, associated (left to right) with vertices of the top row and $\Delta_i^{(i)}$, $i \in [1, n-1]$, associated with vertices of the bottom row to the cluster with the same graph with cluster variables $\Delta_i^{(i)}$, $i \in [1, n]$, associated (left to right) with vertices of the top row and $\Delta_{i+1}^{(i)}$, $i \in [1, n-1]$, associated with vertices of the bottom row. (One has to take into an account that $x_{1n} = 1$ implies that $\Delta_j^{(n)} = \Delta_{n-1}^{(n)}$ for any j.) This means that for $v = u^{-1} = s_{n-1} \cdots s_1$ we can choose $T_{2n-3} \circ \dots \circ T_1$ for $T_{\mathcal{D}}$. Then, for arbitrary pair of Coxeter elements (u, v), $T_{\mathcal{D}}$ can be defined as

$$T_{\mathcal{D}} = T_{w^{-1},w}^{u,v} \circ (T_{2n-3} \circ \dots \circ T_1) \circ T_{u,v}^{w^{-1},w}$$

with $w = s_{n-1} \cdots s_1$. □

10.6. Summary

- Generic elements in a Coxeter double Bruhat cell $G^{u,v}$ are parameterized by $3n-2$ parameters d_i, $i=1,\dots,n$, and c_i^+, c_i^-, $i=1,\dots,n-1$, see Lemma 10.3.
- Parameters d_i and $c_i = c_i^+ c_i^-$ can be restored as monomial expressions in terms of an appropriately chosen collection of Hankel determinants built from the coefficients of the Laurent expansion of the Weyl function, see Theorem 10.9.
- Given a generic element in $G^{u,v}$, we build a planar network in an annulus with one source and one sink such that its boundary measurement coincides with the Weyl function up to a constant factor and the transformation $\lambda \mapsto -\lambda$, see Proposition 10.20. Together with general constructions of Chapter 9, this gives a Poisson bracket on the space $\mathcal{R}_n$ of degree n rational functions in one variable vanishing at infinity and regular at the origin, see Proposition 10.23.
- Based on general techniques developed in Chapter 4, we define a cluster algebra so that the Poisson structure on $\mathcal{R}_n$ becomes compatible with it, see Lemma 10.26. This cluster algebra does not depend on the choice of u, v, and its localization with respect to the stable variables coincides with the ring of regular functions on $\mathcal{R}_n$, see Theorem 10.27.
- Generalized Bäcklund–Darboux transformations between different Coxeter–Toda lattices can be interpreted as cluster transformations in the above cluster algebra, see Theorem 10.36. Several explicit examples of such transformations are given in Example 10.34.

Bibliographical notes

Our exposition in this Chapter is based on [**GSV7**].

10.1. Toda flows for an arbitrary standard semi-simple Poisson–Lie group were studied in [**R**] under the name of characteristic Hamiltonian systems. Factorization method for the exact solution of Toda flows on an arbitrary double Bruhat cell is presented in [**ReST**]. For a discussion of the Weyl function see, e.g., [**DLNT, M, BrF**]. Determinantal formulas for the inversion of the Moser map go back to the work of Stieltjes on continuous fractions [**St**] (see, e.g. [**A**] for details).

Coxeter double Bruhat cells have appeared (for an arbitrary simple Lie group) in [**HKKR**] in the context of integrable systems and in [**BFZ2, YZ**] in connection with cluster algebras of finite type. The term *Coxeter–Toda lattices* was first used in [**HKKR**] in the case $u=v$ for an arbitrary simple Lie group, which generalizes the *relativistic Toda lattice* that corresponds to the choice $u=v=s_{n-1}\cdots s_1$ in GL_n. In [**FG1, FG2**], the corresponding integrable systems for $v=s_{n-1}\cdots s_1$ and an arbitrary Coxeter element u were called *elementary Toda lattices.* Birational transformations between $G^{u,s_{n-1}\cdots s_1}/\mathcal{H}$ and $G^{u',s_{n-1}\cdots s_1}/\mathcal{H}$ for two different Coxeter elements u, u' that serve as generalized Bäcklund–Darboux transformation between the corresponding elementary Toda lattices were first studied in [**FG1**].

A cluster algebra closely related to the one we consider here appeared in [**K**] and was a subject of a detailed combinatorial study in paper [**DK**], where cluster mutations along the edges of a certain subgraph of its exchange graph were shown to describe an evolution of an A_n *type Q-system* – a discrete evolution that arises in the analysis of the XXX-model, which is an example of a *quantum* integrable model. In [**DK**], solutions of the Q-system are represented as Hankel determinants built from coefficients of a certain generating function, that turns out to be rational and can be represented as a matrix element of a resolvent of an appropriate linear operator.

10.2. Recursion operators arising in the theory of orthogonal polynomials on the unit circle, including CMV matrices, and mentioned in Remark 10.6, are studied in [**CMV, Si**].

10.3. The inverse problem studied in Section 10.3 was solved for the case $v = s_{n-1} \cdots s_1$, u arbitrary in [**FG3**]. Relation (10.28) in Lemma 10.14 was proved in the same paper. For the equivalence between coprimality of P and Q and non-vanishing of $\Delta_n^{(n-1)}$ see, e.g. Theorem 8.7.1 in [**Fu**]. The derivation of (10.38) from (10.37) can be found in [**FG2**], Proposition 3.

10.4. The cluster algebra $\mathcal{A}$ built in this section is tightly connected with the cluster algebra studied in [**K, DK**]. To get the latter from $\mathcal{A}$ one has to fix the values of the stable variables at 1. The exchange matrix of this algebra is therefore obtained from $\tilde{B}(\varepsilon)$ by deleting the last two rows. If we take $\varepsilon = (2, 1, \dots, 1, 0)$ and rearrange the cluster variables as $x_{01}, x_{02}, \dots, x_{0n-1}, x_{11}, x_{12}, \dots, x_{1n-1}$, the exchange matrix will be given by

$$\begin{pmatrix} 0 & -C \\ C & 0 \end{pmatrix},$$

where C is the Cartan matrix for A_{n-1}, cp. with Theorem 1.2 in [**DK**].

10.5. For the standard Poisson–Lie structure on double Bruhat cells and its symplectic leaves see [**R, KoZ, Y**]. Conjugation-invariant functions on GL_n are discussed, e.g., in [**ReST**].

We could have also computed brackets (10.59) by specializing general formulas obtained in [**KoZ**] for Poisson brackets for factorization parameters of an arbitrary double Bruhat cell in a standard semisimple Poisson–Lie group.

An explicit solution method to the finite non-periodic Toda lattice was first suggested by Moser [**M**].

Equivalent transformations of perfect networks were introduced in [**Po**].

A connection of the LU-algorithm (as well as similar numerical algorithms, such as QR and Cholesky algorithms) to integrable systems of Toda type is well-documented, see, e.g. [**DLT, W**]. For an arbitrary semisimple Lie group, a restriction of such a transformation to a Coxeter double Bruhat cell of type $G^{u,u}$ was studied, under the name of *factorization dynamics*, in [**HKKR**].

The proof of Proposition 10.38(iii) follows immediately from general results stated in Section 7.1 in [**HKKR**].

The shift $H_i \mapsto H_{i+1}$ plays an important role in the study of Q-systems in [**DK**]. As we mentioned above, the initial cluster in [**DK**] corresponds to $\varepsilon = (2, 1, \dots, 1, 0)$, which results, e. g. from the choice $u = v = s_{n-1} \cdots s_1$. Also, the stable variables x_{0n}, x_{1n} are "frozen" to be equal to 1. The rest of the cluster variables in the initial cluster are realized as Hankel determinants of the form prescribed by (10.40). Therefore, solutions of the corresponding Q-system form a subset of cluster variables in $\mathcal{A}(\varepsilon)$, every element of which is realized as a Hankel determinant with all indices shifted by the same integer.

Bibliography

[A] N. I. Akhiezer, *The classical moment problem and some related questions in analysis.* Hafner Publishing Co., New York 1965.

[A1] V. Arnold, *Arnold's problems.* Springer, Berlin; PHASIS, Moscow, 2004.

[A2] V. Arnold, *Mathematical methods of classical mechanics.* Springer, NY, 1989.

[BaGZ] M. Barot, C. Geiss, and A. Zelevinsky, *Cluster algebras of finite type and positive symmetrizable matrices.* J. London Math. Soc. (2) **73** (2006), 545–564.

[BeD] A. Belavin, V. Drinfeld, *Solutions of the classical Yang–Baxter equation for simple Lie algebras*, Functional Anal. Appl. **16** (1982), 159–180.

[BFZ1] A. Berenstein, S. Fomin, and A. Zelevinsky, *Parametrizations of canonical bases and totally positive matrices.* Adv. Math. **122** (1996), 49–149.

[BFZ2] A. Berenstein, S. Fomin, and A. Zelevinsky, *Cluster algebras. III. Upper bounds and double Bruhat cells.* Duke Math. J. **126** (2005), 1–52.

[BZ1] A. Berenstein and A. Zelevinsky, *Tensor product multiplicities and convex polytopes in partition space.* J. Geom. Phys. **5** (1988), 453–472.

[BZ2] A. Berenstein and A. Zelevinsky, *Total positivity in Schubert varieties.* Comment. Math. Helv. **72** (1997), 128–166.

[BrF] R. W. Brockett, L. Faybusovich, *Toda flows, inverse spectral problems and realization theory*, Systems and Control Letters, **16** (1991), 79-88.

[BGY] K. Brown, K. Goodearl, and M. Yakimov, *Poisson structures on affine spaces and flag varieties. I. Matrix affine Poisson space*, Adv. Math. **206** (2006), 567–629.

[BMRRT] A. Buan, R. Marsh, M. Reineke, I.Reiten, and G. Todorov, *Tilting theory and cluster combinatorics.* Adv. Math. **204** (2006), 572–618.

[BMRT] A. Buan, R. Marsh, I.Reiten, and G. Todorov, *Clusters and seeds in acyclic cluster algebras.* Proc. Amer. Math. Soc. **135** (2007), 3049–3060.

[Bu] Yu. Burman, *Triangulations of surfaces with boundary and the homotopy principle for functions without critical points.* Ann. Global Anal. Geom. **17** (1999), 221–238.

[CMV] M. J. Cantero, L. Moral, L. Velázquez, *Five-diagonal matrices and zeros of orthogonal polynomials on the unit circle*, Linear Algebra Appl. **362** (2003), 29–56.

[CFZ] F. Chapoton, S. Fomin, and A. Zelevinsky, *Polytopal realizations of generalized associahedra.* Canad. Math. Bull. **45** (2002), 537–566.

[CuFl] T. Cusick and M. Flahive, *The Markoff and Lagrange spectra.* Amer. Math. Soc., Providence, 1989.

[DKJM] E. Date, M. Kashivara, M. Jimbo, T. Miwa, *Transformation groups for soliton equations*, Proc. of RIMS Symposium on Non-Linear Integrable Systems, Singapore, World Science Publ. Co., 1983, pp. 39–119.

[DLNT] P. A. Deift., L.-C. Li, T. Nanda and C. Tomei, *The Toda lattice on a generic orbit is integrable*, Comm. Pure Appl. Math. **39** (1986), 183-232.

[DLT] P. A. Deift., L.-C. Li and C. Tomei, *Matrix factorizations and integrable systems*, Comm. Pure Appl. Math. **42** (1989), 443–521.

[DK] P. Di Francesco and R. Kedem, *Q-systems, heaps, paths and cluster positivity*, Comm. Math. Phys. **293** (2010), 727–802.

[ES] P. Etingof and O. Schiffmann, *Lectures on quantum groups.* International Press, Somerville, MA, 2002.

[FT] L. Faddeev, L. Takhtajan, *Hamiltonian methods in the theory of solitons*, Springer, Berlin, 2007.

[FG1] L. Faybusovich, M. I. Gekhtman, *Elementary Toda orbits and integrable lattices*, J. Math. Phys. **41** (2000), 2905–2921.

[FG2] L. Faybusovich, M. I. Gekhtman, *Poisson brackets on rational functions and multi-Hamiltonian structure for integrable lattices*, Phys. Lett. A **272** (2000), 236–244.

[FG3] L. Faybusovich, M. I. Gekhtman, *Inverse moment problem for elementary co-adjoint orbits*, Inverse Problems **17** (2001), 1295–1306.

[Fo] V. Fock, *Dual Teichmüller spaces.* arxiv:dg-ga/9702018.

[FoG] V. Fock and A. Goncharov, *Cluster ensembles, quantization and the dilogarithm*, Ann. Sci. Èc. Norm. Supèr., **42** (2009), 865–930.

[FoR] V. Fock and A. Rosly, *Poisson structure on moduli of flat connections on Riemann surfaces and the r-matrix.* Moscow Seminar in Mathematical Physics, Amer. Math. Soc. Transl. Ser. 2, **191**, Amer. Math. Soc., Providence, RI, 1999, pp. 67–86.

[FRe] S. Fomin and N. Reading, *Generalized cluster complexes and Coxeter combinatorics.* Int. Math. Res. Notices (2005), 2709–2757.

[FST] S. Fomin, M. Shapiro, and D. Thurston, *Cluster algebras and triangulated surfaces. I. Cluster complexes.* Acta Math. **201** (2008), 83–146.

[FZ1] S. Fomin and A. Zelevinsky, *Double Bruhat cells and total positivity.* J. Amer. Math. Soc. **12** (1999), 335–380.

[FZ2] S. Fomin and A. Zelevinsky, *Cluster algebras.I. Foundations.* J. Amer. Math. Soc. **15** (2002), 497–529.

[FZ3] S. Fomin and A. Zelevinsky, *The Laurent phenomenon.* Adv. in Appl. Math. **28** (2002), 119–144.

[FZ4] S. Fomin and A. Zelevinsky, *Cluster algebras. II. Finite type classification.* Invent. Math. **154** (2003), 63–121.

[FZ5] S. Fomin and A. Zelevinsky, *Y-systems and generalized associahedra.* Ann. of Math. (2) **158** (2003), 977–1018.

[FZ6] S. Fomin and A. Zelevinsky, *Cluster algebras: notes for the CDM-03 conference.* Current Developments in Mathematics, Int. Press, Somerville, MA, 2003, pp.1–34.

[FZ7] S. Fomin and A. Zelevinsky, *Cluster algebras. IV. Coefficients.* Compos. Math. **143** (2007), 112–164.

[Fu] P. A. Fuhrmann, *A polynomial approach to linear algebra.* Universitext. Springer-Verlag, New York, 1996.

[GLS] Ch. Geiss, B. Leclerc, and J. Schröer, *Rigid modules over preprojective algebras.* Invent. Math. **165** (2006), 589–632.

[GSV1] M. Gekhtman, M. Shapiro, and A. Vainshtein, *The number of connected components in double Bruhat cells for nonsimply-laced groups.* Proc. Amer. Math. Soc. **131** (2003), 731–739.

[GSV2] M. Gekhtman, M. Shapiro, and A. Vainshtein, *Cluster algebras and Poisson geometry.* Mosc. Math. J. **3** (2003), 899–934.

[GSV3] M. Gekhtman, M. Shapiro, and A. Vainshtein, *Cluster algebras and Weil-Petersson forms.* Duke Math. J. **127** (2005), 291–311; **139** (2007), 407–409.

[GSV4] M. Gekhtman, M. Shapiro, and A. Vainshtein, *On the properties of the exchange graph of a cluster algebra.* Math. Res. Lett. **15** (2008), 321–330.

[GSV5] M. Gekhtman, M. Shapiro, and A. Vainshtein, *Poisson properties of planar networks in a disk.* Selecta Math. **15** (2009), 61–103.

[GSV6] M. Gekhtman, M. Shapiro, and A. Vainshtein, *Poisson properties of planar networks in an annulus.* arxiv:0901.0020.

[GSV7] M. Gekhtman, M. Shapiro, and A. Vainshtein, *Generalized Bäcklund–Darboux transformations of Coxeter–Toda flow from cluster algebra perspective*, Acta Math., *to appear.*

[GrH] P. Griffiths and J. Harris, *Principles of algebraic geometry.* John Wiley & Sons, NY, 1994.

[GrSh] B. Grünbaum and G. Shephard, *Rotation and winding numbers for planar polygons and curves.* Trans. Amer. Math. Soc. **322** (1990), 169–187.

[Ha] A. Hatcher, *On triangulations of surfaces.* Topology Appl. **40** (1991), 189–194.

[HdP] W. Hodge and D. Pedoe, *Methods of algebraic geometry. Vol. II. Book III: General theory of algebraic varieties in projective space. Book IV: Quadrics and Grassmann varieties.* Cambridge University Press, 1994.

[HoL] T. Hodges and T. Levasseur *Primitive ideals of $C_q[SL(3)]$* Comm. Math. Phys. **156** (1993), 581–605.

[HKKR] T. Hoffmann, J. Kellendonk, N. Kutz, and N. Reshetikhin, *Factorization dynamics and Coxeter–Toda lattices.* Comm. Math. Phys. **212** (2000), 297–321.

[HLT] C. Hohlweg, C. Lange, and H. Thomas, *Permutahedra and generalized associahedra.* arXiv:0709.4241.

[Hu1] J. Humphreys, *Arithmetic groups.* Springer, NY, 1980.

[Hu2] J. Humphreys, *Introduction to Lie algebras and representation theory.* Springer, NY, 1978.

[Hu3] J. Humphreys, *Reflection groups and Coxeter groups.* Cambridge University Press, Cambridge, 1990.

[IIKNS] R. Inoue, O. Iyama, A. Kuniba, T. Nakanishi, and J. Suzuki, *Periodicities of T-systems and Y-systems.* arxiv:0812.0667

[Iv] N. Ivanov, *Mapping class groups*, Handbook of geometric topology, North-Holland, Amsterdam, 2002, pp. 523–633.

[Ka] S. Katok, *Fuchsian groups*, Univ. of Chicago Press, 1992.

[K] R. Kedem, *Q-systems as cluster algebras*, J. Phys. A **41** (2008), no. 19, 194011, 14 pp.

[Ke1] B. Keller, *On triangulated orbit categories.* Doc. Math. **10** (2005), 551–581.

[Ke2] B. Keller, *Cluster algebras, quiver representations and triangulated categories*, arxiv0807.1960.

[KoZ] M. Kogan and A. Zelevinsky, *On symplectic leaves and integrable systems in standard complex semisimple Poisson–Lie groups.* Internat. Math. Res. Notices (2002), 1685–1702.

[KR] J. Kung and G.-C. Rota, *The invariant theory of binary forms.* Bull. Amer. Math. Soc. (N.S.) **10** (1984), 27–85.

[Le] C. W. Lee, *The associahedron and triangulations of the n-gon.* European J. Comb. **10** (1989), 551–560.

[Li] B. Lindström, *On the vector representations of induced matroids.* Bull. London Math. Soc. **5** (1973), 85–90.

[MRZ] R. Marsh, M. Reineke, and A. Zelevinsky, *Generalized associahedra via quiver representations.* Trans. Amer. Math. Soc. **355** (2003), 4171–4186.

[Ma] J. Maybee, *Combinatorially symmetric matrices.* Linear Algebra and Appl. **8** (1974), 529–537.

[McMS] P. McMullen and E. Schulte, *Abstract regular polyopes.* Cambridge Univesity Press, Cambridge, 2002.

[M] J. Moser, *Finitely many mass points on the line under the influence of the exponential potential - an integrable system.* Dynamical systems, theory and applications, 467–497, Lecture Notes in Physics **38**, Springer, Berlin, 1975.

[OV] A. Onishchik and E. Vinberg, *Lie groups and algebraic groups.* Springer-Verlag, Berlin, 1990.

[Pe1] R. Penner, *The decorated Teichmller space of punctured surfaces.* Comm. Math. Phys. **113** (1987), 299–339.

[Pe2] R. Penner, *Weil-Petersson volumes.* J. Differential Geom. **35** (1992), 559–608.

[Po] A. Postnikov, *Total positivity, Grassmannians, and networks.* arXiv:math/0609764.

[PY] S. Parter and J. Youngs, *The symmetrization of matrices by diagonal matrices.* J. Math. Anal. Appl. **4** (1962), 102–110.

[R] N. Reshetikhin *Integrability of characteristic Hamiltonian systems on simple Lie groups with standard Poisson Lie structure*, Comm. Mat. Phys. **242** (2003), 1–29.

[ReST] A. Reyman and M. Semenov-Tian-Shansky *Group-theoretical methods in the theory of finite-dimensional integrable systems.* Encyclopaedia of Mathematical Sciences, vol.16, Springer–Verlag, Berlin, 1994, pp. 116–225.

[Sc] J. Scott, *Grassmannians and cluster algebras.* Proc. London Math. Soc. (3) **92** (2006), 345–380.

[Se] A. Seven, *Orbits of groups generated by transvections over $\mathbb{F}_2$.* J. Algebraic Combin. **21** (2005), 449–474.

[SV] B. Shapiro and A. Vainshtein, *Euler characteristics for links of Schubert cells in the space of complete flags*, Theory of singularities and its applications, Adv. Soviet Math., **1**, Amer. Math. Soc., Providence, RI, 1990, pp. 273–286.

[SSV1] B. Shapiro, M. Shapiro, and A. Vainshtein, *Connected components in the intersection of two open opposite Schubert cells in $SL_n(\mathbb{R})/B$.* Int. Math. Res. Not. (1997), 469–493.

[SSV2] B. Shapiro, M. Shapiro, and A. Vainshtein, *Skew-symmetric vanishing lattices and intersections of Schubert cells.* Int. Math. Res. Not. (1998), 563–588.

[SSVZ] B. Shapiro, M. Shapiro, A. Vainshtein, and A. Zelevinsky, *Simply laced Coxeter groups and groups generated by symplectic transvections.* Michigan Math. J. **48** (2000), 531–551.

[ShZ] P. Sherman and A. Zelevinsky, *Positivity and canonical bases in rank 2 cluster algebras of finite and affine types.* Mosc. Math. J. **4** (2004), 947–974.

[Si] B. Simon, *Orthogonal Polynomials on the Unit Circle, Part 1: Classical Theory*, AMS Colloquium Series, American Mathematical Society, Providence, RI, 2005.

[SlTTh] D. Sleator, R. Tarjan, and W. Thurston, *Rotation distance, triangulations, and hyperbolic geometry.* J. Amer. Math. Soc. **1** (1988), 647–681.

[Sp] G. Springer, *Introduction to Riemann surfaces*, 2nd ed., Chelsea Publ. Comp., New York, 1981.

[St] J. Stasheff, *Homotopy associativity of H-spaces. I,II.* Trans. Amer. Math. Soc. **108** (1963), 275–312.

[St] T.J. Stieltjes, *Recherches sur les fractions continues*, in: Oeuvres Complètes de Thomas Jan Stieltjes, Vol. II, P. Noordhoff, Groningen, 1918, pp. 402–566.

[Ta] K. Talaska, *A formula for Plücker coordinates associated with a planar network.* Int. Math. Res. Not. (2008), rnn 081.

[Th] W. Thurston, *Minimal stretch maps between hyperbolic surfaces.* arxiv:math.GT/9801039.

[W] D. S. Watkins, *Isospectral flows*, SIAM Rev. **26** (1984), no. 3, 379–391.

[Y] M. Yakimov, *Symplectic leaves of complex reductive Poisson–Lie groups.* Duke Math. J. **112** (2002), 453–509.

[YZ] S.-W. Yang, A. Zelevinsky, *Cluster algebras of finite type via Coxeter elements and principal minors*, Transform. Groups, **13** (2008), 855–895.

[Zam] Al. Zamolodchikov, *On the thermodynamic Bethe ansatz equations for reflectionless ADE scattering theories.* Phys. Lett. B **253** (1991), 391–394.

[Z] A. Zelevinsky, *Connected components of real double Bruhat cells.* Int. Math. Res. Not. (2000), 1131–1154.

Index